Aquaculture: Aquatic Animals and Plants

Aquaculture: Aquatic Animals and Plants

Editor: Douglas Rodriquez

MURPHY & MOORE
www.murphy-moorepublishing.com

Murphy & Moore Publishing,
1 Rockefeller Plaza,
New York City, NY 10020, USA

Visit us on the World Wide Web at:
www.murphy-moorepublishing.com

ISBN: 978-1-63987-054-7 (Hardback)

Cataloging-in-Publication Data

Aquaculture : aquatic animals and plants / edited by Douglas Rodriquez.
p. cm.
Includes bibliographical references and index.
ISBN 978-1-63987-054-7
1. Aquaculture. 2. Aquatic animals. 3. Aquatic plants. 4. Aquatic organisms.
5. Aquatic biology. 6. Aquatic sciences. I. Rodriquez, Douglas.
SH21 .A68 2022
639.8--dc23

Table of Contents

Preface

The purpose of the book is to provide a glimpse into the dynamics and to present opinions and studies of some of the scientists engaged in the development of new ideas in the field from very different standpoints. This book will prove useful to students and researchers owing to its high content quality.

Aquaculture is concerned with the farming of aquatic plants and animals, such as fish, algae, crustaceans, mollusks, etc. In aquaculture, freshwater and saltwater organisms are cultivated under controlled conditions. Niche industries such as shrimp farming, oyster farming, fish farming, algaculture, and cultivation of ornamental fish fall under aquaculture. Increasingly, practices such as integrated multi-trophic aquaculture and aquaponics are being used in aquaculture today, which integrate both fish as well as aquatic plant farming. Modern research in aquaculture not only examines methods for greater production of farmed organisms but also strives to understand the impacts of modern farming on wild fish, coastal and freshwater ecosystems, as well as the proliferation of fish diseases. This book brings forth some of the most innovative concepts and elucidates the unexplored aspects of aquaculture. The various advancements in aquafarming are glanced at and their applications as well as ramifications are looked at in detail. This book will prove to be immensely beneficial to students and researchers in this field.

At the end, I would like to appreciate all the efforts made by the authors in completing their chapters professionally. I express my deepest gratitude to all of them for contributing to this book by sharing their valuable works. A special thanks to my family and friends for their constant support in this journey.

Editor

1

Growth performance of hybrid striped bass (*Morone chrysops* × *M. saxatilis*) fed with commercial pike perch and trout diets

Stephan S. W. Ende · Vanessa Fuchs · Annabel Schuhn · Christiane von der Marwitz · Andrea Wirtz · Joachim Henjes · Matthew Slater

Abstract Two commercial trout diets (*Oncorhynchus mykiss*) and one commercial pike perch diet (*Sander lucioperca*) were fed to hybrid striped bass (*Morone chrysops* × *M. saxatilis*) (mean initial weight ± SD of 60.7 g ± 12.1; mean initial length SD of 17.2 cm ± 1.1) for 69 days at rations of approximately 1% average body weight. While final body weight (FBW), final length (FBL) and condition factor (Cf) were not significantly influenced by diets, specific growth rate (SGR) in hybrid striped bass fed with the pike perch diet (1.15) was significantly higher than those fed with either of the two trout diets (1.04 and 1.07). The feed conversion ratio (FCR) in hybrid striped bass fed with the pike perch diet (1.0) was significantly lower than the FCR in hybrid striped bass fed with either of the two trout diets (1.1 and 1.2). When hybrid striped bass (mean initial body weight: 65.7 ± 4.5 and 127.7 ± 2.9 g) were fed with the pike perch diet twice per day until satiation for 52 days, the SGR was 1.7 and 1.15% d^{-1} in fishes with an average body weight of 116 and 183 g, respectively. Present results demonstrate that growth performance in hybrid striped bass can be improved when fishes are fed with commercial pike perch diets rather than using commercial trout diets as is the current practice.

Keywords Hybrid striped bass · Commercial diets · Trout · Pike perch · Growth · Nutrition

Introduction

The hybrid striped bass (*Morone chrysops* × *M. saxatilis*) is considered an ideal candidate for aquaculture because of its fast growth, temperature tolerance, flexibility to environmental parameters, disease resistance (Kohler 2004; Morris et al. 1999). Moreover, hybrid striped bass can be produced in basically all production systems such as ponds, tanks (closed or semi-closed recirculating aquaculture systems) or cage systems (FAO 2016; Lougheed and Nelson 2001). The United States is the most significant producer of hybrid striped bass with annual production of around 6000 tons (D'Abramo and Frinsko 2008) though also other countries in Europe and Asia have recently started the production of hybrid striped bass.

Despite their commercial importance, no tailored commercial diets are available for hybrid striped bass. Basic macronutrient requirements of hybrid striped bass have been determined. Reported requirements for protein range from 40 to 41% (Brown et al. 1992; Jaramillo and Gatlin 2004; Liu and Liao 1999; Nematipour et al. 1992; Webster et al. 1995). Energy–protein ratio requirements are between 8 and 9 kcal/g

S. S. W. Ende (✉) · V. Fuchs · A. Schuhn · C. von der Marwitz · A. Wirtz · J. Henjes · M. Slater
Alfred Wegener Institute, Helmholtz Centre for Polar and Marine Research, Am Handelshafen 12, 27570 Bremerhaven, Germany
e-mail: sende@awi.de

(Keembiyehetty and Wilson 1998; Nematipour et al. 1992). Maximum growth with intermediate lipid deposition levels was obtained with dietary lipid levels between 10 and 15% (Gaylord and Gatlin 2000). Though several of the essential amino acids, essential fatty acids, and vitamin and mineral requirements have been determined (Brown et al. 1993; Deng and Wilson 2003; Gaylord and Gatlin 2000; Gaylord et al. 2005; Griffin et al. 1994; Keembiyehetty and Gatlin 1997; Kocabas and Gatlin 1999; Sealey and Gatlin 1999), a complete profile of all amino acid and fatty acid requirements is missing. This knowledge gap is the main reason why commercial feeds formulated for other species are currently used in practice. Diets tailored to salmon were reportedly best suited for hybrid striped bass (Tucker Jr 2012). Growth, feed conversion and physiological indices were near optimum in hybrid striped bass weighing between 38 and 130 g and fed with a commercial salmon feed (45% protein, 12% lipid) at daily rations of 1.0–1.5% (Hung et al. 1993). However, commercial salmon diets generally contain lipid levels in excess of 12% and excess dietary lipid levels (15%) impaired health and product quality in hybrid striped bass (Gaylord and Gatlin 2000). Catfish diets have also been tested but were not suitable for hybrid striped bass (Tucker Jr 2012). Information on performance of hybrid striped bass fed with trout diets is scarce. However, striped bass (*Morone saxatilis*) fed with commercial trout feed was smaller, contained less body lipid and had necrotic liver tissue compared to those fed with a salmon diet (Lemm et al. 1993). These diet-related histological changes in the livers were related to inadequate type and level of dietary essential fatty acids (Lemm et al. 1993). However, this previous observation stating that trout diets are nutritionally inadequate for hybrid striped bass dates back almost 25 years, when nutritional knowledge even for highly commercial aquaculture species like trout was largely incomplete. To our knowledge, the suitability of 'modern' trout diets and alternative 'modern' diets in general for the culture of hybrid striped bass has not been investigated yet. The aim of this study was to evaluate the suitability of a diet formulated for pike perch to culture hybrid striped bass. Suitability was evaluated on the basis of digestibility, feed conversion ratio, condition factor and maximum growth rate.

Materials and methods

Two commercial trout diets were used as reference diets. Experiments were conducted at the Centre for Aquaculture Research (ZAF, Alfred Wegener Institute Bremerhaven, Germany).

Three commercially available diets were compared in the present study: Metabolica, 3 mm (52% Protein, Aller Aqua GmbH) subsequently referred to as 'pike perch' diet; Supreme 15, 3 mm (46% Protein, Coppens International BV) subsequently referred to as 'trout 1′ diet; Efico Alpha 715, 3 mm (44% Protein, Biomar Group) subsequently referred to as 'trout 2' diet. All diets were coated with sunflower oil (100 g kg^{-1}) and Yttrium oxide (Y_2O, 0.1%) as an external marker for subsequent determination of proximate composition and apparent digestibility (Hillestad et al. 1999).

Fish and rearing conditions

Hybrid striped bass ($n = 180$) were obtained from a commercial local fish farm (Fischzucht Hagedorn, Bargstedt-Ohrensen) and were transferred to the ZAF research facility where fishes were acclimatized to the rearing conditions prior to the start of the feeding experiments. The rearing/experimental system consisted of 9 rectangular glass tanks (0.4 $m.^{-2}$ bottom area and a total water volume 250 L per tank). All tanks were connected to a recirculation aquaculture system equipped with a mechanical filter (foam matt), a moving bed filter and a UV unit. Conditions were maintained as follows: (photoperiod 12L:12D; dissolved oxygen > 84%; temperature 21.6 ± 0.7 °C; total ammonia nitrogen < 0.14 mg L^{-1}; nitrite nitrogen < 0.31 mg L^{-1} and nitrate nitrogen < 201.78 mg L^{-1}). The pH was kept above 6.18 except for one day where the pH was at 5.57. The pH was controlled by periodically adding sodium bicarbonate to the system.

Growth experiments

Two controlled feeding (growth) experiments were conducted:

In experiment 1—restricted rations, three commercial diets formulated to meet the nutritional requirements of trout (two diets) and pike perch were fed at restricted rations of approximately 1% average body weight (ABW). Hybrid striped bass were weighed and length was measured individually to the nearest of 0.1 g and 0.1 cm. Fishes (mean initial weight ± SD of 60.7 g ± 12.1; mean initial length SD of 17.2 cm ± 1.1) were randomly assigned to nine experimental tanks at a density of 20 fishes per tank. The nine tanks were randomly assigned to one of the three dietary treatments (three tanks per diet). Fishes were fed twice per day at 09:00 and 16:00 h. Feed was completely consumed, so no uneaten feed needed to be recovered for actual intake utilization calculations. At the end of the experimental period after 69 days, fishes were removed from the tanks and individually weighed and length measured.

In experiment 2—satiation, only the pike perch diet was fed twice per day to apparent satiation. Upon arrival hybrid striped bass were weighed individually to the nearest 0.1 g. In total sixty, 30 smaller fishes (mean initial weight SD of 65.7 g ± 4.5) and 30 larger fishes (mean initial weight SD of 127.7 g ± 2.9) were selected and randomly assigned (only separated by weight) to six experimental tanks at a density of 10 fishes per tank. During the acclimatization period of 4 days, all fishes were fed with the artificial feed at a level of 0.3% average body weight. During the experimental feeding period, fishes were fed twice per day to apparent satiation (at 09:00 and 16:00 h). Previous observations showed no increase in daily feed intake when feed was provided at three meals per day or two meals per day (pers. obs.). Uneaten feed was not recovered from the tanks to avoid stress to the fish (pers. obs.). At the end of the experimental period after 52 days, fishes were removed from the tanks and weighed individually.

Proximate composition and digestibility of the diets

Prior to the digestibility trial, hybrid striped bass (originating from the previous satiation experiment) were weighed individually to the nearest 0.1 g. In total, 48 fishes (mean weight of 312.5 g) were randomly assigned to six experimental tanks at a density of 8 fishes per tank. During the acclimatization period of 5 days, fishes were fed the randomly assigned diet for their treatment tank (without the external marker) at a level of 1.5% average body weight ABW. During the experimental feeding period, fishes were fed twice per day to apparent satiation (at 09:00 and 16:00 h). Feces were siphoned out twice per day, dried and subsequently stored for further analysis. At the end of the experimental period after 28 days, fishes were removed from the tanks and weighed individually.

Analytical methods and calculations

Feed and fecal samples were dried, pooled on tank basis (applied for feces only) and homogenized by grinding before subsequent analyses of dry matter, nitrogen, energy, lipid and yttrium oxide. Samples were analyzed for dry matter according to 'Bundesamt für Verbraucherschutz und Lebensmittelsicherheit'(2004c–07b), crude protein (total nitrogen * 6.25) was determined by the Kjeldahl method according to 'Bundesamt für Verbraucherschutz und Lebensmittelsicherheit'(2004–07a), ash (gravimetrically 520 °C incineration) according to 'Bundesamt für Verbraucherschutz und Lebensmittelsicherheit'(2004–074) and crude fat (gravimetrically after extraction with petroleum ether) according to 'Bundesamt für Verbraucherschutz und Lebensmittelsicherheit'(1980–09). Proximate analysis was done at the laboratory facilities of the University of Applied Sciences (Bremerhaven, Germany). The dietary carbohydrate content (i.e., nitrogen-free extract, NFE) was calculated subtracting the sum of crude protein, crude fat and ash from 1000 in g kg^{-1} DM. Gross energy (GE) was determined using an automated oxygen bomb calorimeter (Parr 6100 calorimeter, Parr Instrument GmbH, Germany) at the facilities of the AWI The digestion indicator was determined using the acid-insoluble ash (AIA) method according to, Bundesamt für Verbraucherschutz und Lebensmittelsicherheit' (2004–07).

Mean initial body weight (IBW) and mean final body weight (FBW) were calculated per tank biomass. The specific growth rate (SGR in %) was calculated as $((\ln(\mathrm{FBW}) - \ln(\mathrm{IBW}))/t$, where t is the experimental period (days) of the growth study. Daily feed intake (DFI; g DM $fish^{-1}$ d^{-1}) was calculated as $\mathrm{TFI}/(t * n)$, where TFI is the total FI per tank over the experimental days and n is the number of fishes per tank. No mortalities occurred during the experiment; hence, calculations were based on eight fishes per tank. Feed conversion ratio (FCR) was calculated as DFI/(FBW−IBW) being expressed on DM basis. Apparent digestibility of the dietary nutrients and energy (ADC) was calculated as100 × (1 – dietary Y2O level/ faecal Y2O Level × faecal nutrient or energy level /dietary nutrient/energy level) (Maynard et al. 1979). The condition factor (C_f) was calculated FBW/(FBL^3), where FBL is the final body length of the fish.

Statistical analysis

Data were log transformed to fulfill the ANOVA requirements of homoscedastic normal distribution of residuals. The effect of diet on performance parameters was analyzed by one-way ANOVA. In case of a significant effect of diet, a *t* test was used for pairwise comparisons. Statistical significance was tested at a 0.05 probability level. All data were analyzed using the statistical software R version 3.3.1. (released 2016-06-21).

Results

The Pike perch diet exhibited markedly higher crude protein and lower nitrogen-free extract than the Trout diets (Table 1). The cumulative feed (restricted rations experiment) given to each tank over the experimental period was 753 g.

Final body weight, final length and condition factor were not significantly affected by diets (Table 2, experiment 1; $P = 0.245$, $P = 0.141$ and $P = 0.869$ for final body weight, final length and condition factor, respectively). The FCR was significantly influenced by diet ($P = 0.010$) with FCR in hybrid striped bass fed the pike perch diet significantly lower than in those fed the trout diets. The SGR was significantly affected by diet ($P = 0.015$). The SGR in hybrid striped bass fed with the pike perch diet was significantly higher than the SGR in hybrid striped bass fed with the trout diets.

Discussion

The present study showed that commercially important hybrid striped bass fed with a commercially available diet formulated for pike perch grew faster (SGR) and showed better FCR values than hybrid striped bass fed with either of two trout diets (experiment 1, Table 2). Previous observations stating that trout diets are nutritionally inadequate for hybrid striped bass date back to 1993 (Lemm et al. 1993), in which nutritional knowledge even for highly commercial aquaculture species like trout was largely incomplete. It may have been expected that hybrid striped bass perform well on advanced 'modern' trout diets.

Table 1 The proximate composition and apparent digestibility coefficients (ADC in %) of commercial pike perch and trout diets after coating with sunflower oil (100 $g.kg^{-1}$) and Yttrium oxide (0.1%)

	Pike perch	Trout 1	Trout 2
Analysis ($g.kg^{-1}$)			
Dry matter (g kg^{-1} WW)	914.64	918.63	928.51
Crude protein (g kg^{-1} DM)	538.83	455.82	478.89
Crude lipid (g kg^{-1} DM)	180.87	191.71	168.66
Ash (g kg^{-1} DM)	65.13	53.43	59.62
NFE (g kg^{-1} DM)[a]	215.17	299.04	292.83
Gross energy (kJ g^{-1} DM)	23.45	22.27	21.7
Apparent digestibility (%)			
Dry matter	72.06	69.87	68.22
Protein	74.19	76.58	73.16
Lipid	88.47	87.59	89.79
Energy	79.22	76.66	73.97

[a]NFE (nitrogen-free extract) calculated subtracting the sum of crude protein, crude lipid and ash from 1000 in g kg^{-1} DM

Table 2 Growth performance in hybrid striped bass (initial body weight: 60.7 ± 12.1 g fed with restricted rations of pike perch and trout diets (two diets) over a period of 69 days (experiment 1), and growth performance in hybrid striped bass (initial body weight: 65.7 ± 4.5 g (a) and 127.7 ± 2.9 g (b)) fed with the pike perch diet twice per day until satiation over a period of 52 days (experiment 2)

Diet	Experiment 1—restricted ration						Experiment 2—satiation			
	Pike Perch	SD	Trout 1	SD	Trout 2	SD	Pike Perch[a]	SD	Pike Perch[b]	SD
Final body weight (g)	134.2	3.67	124.7	2.32	127.2	1.91	167.78	5.85	239.77	3.46
Final total length (cm)	20.7	0.08	19.8	0.15	20	0.52	–	–	–	–
Specific growth rate (%)	1.15^{a}	0.04	1.04^{b}	0.03	1.07^{b}	0.03	1.71	0.08	1.15	0.06
FCR	1.03^{a}	0.05	1.18^{b}	0.04	1.13^{b}	0.04	0.98	0.02	0.88	0.03
Feed intake (g.fish^{-1}·d^{-1})	0.55	0	0.55	0	0.55	0	1.82	0.12	1.79	0.13
Condition factor	1.52	0.04	1.6	0.04	1.59	0.11	–	–	–	–

Values are means of triplicate tanks ($n = 20$ per tank) ± SD (experiment 1) or ($n = 10$ per tank) ± SD (experiment 2). Values without a common superscript letter differ significantly ($P < 0.05$) as calculated by one-way ANOVA and t test significant difference test

The absence of superscript indicates no significant difference between treatments (only applies for experiment 1; no statistical analysis was done in experiment 2)

In the present study, apparent nutrient and energy digestibility coefficients of trout and the pike perch diet were similar (Table 1). Therefore, the higher FCR in hybrid striped bass was not related to poor digestibility of the trout diet. Lemm et al. (1993) suggested that type and level of dietary essential fatty acids of trout diets do not reflect the requirements of hybrid striped bass which may also explain the present observed lower performance of hybrid striped bass fed the present trout diets. The better performance of hybrid striped bass fed with the pike perch diet was, however, somewhat unexpected because the pike perch diet showed the highest protein excess of all present diets (Table 1, 54% compared to the reported requirement of around 40% (Brown et al. 1992; Nematipour et al. 1992; Webster et al. 2001)). In addition, the dietary lipid level of the pike perch diet (Table 1, 18%) was in excess of the species requirement which was estimated between 10 and 15% (Gaylord and Gatlin 2000).

It is known that excess dietary protein and lipid levels can negatively influence the SGR and FCR in fish (Brown et al. 1992; Hatlen et al. 2007; Peres and Oliva-Teles 1999; Saravanan et al. 2012). Excess dietary lipid increases adiposity, which is thought to exert a negative feedback on feed intake via lipostatic regulation mechanisms (Johansen et al. 2002). This should have been reflected by low intake and low growth rates when pike perch diet is fed in unrestricted rations. No intake values and growth rates were found for hybrid striped bass of similar size. However, present SGR values (experiment 2, SGR of 1.7 and 1.15% d^{-1} in fish of 60.7 and 183 g IBW, respectively) are similar to values previously reported in hybrid striped bass of considerably smaller size; Juvenile hybrid striped bass (around 20–40 g IBW) fed diets formulated based on the known nutritional requirements showed SGR of 1.1–1.8 (Bowzer et al. 2016; Thompson et al. 2000) Similar sized juvenile hybrid striped bass fed a commercial salmonid diet showed SGR of 1.5% (Hung et al. 1993). In addition, the moderate excess of dietary lipid in the pike perch diet (18%, Table 2) did not seem to negatively influence FCR. In fact, present FCR values (Table 2, 0.88–0.98) compared favorably to FCR values previously reported in hybrid striped bass fed with diets of optimized dietary lipids levels of 12% (FCR of 1.1–1.5% (Bowzer and Trushenski 2015; Thompson et al. 2000)). The absence of negative effects on FCR fed diet moderately in excess of dietary lipids requirements is in line with previous observations. Higher dietary lipid levels > 15% did not negatively influence feed efficiency in hybrid striped bass; however, the deposition of lipid in the liver and peritoneal cavity increased whereas the muscle yield decreased (Gaylord and Gatlin 2000). This was also observed in the present study. In this study, random individuals were sampled and showed excessive body fat depositions (unpubl. observation). The ADC values (Table 1) obtained in hybrid striped bass fed with the pike perch diet (and in those fed the two trout diets) are in the lower range of ADC values for feedstuff of animal origin previously reported for hybrid striped bass (Sullivan and Reigh 1995). However, considering the adequate FCR values, present ADC values may rather be a result of an inadequate fecal sampling method than related to low ingredient quality. In this study, settled feces were recovered twice per week. The recovery of settled feces may facilitate under estimation of digestibility due to disintegration/

separation of feces, or leaching of nutrients and/or marker from the fecal matter (Cho and Kaushik 1990). Disintegrated feces (as in the present study) are more susceptible to nutrient loss by leaching (Cho et al. 1985).

In conclusion, present results demonstrate that growth performance in hybrid striped bass can be improved when fishes are fed with commercial pike perch diets rather than using commercial trout diets as is the current practice. Key performance data in terms of growth (SGR) and growth by feed input (FCR) were better in hybrid striped bass fed with the pike perch diet compared to two trout diets. Acceptable growth rates were obtained when hybrid striped bass were fed with unrestricted rations of pike perch diet.

Acknowledgements This work was supported by 'The Central Innovation Programme for SMEs' (ZIM) from the Federal Ministry of Economic Affairs and Energy, BMWi (grand number Germany, ZF4010501SK5). The authors thank Mrs. Märkle (University of Applied Sciences Bremerhaven) for her support in the laboratory as well as staff from the 'Centre for Aquaculture Research' (ZAF) for conducting water quality analyses and helping with daily feeding and system management activities.

References

Bowzer J, Trushenski J (2015) Growth performance of hybrid striped bass, rainbow trout, and cobia utilizing Asian carp meal-based aquafeeds. N Am J Aquac 77:59–67. https://doi.org/10.1080/15222055.2014.960117

Bowzer J, Jackson C, Trushenski J (2016) Hybrid striped bass feeds based on fish oil, beef tallow, and eicosapentaenoic acid/docosahexaenoic acid supplements: Insight regarding fish oil sparing and demand for n-3 long-chain polyunsaturated fatty acids. J Anim Sci 94:978–988. https://doi.org/10.2527/jas.2015-9199

Brown ML, Nematipour GR, Gatlin DM (1992) Dietary Protein Requirement of Juvenile Sunshine Bass at Different Salinities Prog Fish Cul 54:148-156. doi:10.1577/1548-8640(1992)054<0148:dprojs>2.3.co;2

Brown ML, Jaramillo F, Gatlin DM (1993) Dietary phosphorus requirement of juvenile sunshine bass, *Morone chrysops* × *M. saxatilis*. Aquaculture 113:355–363

Bundesamt für Verbraucherschutz und Lebensmittelsicherheit (1980–09) Amtliche Sammlung von Untersuchungsverfahren nach § 64 LFGB, § 35 Vorläufiges Tabakgesetz, § 28b GenTG 1 (L), Verfahren zur Probenahme und Untersuchung von Lebensmitteln, Bestimmung des Gesamtfettgehaltes in Fleisch und Fleischerzeugnissen, ASU L0600-3

Bundesamt für Verbraucherschutz und Lebensmittelsicherheit (2004–07) Amtliche Sammlung von Untersuchungsverfahren nach § 64 LFGB, § 35 Vorläufiges Tabakgesetz, § 28b GenTG 1 (L), Verfahren zur Probenahme und Untersuchung von Lebensmitteln, Bestimmung der säureunlöslichen Asche in Fleisch und Fleischerzeugnissen, ASU L4700-5

Bundesamt für Verbraucherschutz und Lebensmittelsicherheit (2004–07a) Amtliche Sammlung von Untersuchungsverfahren nach § 64 LFGB, § 35 Vorläufiges Tabakgesetz, § 28b GenTG 1 (L), Verfahren zur Probenahme und Untersuchung von Lebensmitteln, Bestimmung der Rohprotein Gehaltes in Fleisch und Fleischerzeugnissen, ASU L1500-3

Bundesamt für Verbraucherschutz und Lebensmittelsicherheit (2004–07b) Amtliche Sammlung von Untersuchungsverfahren nach § 64 LFGB, § 35 Vorläufiges Tabakgesetz, § 28b GenTG 1 (L), Verfahren zur Probenahme und Untersuchung von Lebensmitteln, Bestimmung der Trockenmasse in Fleisch und Fleischerzeugnissen, ASU L0600-3

Bundesamt für Verbraucherschutz und Lebensmittelsicherheit (2004–074) Amtliche Sammlung von Untersuchungsverfahren nach § 64 LFGB, § 35 Vorläufiges Tabakgesetz, § 28b GenTG 1 (L), Verfahren zur Probenahme und Untersuchung von Lebensmitteln, Bestimmung der Asche in Fleisch und Fleischerzeugnissen, ASU L0600-4

Cho CY, Kaushik SJ (1990) Nutritional energetics in fish: energy and protein utilization in rainbow trout (*Salmo gairdneri*). In: Aspects of food production, consumption and energy values. Karger Publishers, pp 132–172

Cho CY, Cowey CB, Watanabe T (1985) Finfish nutrition in Asia: methodological approaches to research and development. International Development Research Centre, Manila

D'Abramo LR, Frinsko MO (2008) Hybrid Striped bass: pond production of food fish. SRAC publication, Stoneville

Deng DF, Wilson RP (2003) Dietary riboflavin requirement of juvenile sunshine bass (*Morone chrysops* × *Morone saxatilis*). Aquaculture 218:695–701

FAO (2016) Cultured aquatic species information programme. *Morone* hybrid. FAO Fisheries and Aquaculture Department, Rome

Gaylord TG, Gatlin DM (2000) Dietary lipid level but not l-carnitine affects growth performance of hybrid striped bass (*Morone chrysops* × *M. saxatilis*). Aquaculture 190:237–246

Gaylord TG, Rawles SD, Davis KB (2005) Dietary tryptophan requirement of hybrid striped bass (*Morone chrysops* × *M. saxatilis*). Aquac Nutr 11:367–374

Griffin ME, Wilson KA, Brown PB (1994) Dietary arginine requirement of juvenile hybrid striped bass *J Nutr* 124:888–893

Hatlen B, Helland SJ, Grisdale-Helland B (2007) Energy and nitrogen partitioning in 250 g Atlantic cod (*Gadus morhua* L.) given graded levels of feed with different protein and lipid content. Aquaculture 270:167–177

Hillestad M, Åsgård T, Berge GM (1999) Determination of digestibility of commercial salmon feeds. Aquaculture 179:81–94

Hung SSO, Conte FS, Hallen EF (1993) Effects of feeding rates on growth, body composition and nutrient metabolism in striped bass (*Morone saxatilis*) fingerlings. Aquaculture 112:349–361

Jaramillo F, Gatlin DM (2004) Comparison of purified and practical diets supplemented with or without β-glucan and selenium on resistance of hybrid striped bass *Morone chrysops* ♀ × *M. saxatilis* ♂ to *Streptococcus iniae* infection. J World Aquac Soc 35:245–252 https://doi.org/10.1111/j.1749-7345.2004.tb01081.x

Johansen SJS, Ekli M, Jobling M (2002) Is there lipostatic regulation of feed intake in Atlantic salmon *Salmo salar* L.? Aquacult Res 33:515–524. https://doi.org/10.1046/j.1365-2109.2002.00736.x

Keembiyehetty CN, Gatlin DM (1997) Dietary threonine requirement of juvenile hybrid striped bass (*Morone chrysops* × *M saxatilis*). Aquac Nutr 3:217–221

Keembiyehetty CN, Wilson RP (1998) Effect of water temperature on growth and nutrient utilization of sunshine bass (*Morone chrysops*♀ × *Morone saxatilis*♂) fed diets containing different energy/protein ratios. Aquaculture 166:151–162

Kocabas AM, Gatlin DM (1999) Dietary vitamin E requirement of hybrid striped bass (*Morone chrysops* female × *M. saxatilis* male). Aquac Nutr 5:3–8

Kohler CC (2004) A white paper on the status and needs of hybrid bass aquaculture in the north central region. NCRAC White Papers, Paper, p 2

Lemm CA, Herman RL, Lemarie DP, Arzapalo A (1993) Effects of diet and environmental salinity on the growth, mortality, and tissue structure of juvenile striped bass. J Aquat Anim Health 5:294–305

Liu FG, Liao IC (1999) Effect of feeding regimen on the food consumption, growth, and body composition in hybrid striped bass *Morone saxatilis* × *M. chrysops*. Fish Sci 65:513–519. https://doi.org/10.2331/fishsci.65.513

Lougheed M, Nelson B (2001) Hybrid striped bass production, markets and marketing. Michigan State University, East Lansing

Maynard LA, Loosli JK, Hintz HF, Warner RG (1979) Animal Nutrition. 7th edn. New York

Morris JE, Kohler CC, Mischke CC (1999) Pond culture of hybrid striped bass in the North Central Region

Nematipour GR, Brown ML, Gatlin DM (1992) Effects of dietary energy: protein ratio on growth characteristics and body composition of hybrid striped bass, *Morone chrysops* x *M. saxatilis*. Aquaculture 107:359–368

Peres H, Oliva-Teles A (1999) Effect of dietary lipid level on growth performance and feed utilization by European sea bass juveniles (*Dicentrarchus labrax*). Aquaculture 179:325–334

Saravanan S, Geurden I, Figueiredo-Silva AC, Kaushik SJ, Haidar MN, Verreth JA, Schrama JW (2012) Control of voluntary feed intake in fish: a role for dietary oxygen demand in Nile tilapia (*Oreochromis niloticus*) fed diets with different macronutrient profiles. Br J Nutr 108:1519–1529. https://doi.org/10.1017/s0007114511006842

Sealey WM, Gatlin DM (1999) Dietary vitamin C requirement of hybrid striped bass *Morone chrysops* × *M. saxatilis*. J World Aquac Soc 30:297–301

Sullivan JA, Reigh RC (1995) Apparent digestibility of selected feedstuffs in diets for hybrid striped bass (*Morone saxatilis* × *Morone chrysops*). Aquaculture 138:313–322

Thompson KR, Webster CD, Morgan AM, Grisby EJ (2000) Effects of different feeding frequencies on growth, body composition, and fillet composition of juvenile sunshine bass, *Morone chrysops* × *M. saxatilis* grown indoors. J Appl Aquacult 10:55–65. https://doi.org/10.1300/J028v10n02_05

Tucker Jr JW (2012) Marine fish culture. Springer Science & Business Media

Webster CD, Tiu LG, Tidwell JH, Van Wyk P, Howerton RD (1995) Effects of dietary protein and lipid levels on growth and body composition of sunshine bass (*Morone chrysops* × *M. saxatilis*) reared in cages. Aquaculture 131:291–301

Webster CD, Thompson KR, Morgan AM, Grisby EJ, Dasgupta S (2001) Feeding frequency affects growth, not fillet composition, of juvenile sunshine bass *Morone chrysops* × *M. saxatilis* grown in cages. J World Aquac Soc 32:79–88. https://doi.org/10.1111/j.1749-7345.2001.tb00925.x

2

Oxidative effect of L-carnitine on energy metabolism in diploid and triploid rainbow trout (*Oncorhynchus mykiss*): impact on metabolites

José Fernando M. Gonçalves · Mariana Hinzmann · Jorge Machado · Sandra Araújo · José Carlos Oliveira · Rodrigo O. A. Ozório

Abstract L-Carnitine plays a key role in the regulation of energy metabolism and growth in fish. The aim of this study was to determine the effects of dietary L-carnitine levels on carnitine homeostasis and energy metabolism in diploid and triploid trout (*Oncorhynchus mykiss*). Diploid and triploid trout (29.5 ± 0.6 and 31.8 ± 0.5 g, respectively) were fed with three diets supplemented with 15, 200 or 530 mg L-carnitine/kg, respectively, for 56 days. Compared to 15 mg L-carnitine, total carnitine content increased in the liver by 182% in diploid and by 154% in triploid trout fed 530 mg L-carnitine. In muscle, total carnitine content increased by 60 and 73% in diploid and by 34 and 75% in triploid with the increase in dietary L-carnitine levels. L-Carnitine caused no significant change in the plasma concentrations of metabolites like proteins (ammonia, urea and protein), lipid (triglycerides), carbohydrate (glucose, lactate) and enzyme activities (lipase, lactate, alanine transaminase, lactic acid dehydrogenase). Triploid trout had significantly lower plasma ammonia ($P = 0.003$), lipase ($P = 0.005$) and triiodothyronine (T_3; $P = 0.003$) levels than diploid trout. In conclusion, ploidy significantly affected the energy metabolism in rainbow trout, dietary L-carnitine levels altered the L-carnitine homeostasis, but not influence nutritional metabolism.

Keywords L-Carnitine · Carnitine metabolism · Plasma metabolites · Trout ploidy

Abbreviations

ABW	Average body weight
ALT	Alanine aminotransferase
ATP	Adenosine triphosphate
BW	Body weight
DGI	Daily growth index

J. F. M. Gonçalves · M. Hinzmann · J. Machado · S. Araújo · R. O. A. Ozório
Interdisciplinary Centre of Marine and Environmental Research, CIIMAR-UP, University of Porto, Terminal de Cruzeiros do Porto de Leixões, Av. General Norton de Matos s/n, 4450-208 Matosinhos, Portugal

J. F. M. Gonçalves · M. Hinzmann (✉) · J. Machado · S. Araújo
ICBAS-UP, Abel Salazar Institute of Biomedical Sciences, University of Porto, Rua Jorge Viterbo Ferreira n.° 228, 4050-313 Porto, Portugal
e-mail: mhinzmann@icbas.up.pt

J. C. Oliveira
CHO-Hospital Centre of Porto, Lg. Prof. Abel Salazar, 4099-001 Porto, Portugal

FBW	Final body weight
FCR	Feed conversion ratio
GLDH	Glutamate dehydrogenase
IBW	Initial body weight
IFCC	International Federation of Clinical Chemistry and Laboratory Medicine
LDH	Lactate dehydrogenase
VFI	Voluntary feed intake

Introduction

During the past three decades, aquaculture production has increased steadily, an average of 9% year, reaching a staggering 60 million tones (excluding aquatic plants and non-food products) in 2010 (FAO 2012). Fish meal and fish oil derived from wild catches are dominant ingredients in aqua feed. It is estimated, at the present annual growth rates, that fishmeal supplies will be entirely consumed by the aquaculture sector by 2025. Plant protein ingredients have increasingly been used in diets of aquaculture species (Glencross Booth and Allan 2007), including carnivore marine species (Oliva-Teles and Gonçalves 2001; Colburn et al. 2012), is crucial the reduction of aquaculture farming dependent on fish meal. The replacement of fish meal may trigger, however, specific deficiency symptoms, including L-carnitine deficiency. Fishmeal may contain 10–20 times more L-carnitine than plant-based feeds (Ozório et al. 2001).

L-Carnitine, a multi-physiological and bioactive vitamin-like nutrient, plays an important role in cellular nutritional metabolic status, including the regulation of fish energy metabolism and growth (Harpaz 2005; Mohseni et al. 2008). Fish fed high-fat diets supplemented with 600 mg L-carnitine showed a reduction in plasma glucose and lactate dehydrogenase (LDH) (Ozório et al. 2003) suggesting an inhibitory effect on pyruvate dehydrogenase complex (PDH), the key rate-limiting step in carbohydrate oxidation (Randle 1998; Rasmussen and Wolfe 1999) or enhancement of glucose flux toward lipogenesis. A decrease in LDH levels may also indicate an increase protection of L-carnitine against oxidative stress (Ronca et al. 1992).

According to Rasmussen and Wolfe (1999) growth is impaired when a dietary L-carnitine level is limited. In fish, an increase in energy demand is followed by the increase in lipid metabolism and body L-carnitine turnover (Ozório et al. 2001). Carnitine reserves are mainly replenished by the diet since carnitine biosynthesis is not sufficient, especially in very young animals (Harpaz 2005). Ozório et al. (2008) observed an improvement in survival, cell viability and ventilation frequency of turbot *Psetta maxima* L. juveniles fed L-carnitine supplements following acute exposure to toxic ammonia levels and to pathogen agent (*S. parauberis*). In case of fishmeal replacement, body L-carnitine reserves may become limited, thus dietary supplementation is required.

In salmonid production, sterile triploids have been increasingly used. After the onset of sexual maturation, triploid fish often grow faster and have better flesh quality than their diploid counterparts. In addition, triploid fish may display greater hypertrophic growth and reduced energy costs for gamete production (Thorgaard 1986; Qin et al. 1998; Oppedal et al. 2003; Maxime 2008). Moreover, its use in aquaculture may minimize the risks of genetic losses from wild populations (O'Flynn et al. 1997; Benfey 1999; Atkins and Benfey 2008).

The comparative results between diploid and triploid fish are often contradictory (Ihssen et al. 1990; Galbreath et al. 1994; Li et al. 2017). The general consensus is that the survival and growth are substantially lower in the early life stages in triploids compared to diploids (Thorgaard 1986; Sutterlin et al. 1987; Malison et al. 1993; Withler et al. 1995). In addition, triploids appear to be less resilient and less tolerant to poor water quality than their diploid counterpart (Benfey 1999; Benfey and Biron 2000; Altimiras et al. 2002; Segato et al. 2006; Atkins and Benfey 2008; Maxime 2008).

The aim of this study was to assess the effects of dietary L-carnitine levels on energy metabolism in diploid and triploid rainbow trout (*Oncorhynchus mykiss*).

Materials and methods

Fish and husbandry conditions

The current study is an extension of a published study (Ozório et al. 2012) in which the results of growth performance and body composition has been presented. The growth trial was carried out in accordance with European Directive 86/609/UE by a certified researchers under National and International animal welfare laws, according to the FELASA requirements (level C).

Two groups of 5-month-old rainbow trout were used, one of all-female diploid trout (initial weight 29.5 ± 0.6 g) and another of triploid trout (initial weight 31.8 ± 0.5 g). Fish were outbred siblings and had identical nutritional and husbandry history and were kept in standard culture conditions at a density of 15–20 kg/m^3. After an acclimatization period of 2 weeks, 612 rainbow trout were individually weighed and randomly distributed in 18 fiberglass tanks of 50 L. The water supply system consisted of a closed circuit (TMC-5000, TMC Iberia, Lisbon, Portugal), which included mechanical and biological filtration, a skimmer, an ultraviolet sterilization unit and a refrigerator. The water quality parameters such as temperature (16.9 ± 0.5 °C), pH (7.3 ± 0.1) dissolved oxygen (8.0 ± 0.4 mg/L), ammonia (0.6 ± 0.8 mg/L) and nitrite (1.2 ± 0.5 mg/L) were checked on a daily basis. Photoperiod (12 h L/12 h D) and light intensity (200 Lux) was kept constant.

Diets, sampling and experimental design

The experimental diets were formulated to contain three dietary L-carnitine levels (15, 200 or 530 mg/kg diet) (Table 1) and fed in triplicate to diploid and triploid trout groups for a period of 56 days. All diets were isoproteic (44% crude protein) and isocaloric (20 MJ/kg) and had identical formulation, except for L-carnitine content. Since the concentration of 200 mg L-carnitine per kilogram diet is the generally obtained in commercial aquafeed, the low L-carnitine level (15 mg/kg diet) was possible using several plant protein

Table 1 Formulation and chemical composition of the diets used in L-carnitine trials

	L-Carnitine (mg/kg)		
	15	200	530
Ingredients			
Fishmeal (60% crude protein)	190	190	190
Soymeal (65% crude protein)	164	164	164
Soya bran	145	145	145
Fish oil	145	145	145
Maize gluten	120	120	120
Wheat gluten	100	100	100
Wheat	100	100	100
Lecithin	30	30	30
Premix[a]	5	5	5
L-Carnitine (mg/kg diet)[b]	–	200	550
Chemical composition (analyzed)			
Dry matter	903	889	916
Crude protein	445	440	450
Crude lipid	210	200	213
Ash	62	58	62
Gross energy (MJ/kg diet)	19.9	19.9	19.9
L-Carnitine (mg/kg diet)[b]	15	200	530

[a]Premix was prepared with: vitamins: 774 mg vitamin A; 6.25 mg vitamin D; 20,000 mg vitamin E. Minerals: 501 mg $CuSO_4$; 15,000 mg $ZnSO_4$; 0.01 mg $MnSO_4$; 500 $CoSO_4$; 500 mg KI; 35 mg Na_2SeO_3 per kilogram of feed

[b]L-Carnitine was supplemented by the use of Carniking®

ingredients know to have very low levels or deficient in L-carnitine in replacement of animal ingredients, rich in L-carnitine. The 530 mg L-carnitine concentration was selected based on previous studies, which have showed that L-carnitine improved performance until a dietary level of 600 mg/kg, with best performance around 500 mg L-carnitine/kg diet (Ozório 2001; Ozório et al. 2001, 2012). Fish were fed twice daily (10:00 and 15:00 h), to apparent satiation.

At the end of the trial all fish were starved for 24 h prior to sampling. Before sampling, fish were euthanized by an overdose bath of 0.4 mL ethylene-glycol–monophenol–ether/L water (Merck, Darmstadt, Germany). A total of 12 fish per group were sampled for muscle, liver and blood plasma analyses. The samples collected were stored at − 80 °C, until further processing.

Growth performance

The results of growth performance parameters are presented solely in the text of the Result section. For detailed information about the growth performance, see Ozório et al. (2012). The voluntary feed intake (VFI), feed conversion ratio (FCR) and growth rate (daily growth index, DGI) was calculated as follows:

$$\text{VFI (\%BW/day)} = \text{feed } [(100)\,(\text{feed intake/average body weight (ABW)}/(t))]$$

$$\text{where ABW} = (\text{initial body} + \text{final body weight})/2.$$

$$\text{FCR (g/g)} = \text{total dry feed intake (g)/total wet weight gain (g)}.$$

$$\text{DGI (\% BW/day)} = \left((\text{FBW}^{1/3})\ \text{IBW}^{1/3}/t\right),$$

where FBW - final body weight, IBW - initial body weight and t - trial duration.

Carnitine analyses

Diets, liver, and muscle tissues were analyzed for the free- and acyl L-carnitine content. All samples, prior to L-carnitine analysis, were deproteinated with perchloric acid, subsequently neutralized with potassium hydroxide, centrifuged and the supernatant collected.

L-Carnitine was extracted by stepwise heating, ultrasonic treatment or extraction with various detergents. Additionally, alkaline hydrolysis was performed for the determination of L-carnitine esters. The L-carnitine determination in diets and tissues was performed by radiometric detection of free and L-carnitine esters. The assay is based on the reaction of free L-carnitine with acetyl-CoA catalyzed by carnitine acetyltransferase with the production of acetyl carnitine and coenzyme (Christiansen and Bremer 1978).

Plasma analyses

Blood was collected by caudal puncture with lithium-heparin coated syringes. Blood was immediately centrifuged (10 min at 1500 g at 4 °C) and plasma stored at − 80 °C until subsequent analysis. All plasma tests were performed in triplicate. Plasma cortisol (ng/mL) was quantified by Radioimmunoassay (RIA) technique, using commercial kits (Cortisol Coat-A-Count, DPC, Los Angeles, CA, USA) validated for salmonids.

The total ammonia concentration in plasma was determined using the method described by Bergmeyer and Beutler (1983) and plasma osmotic pressure (mOsm) was determined with an osmometer (model 15—Löser Messtechnik, Berlin, Germany). Plasma glucose concentrations were measured by the glucose oxidase method using a Beckman glucose analyzer 2 (Beckman Instruments, Brea, CA, USA).

Urea was quantified colourimetrically with the Glutamate dehydrogenase (GLDH) enzymatic method. Lipase was quantified colourimetrically with an enzymatic method using 1,2-0-dilauryl-rac-glycero-3-glutaric acid-(6′-methyl resorufin)-ester as subtracting. Alanine aminotransferase (ALT) was quantified by an enzymatic method in accordance with IFCC. Lactate dehydrogenase (LDH) was quantified by an enzymatic method in accordance with IFCC. Lactate was quantified colourimetrically with the lactate oxidase/peroxidase enzymatic method and protein was quantified colourimetrically with the biuret method. Triglyceride was quantified colourimetrically with the glycerol phosphate oxidase/peroxidase enzymatic method. All the methods were executed in a Roche Cobas Integra 800 Chemistry Analyser (Roche® Sistemas de Diagnóstico, Amadora, Portugal), with Roche reagents. Triiodothyronine (T3) and thyroxine (T4) were quantified by electrochemiluminescence immunoassays in a Roche Cobas e 411 Analyser (Roche® Sistemas de Diagnóstico, Amadora, Portugal), with Roche reagents.

Statistical analyses

All data were tested for normality using Kolmogorov–Smirnov and Levene tests and analyzed using two-way analyses of variance (ANOVA). The model included the effect of ploidy (two levels) and dietary L-carnitine (three levels) and their interaction. Data were analyzed using the General Linear Model procedure (SAS Institute Inc., Cary, NC, USA). The results (mean ± standard deviation) are reported as significant at $P < 0.05$.

Results

Voluntary feed intake (VFI: 1.77–1.99% BW/day) was significantly affected by ploidy and L-carnitine levels ($P < 0.001$). Feed conversion ratio (FCR: 0.95–1.07 g/g) was significantly affect by dietary L-carnitine levels ($P < 0.05$). Growth rate (2.52–2.65% BW/day) was not affected by diet or ploidy.

The concentration of L-carnitine fractions (free and acyl L-carnitine) significantly increased in muscle and liver (Fig. 1) with the dietary L-carnitine intake (Fig. 2). Muscle acyl L-carnitine content of fish fed 15, 200 and 530 mg L-carnitine was significantly higher ($P = 0.003$) in triploid (19.4, 26.6 and 33.6 mg/kg) than in diploid trout (15.6, 25.3 and 26.5 mg/kg). Free carnitine content was significantly higher than acyl L-carnitine content in both muscle and liver tissues.

Ploidy and L-carnitine supplementation did not affect plasma glucose, urea, protein, and triglyceride levels (Table 2). Diploid trout fed 200 and 530 mg L-carnitine had the highest (513 ± 67.78 U/L) and the lowest LDH levels (404.0 ± 10.0 U/L), respectively. Triploid fish had significantly lower plasma ammonia ($P = 0.003$), lipase ($P = 0.005$) and triiodothyronine (T_3, $P = 0.003$) levels than diploid trout. Trout fed 200 mg L-carnitine had the lowest ammonia (8.89 ± 0.64 μg/mL) and T_3 (1.84 ± 0.31 ng/mL) levels, while trout fed 530 mg L-carnitine had the lowest lipase level (2.80 ± 0.51 U/L). Plasma cortisol ranged from 4.8 ± 1.6 to 7.8 ± 3.3 ng/mL and was not significantly affected by the different treatments. Plasma osmolality was significantly affected by ploidy ($P = 0.008$), with triploid trout having lower osmolality (298.2 ± 7.1 mOsm) than diploid trout (303.5 ± 3.3 mOsm).

Discussion

L-Carnitine supplementation improved growth performance when fed to fish larvae and early juvenile fish species (Santulli and D'Amello 1986; Torreele et al. 1993; Chatzifotis et al. 1995; Keshavanath and Renuka 1998). In studies with larger animals, the positive effects of L-carnitine become less evident, as confirmed by Chatzifotis et al. (1996). In the current study using juveniles, with exception of FCR, L-carnitine supplementation did not improve growth performance. The ambiguity among many dose–response studies on L-carnitine may be caused by different L-carnitine concentrations, the husbandry conditions or the duration of the dietary treatment (Ozório et al. 2012).

The carnitine pool in animals is maintained by a combination of absorption of carnitine from supplemental sources, a modest rate of biosynthesis, and an efficient reabsorption of carnitine (Rebouche and Seim 1998). The body distribution of carnitine is determined by a series of systems that transport carnitine into cells against a concentration gradient. The liver plays a unique role in the whole-body carnitine homeostasis. The liver transport system has a higher affinity for butyrobetaine ($K_m \approx 0.5$ mM) (Christiansen and Bremer 1978; Kerner and Hoppel 1998) than for carnitine ($K_m \approx 5.0$ mM), this high affinity for butyrobetaine favors its transport into the liver and the efflux system acyl L-carnitine over L-carnitine (Hokland 1988).

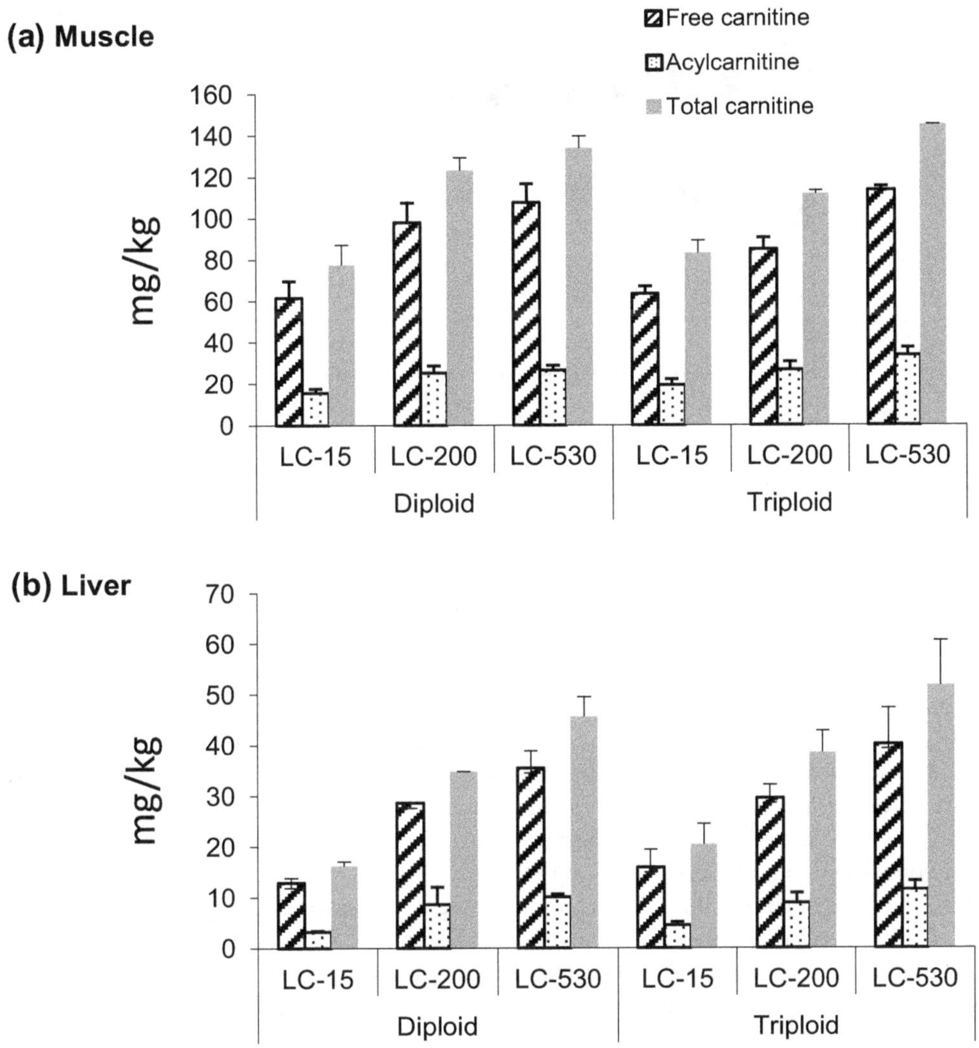

ANOVA (*P*-value)		Carnitine	Ploidy	Carnitine × Ploidy
Liver	Free-carnitine	<0.0001	0.08	0.93
	Acyl-carnitine	<0.0001	0.03	0.95
Muscle	Free-carnitine	<0.0001	0.48	0.53
	Acyl-carnitine	<0.0001	0.002	0.67

Fig. 1 Acyl L-carnitine, free L-carnitine and total carnitine content in muscle (**a**) and liver (**b**) of diploid and triploid rainbow trout fed the experimental diets

Free and acyl L-carnitine content increased with the dietary L-carnitine supplementation, whereas higher levels were observed in muscle than in liver, indicating a direct relationship between the dietary and body L-carnitine. The muscle may store a large amount of L-carnitine against a concentration gradient by an active transport process, which could be to ensure an immediate and efficient energy for the muscle during exercise and recovery. Ozório et al. (2001) working with African catfish *Clarias gariepinus* observed that the muscle L-carnitine concentration was about 12 times higher than plasma L-carnitine. The L-carnitine concentration gradient was even more pronounced between muscle and liver tissues (Ozório et al. 2001).

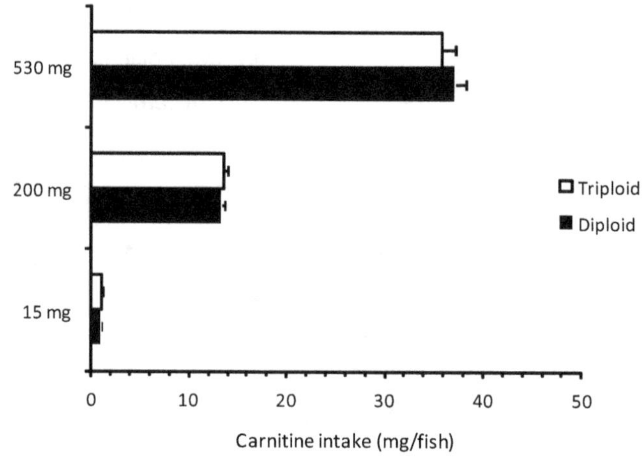

Fig. 2 Comparison of total L-carnitine intake (mg/fish) between diploid and triploid rainbow trout

The ability to take up and retain dietary carnitine was observed in other species, such as Atlantic salmon *Salmo salar* L. (Ji et al. 1996), red seabream *Pagrus major* (Chatzifotis and Takeuchi 1997), rabbits (Bell et al. 1992), piglets (Böhles et al. 1984), rats (Tsai et al. 1974), among other species (Li et al. 2017).

The increase in the acyl L-carnitine fraction in fish tissues may indicate an increase in lipid catabolism and that the diet quality and quantity may affect the body L-carnitine level. These results are consistent with the study described by Ozório et al. (2001), which demonstrated that the concentration of L-carnitine in muscle differs between fish fed with low and high L-carnitine levels. In fact, muscle L-carnitine levels increased moderately when African catfish were fed between 40 and 600 mg L-carnitine. Thereafter, a sharp increase in muscle L-carnitine level was observed. The latter suggests that L-carnitine requirement in catfish, under a normal physiological condition, would be below 600 mg/kg diet.

In the current study, the ploidy affected the acyl L-carnitine content in muscle, with higher values observed in triploid fish. This could indicate that triploid trout have higher L-carnitine requirement than diploid fish. In fact, triploid fish have a lower respiratory efficiency and are more susceptible to stress than diploid fish. Triploid fish have fewer red blood cells which probably affect the exchange of oxygen and decrease the ability to transport oxygen in the blood (Benfey 1999; Benfey and Biron 2000; Tiwary and Ray 2004) and, therefore, spend more energy to maintain internal balance than their diploid counterpart, possibly mobilizing more body L-carnitine for energy production. Plasma lipase was higher in diploid fish and is in agreement with the slight increase in plasma triglyceride. Lipase is the rate-limiting enzyme for the hydrolysis of triglyceride in plasma (Albalat et al. 2006).

Plasma cortisol levels were similar to the basal levels observed for this species (< 20 ng/mL) (Taylor et al. 2007), with no effect of diet and ploidy. In the current study, triploid trout had lower plasma ammonia concentrations, which may reflect on a higher capacity to excrete ammonia than diploid trout. Considering that ammonia excretion rates are linked to acid–base regulation (Kieffer and Tufts 1996), triploid rainbow trout may have a greater clearance capacity of metabolic protons. This would allow for faster recovery from acidosis and energy (ATP) depletion after exercise, which is often the case during handling stress. A faster recovery of the blood acid–base status may also reflect a faster recovery of ion and osmoregulatory imbalance, which was also shown by the significant differences in plasma osmolality between triploid and diploid trout. Plasma triiodothyronine hormone (T_3) was significantly higher in diploid than triploid trout, as well as the feed intake (VFI). Feed intake, quality and quantity, is known to influence thyroid hormone levels in rainbow trout (Leatherland et al. 1977; Himick et al. 1991). Thyroid hormones interact closely with the energy metabolism, with T_3 playing a role in the regulation of growth and nutrient partitioning in teleosts (Peter and Marchant 1995; Korytko and Cuttler 1997).

In the current study triploid trout had lower plasma LDH level. As mentioned earlier, triploid fish have lower oxygen transport ability, which may explain their lower activity when compared to diploid trout. According to Sullivan and Somero (1980), fish with different locomotors activity would have different LDH

Table 2 Comparison of plasma parameters (mean ± SD) of diploid and triploid rainbow trout fed with three different L-carnitine levels, *P* values of are indicated in the case of significant difference

	Diploid			Triploid			ANOVA (*P* value)		
	15	200	530	15	200	530	*P*	*C*	*P* × *C*
Glucose (mg/dL)	73.33 ± 1.25	75.67 ± 4.64	74.67 ± 2.49	73 ± 3.56	84 ± 10.23	70 ± 3.27	●	●	●
Urea (mg/dL)	2.33 ± 0.47	2.67 ± 0.94	2.67 ± 0.47	2.00 ± 0.82	2.00 ± 0.00	1.67 ± 0.47	●	●	●
Lipase (U/L)	3.17 ± 0.09	3.93 ± 0.09	3.83 ± 0.25	2.97 ± 0.5	2.90 ± 0.51	2.80 ± 0.51	0.005	●	●
Alanine aminotransferase (ALT; U/L)	3.33 ± 0.47	4.33 ± 1.25	2.67 ± 0.47	3.00 ± 0.82	2.67 ± 0.47	2.67 ± 0.47	●	●	●
Lactate dehydrogenase (LDH; U/L)	494.7 ± 62.9	513.0 ± 67.8	404.0 ± 10.0	416.0 ± 3.7	407.7 ± 48.6	412.0 ± 41.4	0.05	●	●
Lactate (mmol/L)	2.69 ± 0.18	3.20 ± 0.31	2.97 ± 0.35	3.28 ± 0.22	3.54 ± 0.03	2.71 ± 0.33	●	t	●
Protein (g/dL)	3.36 ± 0.16	3.69 ± 0.23	3.58 ± 0.06	3.40 ± 0.22	3.40 ± 0.30	3.19 ± 0.15	●	●	●
Triglyceride (mg/dL)	517.0 ± 73.3	565.5 ± 67.5	511.7 ± 37.9	393.0 ± 14.9	577.5 ± 9.5	467 ± 108.9	●	●	●
Triiodothyronine (T_3; ng/mL)	2.2 ± 0.41	2.17 ± 0.11	2.34 ± 0.46	1.42 ± 0.14	1.41 ± 0.29	1.84 ± 0.31	0.003	●	●
Thyroxine (T_4; µg/dL)	0.67 ± 0.06	0.62 ± 0.06	0.67 ± 0.05	0.66 ± 0.04	0.59 ± 0.02	0.62 ± 0.03	●	●	●
Ammonia (µg/mL)	12.26 ± 1.99	12.15 ± 1.54	12.33 ± 0.58	10.04 ± 1.78	8.89 ± 0.64	11.03 ± 0.92	0.003	●	●
Cortisol (ng/mL)	7.44 ± 2.32	5.08 ± 0.62	7.12 ± 3.04	7.82 ± 3.33	6.73 ± 2.30	6.62 ± 0.85	●	●	●
Osmolality (mOsm)	303.28 ± 4.07	298.19 ± 7.07	303.46 ± 3.28	290.81 ± 12.92	285.39 ± 13.52	287.36 ± 15.95	0.001	●	●

● not significant, *P* ploidy, *C* carnitine, *P* × *C* interaction effect

activity to compensate for the amount of lactate produced. Wardle (1978) reported lower lactate ion concentration in flatfish than in rainbow trout.

Conclusion

Although a multitude of questions concerning the regulation of dietary carnitine requirements remain to be elucidated, our study suggested that ploidy significantly affected the energy metabolism, whereas dietary L-carnitine levels altered the carnitine homeostasis, but did not influence energy metabolism in diploid and triploid rainbow trout.

Acknowledgements This research was partially supported by the European Regional Development Fund (ERDF) through the COMPETE—Operational Competitiveness Programme and national funds through FCT—Foundation for Science and Technology, under the project "PEst-C/MAR/LA0015/2011.

Author contributions All authors have contributed intellectually in this study. GJFM gave substantial contributions in the concept, design and interpretation of data, MJ was involved in revising the manuscript, HM and AS were involved in interpretation of data, drafting the manuscript and revising, OJC was more involved in acquisition of data, analysis and interpretation and OROA was involved in all phases of the study, supervising and giving intellectual feedback.

References

Albalat A, Sánchez-Gurmaches J, Gutiérrez J, Navarro I (2006) Regulation of lipoprotein lipase activity in rainbow trout (*Oncorhynchus mykiss*) tissues. Gen Comp Endocrinol 146:226–235

Altimiras J, Axelsson M, Claireaux G, Lefrançois C, Mercier C, Farrell AP (2002) Cardiorespiratory status of triploid brown trout during swimming at two acclimation temperatures. J Fish Biol 60:102–116

Atkins ME, Benfey TJ (2008) Effect of acclimation temperature on routine metabolic rate in triploid salmonids. Comp Biochem Physiol A Mol Integr Physiol 149:157–161

Bell FP, Vidmar TJ, Raymond TL (1992) L-Carnitine administration and withdrawal affect plasma and hepatic carnitine concentrations, plasma lipid and lipoprotein composition, and in vitro hepatic lipogenesis from labeled mevalonate and oleate in normal rabbits. J Nutr 122:959–966

Benfey TJ (1999) The physiology and behavior of triploid fishes. Rev Fish Sci 7:39–67

Benfey T, Biron M (2000) Acute stress response in triploid rainbow trout (*Oncorhynchus mykiss*) and brook trout (*Salvelinus fontinalis*). Aquaculture 184:167–176

Bergmeyer HU, Beutler HO (1983) Ammonia. Methods Enzymat Anal 8:454–461

Böhles H, Michalk D, Brandl U, Fekl W, Börresen HC, Stehr K (1984) The effect of L-carnitine-supplemented total parenteral nutrition on tissue amino acid concentrations in piglets. J Nutr 114:671–676

Chatzifotis S, Takeuchi T (1997) Effect of supplemental carnitine on body weight loss, proximate and lipid compositions and carnitine content of red sea bream (*Pagrus major*) during starvation. Aquaculture 158:129–140

Chatzifotis S, Takeuchi T, Seikai T (1995) The effect of dietary L-carnitine on growth performance and lipid composition in red sea bream fingerlings. Fish Sci 61:1004–1008

Chatzifotis S, Takeuchi T, Seikai T (1996) The effect of dietary carnitine supplementation on growth of red sea bream (*Pagrus major*) fingerlings at two levels of dietary lysine. Aquaculture 147:235–248

Christiansen RZ, Bremer J (1978) Acetylation of tris(hydroxymethyl)aminomethane (tris) and tris derivatives by carnitine acetyltransferase. FEBS Lett 86:99–102

Colburn HR, Walker AB, Breton TS, Stilwell JM, Sidor IF, Gannam AL, Berlinsky DL (2012) Partial replacement of fishmeal with soybean meal and soy protein concentrate in diets of atlantic cod. N Am J Aquacult 74:330–337

FAO (2012) The state of world fisheries and aquaculture. FAO, Rome. www.fao.org/docrep/016/i2727e/i2727e00.htm

Galbreath PF, St Jean W, Anderson V, Thorgaard GH (1994) Freshwater performance of all-female diploid and triploid atlantic salmon. Aquaculture 128:41–49

Glencross BD, Booth M, Allan GL (2007) A feed is only as good as its ingredients—a review of ingredient evaluation strategies for aquaculture feeds. Aquacult Nutr 13:17–34

Harpaz S (2005) L-Carnitine and its attributed functions in fish culture and nutrition—a review. Aquaculture 249:3–21

Himick BA, Higgs DA, Eales JG (1991) The acute effects of alteration in the dietary concentrations of carbohydrate, protein, and lipid on plasma t4, t3, and glucose levels in rainbow trout, *Oncorhynchus mykiss*. Gen Comp Endocrinol 82:451–458

Hokland BM (1988) Uptake, metabolism and release of carnitine and acylcarnitines in the perfused rat liver. Biochim Biophys Acta 961:234–241

Ihssen PE, McKay LR, McMillan I, Phillips RB (1990) Ploidy manipulation and gynogenesis in fishes: cytogenetic and fisheries applications. Trans Am Fish Soc 119:698–717

Ji H, Bradley TM, Tremblay GC (1996) Atlantic salmon (*Salmo salar*) fed L-carnitine exhibit altered intermediary metabolism and reduced tissue lipid, but no change in growth rate. J Nutr 126:1937–1950

Kerner J, Hoppel C (1998) Genetic disorders of carnitine metabolism and their nutritional management. Annu Rev Nutr 18:179–206

Keshavanath P, Renuka P (1998) Effect of dietary L-carnitine on growth and body composition of fingerling rohu, *Labeo rohita* (Hamilton). Aquacult Nutr 4:83–87

Kieffer JD, Tufts BL (1996) The influence of environmental temperature on the role of the rainbow trout gill in correcting the acid–base disturbance following exhaustive exercise. Physiol Zool 69:1301–1323

Korytko AI, Cuttler L (1997) Thyroid hormone and glucocorticoid regulation of pituitary growth hormone-releasing hormone receptor gene expression. J Endocrinol 152:R13–R17

Leatherland FJ, Cho CY, Slinger SJ (1977) Effects of diet, ambient temperature, and holding conditions on plasma thyroxine levels in rainbow trout *Salmo gairdneri*. J Fish Res Board Can 34:677–682

Li JM, Li LY, Qin X, Ning LJ, Lu DL, Li DL, Zhang ML, Wang X, Du ZY (2017) Systematic regulation of L-carnitine in nutritional metabolism in zebra-fish, Danio rerio. Sci Rep 7:40815

Malison JA, Procarione LS, Held JA, Kayes TB, Amundson CH (1993) The influence of triploidy and heat and hydrostatic pressure shocks on the growth and reproductive development of juvenile yellow perch (*Perca jlavescens*). Aquaculture 116:121–133

Maxime V (2008) The physiology of triploid fish: current knowledge and comparisons with diploid fish. Fish Fish 9:67–78

Mohseni M, Ozório ROA, Pourkazemi M, Bai SC (2008) Effects of dietary L-carnitine supplements on growth and body composition in beluga sturgeon (*Huso huso*) juveniles. J Appl Ichthyol 24:646–649

O'Flynn MF, McGeachy SA, Friars GW, Benfey TJ, Bailey JK (1997) Comparisons of cultured triploid and diploid atlantic salmon, *Salmo salar* L. ICES J Mar Sci 54:1160–1165

Oliva-Teles A, Gonçalves P (2001) Partial replacement of fishmeal by brewers yeast (*Saccharomyces cerevisiae*) in diets for sea bass (*Dicentrarchus labrax*) juveniles. Aquaculture 202:269–278

Oppedal F, Taranger GL, Hansen T (2003) Growth performance and sexual maturation in diploid and triploid atlantic salmon (*Salmo salar L.*) in seawater tanks exposed to continuous light or simulated natural photoperiod. Aquaculture 215:145–162

Ozório ROA (2001) Dietary L-carnitine and energy and lipid metabolism in African catfish (*Clarias gariepinus*) juveniles. 1 vols, p. 136. Wageningen: Ponsen and Looijen. (ISBN 90-5808-493-0), PhD Thesis

Ozório ROA, Uktoseja JL, Huisman EA, Verreth JA (2001) Changes in fatty acid concentrations in tissues of african catfish, *Clarias gariepinus* (Burchell), as a consequence of dietary carnitine, fat and lysine supplementation. Br J Nutr 86:623–636

Ozório ROA, Verreth JA, Aragão CR, Vermeulen CJ, Schrama JW, Verstegen MWA (2003) Dietary carnitine supplements increased lipid metabolism and decreased protein oxidation in african catfish (*Clarias gariepinus*) juveniles fed high fat levels. J Aquacult Trop 18:225–238

Ozório ROA, Gonçalves R, Dias S, Correia AD, Gonçalves J, Wilson J (2008) Protective effects of dietary L-carnitine on turbot, *Psetta maxima*, challenged with *S. parauberis* and sublethal ammonia concentration. In: Proceedings 13th international symposium on nutrition and feeding of fish, Florianópolis, Brazil. Aquacult Res 41(3): 168–168

Ozório ROA, Escorcio C, Bessa RJB, Ramos B, Gonçalves JFM (2012) Comparative effects of dietary L-carnitine supplementation on diploid and triploid rainbow trout (*Oncorhynchus mykiss*). Aquacult Nutr 18:189–201

Peter RE, Marchant TA (1995) The endocrinology of growth in carp and related species. Aquaculture 129:299–321

Qin JG, Fast AW, Ako H (1998) Growout performance of diploid and triploid chinese catfish *Clarias fuscus*. Aquaculture 166:247–258

Randle PJ (1998) Regulatory interactions between lipids and carbohydrates: the glucose fatty acid cycle after 35 years. Diabetes Metab Res Rev 14:263–283

Rasmussen BB, Wolfe RR (1999) Regulation of fatty acid oxidation in skeletal muscle. Annu Rev Nutr 19:463–484

Rebouche CJ, Seim H (1998) Carnitine metabolism and its regulation in microorganisms and mammals. Annu Rev Nutr 18:39–61

Ronca G, Ronca F, Yu G, Zucchi R, Bertelli A (1992) Protection of isolated perfused working rat heart from oxidative stress by exogenous L-propionyl carnitine. Drugs Exp Clin Res 18:475–480

Santulli A, D'Amello V (1986) The effects of carnitine on the growth of sea bass, *Dicentrarchus labrax L.* fry. J Fish Biol 28:81–86

Segato S, Bertotto D, Fasolato L, Francescon A, Barbaro A, Corato A (2006) Effects of triploidy on feed efficiency, morphometric indexes and chemical composition of shi drum, *Umbrina cirrosa* L. Aquacult Res 37:71–77

Sullivan KM, Somero GN (1980) Enzyme activities of fish skeletal muscle and brain as influenced by depth of occurrence and habits of feeding and locomotion. Mar Biol 60:91–99

Sutterlin AM, Holder J, Benfey TJ (1987) Early survival rates and subsequent morphological abnormalities in landlocked, anadromous and hybrid (landlocked × anadromous) diploid and triploid atlantic salmon. Aquaculture 64:157–164

Taylor JF, Needham MP, North BP, Morgan A, Thompson K, Migaud H (2007) The influence of ploidy on saltwater adaptation, acute stress response and immune function following seawater transfer in non-smolting rainbow trout. Gen Comp Endocrinol 152:314–325

Thorgaard GH (1986) Ploidy manipulation and performance. Aquaculture 57:57–64

Tiwary KB, Ray AK (2004) Alterations in air-sac and skeleton of triploid heteropneustes fossilis. J Fish Biol 64:268–272

Torreele E, van der Sluiszen A, Verreth J (1993) The effect of dietary L-carnitine on the growth performance in fingerlings of the african catfish (*Clarias gariepinus*) in relation to dietary lipid. Br J Nutr 69:289–299

Tsai CA, Romsos DR, Leveille GA (1974) Significance of dietary carnitine for growth and carnitine turnover in rats. J Nutr 104:782–792

Wardle CS (1978) Non-release of lactic acid from anaerobic swimming muscle of plaice Pleuronectes platessa L.: a stress reaction. J Exp Biol 77:141–155

Withler RE, Beacham TD, Solar II, Donaldson EM (1995) Freshwater growth, smolting, and marine survival and growth of diploid and triploid coho salmon (*Oncorhynchus kisutch*). Aquaculture 136:91–107

3

Effect of dietary *Sargassum wightii* and its fucoidan-rich extract on growth, immunity, disease resistance and antimicrobial peptide gene expression in *Labeo rohita*

Adnan Hussain Gora · Narottam Prasad Sahu · Sujata Sahoo · Saima Rehman · Showkat Ahmad Dar · Irshad Ahmad · Deepak Agarwal

Abstract This work aimed to appraise growth performance, immune responsiveness, bacterial disease resistance and mRNA expression of two antimicrobial peptides in *Labeo rohita* fingerlings fed diets supplemented with seaweed *Sargassum wightii* and its fucoidan-rich extract. Two hundred and twenty-five fingerlings were divided into five experimental groups in triplicates, each replicate having 15 fish in a 90-L tank. Different groups were fed with isonitrogenous (35% crude protein) and isoenergetic (1408.16 kJ/100 g) diets containing either 0% fucoidan-rich seaweed extract (FRSE), 1% FRSE, 2% FRSE, 3% seaweed powder and 6% seaweed powder for 60 days. Partial nucleotide sequences of the antimicrobial peptides, hepcidin and defensin, were characterized in *Labeo rohita* fingerlings, and gene expression in the different tissues of experimental fish was studied. The fish-fed FRSE-supplemented diets showed significant increase ($P < 0.05$) in the mRNA expression of antimicrobial peptides in liver, skin and intestine tissues. Bacterial challenge study using *Aeromonas hydrophila* showed higher survival in the FRSE-fed groups. Fish-fed FRSE-supplemented diets also showed better non-specific immune response (NBT reduction, serum lysozyme activity, serum albumin: globulin ratio and phagocytic activity) in pre-challenge and post-challenge periods. No significant difference was observed in the growth performance in the different experimental groups. The study indicates that although dietary FRSE is a potential immunostimulant in *Labeo rohita* fingerlings, it does not improve growth performance.

Keywords Antimicrobial peptides · Nutraceutical · Immunomodulation · Fish

---

A. H. Gora · S. Rehman
Division of Marine Biotechnology, Chennai Research Centre of ICAR-Central Marine Fisheries Research Institute, Chennai, India

A. H. Gora (✉)
Chennai Research Centre of ICAR-Central Marine Fisheries Research Institute, 75, Santhome High Road, R.A. Puram, Chennai 600028, India
e-mail: goraadnan@gmail.com

N. P. Sahu · S. Sahoo · S. A. Dar
Division of Fish Nutrition, Biochemistry and Physiology, Central Institute of Fisheries Education, Mumbai, India

I. Ahmad
Division of Aquaculture, Central Institute of Fisheries Education, Mumbai, India

D. Agarwal
Division of Fish Genetics and Biotechnology, Central Institute of Fisheries Education, Mumbai, India

Introduction

Disease outbreak is one of the main limiting factors for sustainable aquaculture production. Tropical countries suffer proportionally greater losses during disease outbreaks (Robar et al. 2010). Motile aeromonas septicaemia, caused by *Aeromonas hydrophila* (Plumb 1999), is one of the most deadly bacterial infections encountered in aquaculture practices. The disease is highly contagious among the fish and results in high mortalities within few days (Roberts 1993) with ulceration, exophthalmia and abdominal distension as symptoms (Camus et al. 1998). There is no commercially available vaccine that is effective against all strains of *A. hydrophila* (Poobalane et al. 2008). Hence, various chemotherapeutants may be used for prevention of this disease in aquaculture. However, their use does not come without the risk of antibiotic resistance in pathogenic bacteria (Huys et al. 2000), tissue antibiotic residues (Angulo et al. 2004), aquatic environment pollution (Hektoen et al. 1995) and transfer of resistance to human pathogenic bacteria (Sørum 2006). Therefore, novel strategies to control bacterial infections are needed (Defoirdt et al. 2011). These alternative measures involve the use of natural bioactive compounds that stimulate the immune system of fish. Antimicrobial peptides (AMPs) are components of the innate immune system of fish that provide protection against bacteria, fungi and virus (Cuesta et al. 2011; Zasloff 2002). Most antimicrobial peptides prevent the colonization of pathogens by lytic or ionophoric mechanism (Smith et al. 2010). Two important and well-known classes of antimicrobial peptides are defensins and hepcidins. Defensins and hepcidins characterized so far in fishes have shown strong antimicrobial activity against pathogens (Falco et al. 2008; Xu and Faisal 2010; Álvarez et al. 2014; Xu et al. 2014). Furthermore, these antimicrobial peptides exhibit increased expression at the transcriptomic level in fishes fed with dietary immunostimulants (Marel et al. 2012; Gong et al. 2014).

An immunostimulant is a chemical, drug, or stressor that enhances the immune response by interacting directly with cells of the system activating them. Non-specific immunostimulants act irrespective of antigenic specificity to augment immune response and stimulate components of the immune system. Seaweeds are cheap and easily available resource of bioactive compounds that can be used as immunostimulants in aquaculture (Deivasigamani and Subramanian 2016). *Sargassum*, a genus of brown seaweed (*Phaeophyceae*), contains a conglomerate of bioactive compounds such as steroids, alkaloids, phenolic compounds, saponins and flavonoids rendering it a number of different bioactive properties (Arunkumar et al. 2005; Zubia and Deslandes 2008). *Sargassum wightii* also contains fucoidan, a water-soluble, sulphated fibrous polysaccharide. Fucoidan is bestowed with many biological properties that include immunostimulation (Kitikiew et al. 2013; Prabu et al. 2016) disease resistance (Chotigeat et al. 2004; Immanuel et al. 2012), growth stimulation in shrimps (Traifalgar et al. 2010) and finfish (Tuller et al. 2014; Prabu et al. 2016). The effect of dietary *Sargassum wightii* and its fucoidan-rich extract on the immuno-physiological responses and growth performance of Indian major carps has not been studied in detail. In this context, the present work was designed with the objective to study the dietary effects of *Sargassum wightii* and its fucoidan-rich seaweed extract (FRSE) on the non-specific immune status, gene expression level of β-defensin 1 and hepcidin and survival of *L. rohita* fingerlings challenged with *A. hydrophila*.

Materials and methods

Preparation of fucoidan-rich seaweed extract

The brown seaweed, *Sargassum wightii*, required for the experiment was collected from Mandapam, Tamil Nadu (India) with the help of fishermen. Seaweeds were subsequently sorted out, washed thoroughly with water, shade dried and finally ground into fine powder. This powder was used for feed preparation and extraction of crude fucoidan. The crude FRSE was obtained from seaweed powder by the alcohol–hot water method of Prabu et al. (2016). To quantify the fucoidan content in crude FRSE, it was dialyzed in a cellulose tubing having molecular weight cut off of 5000. The L-Fucose content in the dialyzed sample was calculated according to Dubois et al. (1956) and from this the fucoidan content was obtained by multiplying the fucose values with 1.75.

Experimental fish and diets

Three hundred five fingerlings of *L. rohita* with an average weight of 8.852 ± 0.2 g were procured from R.A. Farm, Mumbai, India. Fish were acclimatized for 15 days in laboratory condition in 1000 L tanks at 30 °C, under continuous aeration before starting the experiment. From this stock, eighty fingerlings were used for determination LD_{50} prior to bacterial challenge study. Two hundred and twenty-five fingerlings were used for the feeding experiment and bacterial challenge study. Semipurified ingredients were used for feed formulation (Table 1). Five experimental diets (crude protein and lipid levels at 35 and 8%, respectively) were fed to different experimental groups. Continuous aeration was provided to all the tubs throughout the experimental period. The fish were fed to satiation daily at 10.00 h in the morning and 18.00 h in the evening under normal light regime.

Experimental design and in vivo study

Two hundred and twenty-five fingerlings were randomly transferred to plastic tanks of 100 L capacity with 15 fingerlings per tank in triplicates following a completely randomized design (CRD). The first group was maintained as control in which fish were given no dietary FRSE or seaweed. In second and third groups, fish were given 1 and 2% dietary FRSE, respectively. In the fourth and fifth groups, the fish were provided finely ground seaweed powder at the levels of 3 and 6% of the diet, respectively. After 60 days of feeding trial, five fish from each replicate was selected for blood collection (pre-challenge) by caudal vein puncture using a 1-ml 26G syringe. For analysis, fresh blood was used. For obtaining the serum, part of the blood was allowed to stand for 4 h and then centrifuged at 1006×*g* for 5 min. The straw color serum was pipetted out as supernatant and stored at − 80 °C. These fish were killed and liver, gill, spleen, kidney, intestine and skin tissues were

Table 1 Composition of experimental diets (g%)

Ingredients	Control (*C*)	1% fucoidan (T1F)	2% fucoidan (T2F)	3% dry seaweed (T3S)	6% dry seaweed (T4S)
Casein	30	30	30	30	30
Gelatin	9.5	9.5	9.5	9.5	9.5
Starch	17	17	17	17	17
Dextrin	20	20	20	20	20
Cellulose	11	10	9	8	5
Cod liver oil	4	4	4	4	4
Sunflower oil	4	4	4	4	4
Vit-min premix	2	2	2	2	2
Carboxymethyl cellulose	2	2	2	2	2
Choline chloride	0.4	0.4	0.4	0.4	0.4
Butylated hydroxy toluene	0.1	0.1	0.1	0.1	0.1
FRSE/seaweed powder	0	1	2	3	6
Total	100	100	100	100	100
Proximate analysis (g%)					
Moisture	7.6 ± 0.30	7.2 ± 0.19	8.0 ± 0.10	7.7 ± 0.10	7.9 ± 0.15
Crude protein	35.7 ± 0.12	34.5 ± 0.03	34.9 ± 0.08	35.0 ± 0.1	35.6 ± 0.15
Crude lipid	7.9 ± 0.03	8.1 ± 0.12	7.8 ± 0.27	7.7 ± 0.02	8.0 ± 0.19
Crude fiber	13.1 ± 0.2	12.8 ± 0.56	13.5 ± 0.34	12.6 ± 1.8	13.6 ± 2.1
Ash	9.3 ± 0.7	9.2 ± 0.2	9.6 ± 0.34	9.8 ± 0.23	10.3 ± 1.2

Composition of vitamin mineral mix (PREMIX PLUS) (quantity/2.5 kg): vitamin A, 5,500,000 IU; vitamin D3, 1,100,000 IU; vitamin B2, 2000 mg; vitamin E, 750 mg; vitamin K, 1000 mg; vitamin B6, 1000 mg; vitamin B12, 6 μg; calcium pantothenate, 2500 mg; nicotinamide, 10 g; choline chloride, 150 g; Mn, 27,000 mg; I, 1000 mg; Fe, 7500 mg; Zn, 5000 mg; Cu, 2000 mg; Co, 450 L-lysine, 10 g; DL-methionine, 10 g; selenium, 50 ppm

taken for mRNA expression studies. All the sampled tissues were kept in RNA later and stored in − 80 °C for RNA extraction.

Median lethal dose (LD_{50})

For determination of (LD_{50}) value, 80 fishes from the same stock as the experimental fish were selected which were being reared in similar physicochemical conditions and were fed the experimental control diet (Table 1). A rough idea of the doses to be injected was obtained from the study of Sahu et al. (2011). The bacterial culture was obtained as glycerol and reactivated by immersing in Müeller–Hinton Broth and incubation for 24 h at 30 °C. The bacteria were plated on Müeller–Hinton Agar by the spread-plate technique to obtain single colony. A single colony was taken and again regrown in Müeller–Hinton Broth for 24 h at 30 °C. The culture broth was centrifuged at 1006×*g* for 10 min. The supernatant was discarded and the pellets were resuspended in sterile phosphate buffer saline (PBS, pH 7.4) and the OD of the solution was adjusted to 0.50–0.48 nm. This corresponded to a count of 1.79×10^7, confirmed by total plate count. Then the bacterial cells were diluted to get different concentrations required for the experiment. Eighty fish were distributed randomly into eight groups. The fish of seven groups were injected intraperitoneally with 0.1 mL/g of American Type Culture Collection (ATCC) 7966 *Aeromonas hydrophila* bacterial cell suspension at concentrations of 1.1×10^1, 1.1×10^2, 1.1×10^3, 1.1×10^4, 1.1×10^5, 1.1×10^6 and 1.1×10^7 cfu/mL, respectively. The control group was injected 0.1 mL/g of PBS. The fish were maintained for another 96 h and were observed regularly for any overt signs of disease including behavioral abnormalities and mortality. In case of mortality, the causative agent was confirmed by re-isolating *A. hydrophila* from the moribund fish. LD_{50} value was calculated as per the trimmed Spearman–Karber method (Hamilton et al. 1977).

Bacterial challenge study

Following the feeding trial, 30 fish from each treatment group (10 from each replicate) were anesthetized using 2-phenoxy ethanol and subjected to intraperitoneal injection 0.1 mL/g of PBS containing *A. hydrophila* (ATCC 7966) having 1.79×10^4 cfu/mL. The fish were maintained for another 10 days and were observed regularly for any overt signs of disease including behavioral abnormalities and mortality. In case of mortality, the causative agent was confirmed by re-isolating *A. hydrophila* from the moribund fish. Blood was collected after challenge of *A. hydrophila* (post-challenge) for immunological assays.

Growth study

On the 1st and 60th day of feeding trial, weight measurement was taken using an electronic balance (Citizen, India). The growth performance of fish was evaluated in terms of percentage weight gain (%), feed conversion ratio (FCR), feed efficiency ratio (FER), specific growth rate (SGR), and protein efficiency ratio (PER) using the following formulae:

$$\text{Weight gain (\%)} = [(\text{FW} - \text{IW}) \times 100]/\text{IW},$$
$$\text{FCR} = \text{Feed given (DW)}/\text{body weight gain (WW)},$$
$$\text{SGR (\%)} = \{[\ln(\text{FW}) - \ln(\text{IW})]/N\} \times 100,$$
$$\text{PER} = \text{body weight gain (WW)}/\text{crude protein fed (DW)},$$

(where FW is the final weight, IW is the initial weight, DW is the dry weight, WW is the wet weight gain, ln is the natural log, and *N* is the number of cultural days).

RNA extraction and preparation of cDNA

Total *RNA* from the tissues was extracted using Trizol™ reagent (Invitrogen Waltham, USA). *DNaseI* treatment was given to extract *RNA* using DNaseI, RNase-free kit (Thermo scientific, Waltham, MA, USA) to remove any co-purified genomic *DNA*. The DNase I treated RNA isolated from different tissues of *L. rohita* was analyzed by agarose gel electrophoresis and displayed three distinct bands. The ratio of 28S:18S RNA was similar to 2:1, which indicated that total RNA had good integrity. The OD 260/280 value was found to be 2.0, which confirmed that RNA was highly pure and met the requirements for qRT-PCR. *cDNA* was synthesized from total *RNA* (5 μg) using RevertAid^{TM} reverse transcriptase First Strand cDNA Synthesis kit (Thermo scientific, USA) as per the manufacturer's instructions. The OD 260/280 value was found to be 1.8, which confirmed that cDNA was highly pure and met the requirements for qPCR. Successful amplification of house keeping genes using cDNA as the template further confirmed the quality of cDNA.

Molecular cloning of β-defensin 1 and hepcidin genes

Primers for β-defensin 1 and hepcidin (Table 2) were designed based on highly conserved regions of β-defensin 1 and hepcidin gene from various teleost species available in NCBI using Gene runner software v3.5 (http://www.generunner.net/). PCR amplification was performed in a total volume of 20 μL containing 2.5 μL Taq buffer (10×), 0.5 μL dNTP mix (10 mM), 2.5 μL $MgCl_2$ (25 mM), 1 μL each of sense and antisense primers (10 pmol), 1 μL of *cDNA*, 0.25 μL of TaqDNA polymerase (5 U/μL) and 12.25 μL of nuclease-free water. Amplification was performed in Takara PCR System (Takara, Japan). The PCR conditions included an initial denaturation at 94 °C for 4 min, followed by 35 cycles of 94 °C for 30 s, 55 °C for 45 s (β-defensin 1) or 56 °C for 45 s (hepcidin) and 72 °C for 1 min and a final extension of 10 min at 72 °C.

The PCR products were separated on 1.5% agarose gels, the desired DNA fragments were sliced carefully and eluted using QIAquick Gel Extraction Kit (Qiagen) following manufacturer's instructions. The desired DNA fragments were cloned into pTZ57R/T vector and transformed into *E. coli* DH5α strain using InsTAclone^{TM} PCR Cloning Kit (Fermentas, USA) following manufacturer's instructions. After confirmation of positive clones by colony PCR, plasmid DNA was isolated from positive colonies using QIAprep Spin Miniprep Kit (Qiagen, USA), following manufacturer's instructions. The plasmid was quantified using Nanodrop spectrophotometer and visualized on 1% agarose gel. The samples were sequenced using universal M13 primers and Sanger's sequencing method. The sequences obtained were confirmed using NCBI nBLAST homology tool (http://www.ncbi.nlm.nih.gov/blast).

Non-specific immune parameters

The lysozyme activity level was measured using the turbidimetric assay (Sankaran and Gurnani 1972). In the 96-well U-bottom microtitre plate, 150 μL of *Micrococcus luteus* suspension in phosphate buffer (A450 = 0.5–0.7) was taken and 15 μL of serum samples was added. The absorbance was taken immediately at 450 nm in ELISA reader (BioTek Power Wave 340, India) and the plate was incubated at 25 °C for 1 h and final absorbance was taken. This absorbance was compared with standard hen egg white lysozyme of a known activity following the same procedure as above. The activity was expressed as U/min/mg protein. The respiratory burst activity of neutrophils was measured by nitroblue tetrazolium (NBT) assay (Secombes 1990). Fifty microliter of blood was placed into the wells of flat bottom microtitre plates and incubated at 37 °C for

Table 2 List of primers used for RT-PCR and qPCR expression of defensin and hepcidin genes in *Labeo rohita*

Primer name	Primer sequence 5′–3′
Carp defensin F	ACTTGTCTTGCTTGTCCT
Carp defensin R	ACAGCCTAAGGGTCC
Carp hepcidin F	TTCTCACGTGTGGCTCTCG
Carp hepcidin R	CAGAATTTGCAGCAGTA
Defensin qPCR F	TCTTGCTTGTCCTTGTCGTCTTAG
Defensin qPCR R	CCTTGTCTGCAAACTCCAGTGA
Hepcidin qPCR F	CGCTTGCGTCCTGAACAATG
Hepcidin qPCR F	TTCTCAACGTGTGGCTCTCG
Beta-actin F	GCCGAGAGGGAAATTGTCCGTGAC
Beta-acin R	TTGCCAATGGTGATGACCTGTCCG

1 h to facilitate adhesion of cells. Then the supernatant was removed and the loaded wells were washed thrice in phosphate buffer saline (PBS). After washing, 50 μl of 0.2% NBT was added and was incubated for 1 h. The cell was then fixed with 100% methanol for 2–3 min and was again washed thrice with 30% methanol. The plates were then air dried. Sixty microlitres of 2 N potassium hydroxide and 70 μL dimethyl sulphoxide were added into each well to dissolve the formazan blue precipitate formed. The OD of the turquoise blue colored solution was then read in ELISA reader (BioTek Power Wave 340, India) at 620 nm. Serum protein was estimated by biuret and BCG dye binding method (Reinhold 1953) using total protein kit (Merck, India). Albumin was estimated by bromocresol green binding method (Doumas et al. 1971) using albumin kit (Merck, India). Globulin was calculated by subtracting albumin values from total serum protein values. A:G ratio was calculated by dividing albumin values by globulin values. Phagocytic activity was determined following the method of Anderson and Siwicki (1995). A sample of 0.1 mL of serum was taken in wells of microtiter plate and 0.1 mL of *Staphylococcus aureus* (1×10^7 cells) suspended in phosphate buffer saline (pH 7.2), was added and shaken sufficiently. This bacterial serum mixture was incubated at room temperature for 20 min. From this mixture, 5 μL was taken on a clean glass slide and a smear was prepared. The smear was air dried and fixed with 95% of ethanol for 5 min and air dried once again. The smear was stained with Giemsa stain for 10 min. A total of two hundred haemocytes from each smear were observed under the light microscope and the number of phagocytizing cells and bacteria engulfed by the phagocytes were counted. Phagocytic activity was expressed using the formula:

$$\text{Phagocytic activity} = (\text{Number of phagocytizing cells}/\text{Total number of phagocyte cells counted}) \times 100.$$

Gene expression study using qPCR

Real-time PCR was conducted to determine the mRNA expression levels of β-defensin 1 and hepcidin. For gene expression studies, tissue samples from three fish from each replicate were selected and pooled. Investigation of gene expression was done using qRT-PCR gene-specific primers with β-actin gene used as an internal control. Real-time PCR amplifications were carried out in LightCycler® Real-time PCR detection system (Roche, USA). The 25 μL reaction mix volume was prepared containing 12.5 μL of Maxima™ SYBR Green qPCR master mix (Thermo scientific, USA), 0.5 μL of (0.3 pM) each gene-specific primer and 2 μL (20 ng) of cDNA and 9.5 μL of nuclease-free water. The default thermal profile was used for PCR amplification and it consisted of initial denaturation at 95 °C for 10 min, followed by 40 cycles of denaturation at 95 °C for 15 s and annealing and extension at 60 °C for 1 min. Cycle threshold (C_t) values were obtained from the exponential phase of PCR amplification, and the expressions of genes were normalized to the expression of β-actin generating a ΔC_t value [(C_t of target gene) − (C_t of β-actin)]. The relative expression level of the target gene in the treatment group to target gene in the control group was determined using $2^{-\Delta\Delta C_t}$ method proposed by Livak and Schmittgen (2001), which gives the value that corresponds to the *n*-fold difference relative to the control.

Calculation of relative percentage survival

Relative percentage of survival (RPS) versus control was calculated at the end of 14 days of infection using the following formula (Amend 1981):

$$\text{RPS} = 1 - (S_T/S_C) \times 100,$$

where S_T is the percentage of surviving fish in treatment group and S_C is the percentage of surviving fish in control group.

Statistical analysis

The results of immunological parameters, growth performance and gene expression studies obtained from treatments were tested for statistical significance using one-way analysis of variance (ANOVA) followed by

Duncan's new multiple range tests using the statistical package, SPSS 16.0. $P < 0.05$ was considered as statistically significant.

Results

Quantification of fucoidan

The L-fucose content of the dialyzed extract was 32 mg/g dried seaweed powder. This was multiplied by a factor of 1.75 to get the total fucoidan content of 5.6 g/100 g dried seaweed powder.

Growth parameters

There was no significant difference in the percentage weight gain, specific growth rate, protein efficiency ratio and feed conversion ratio among the different treatment groups (Table 3).

Non-specific immune parameters

In the pre-challenge period, the 2% FRSE (T2F) group revealed a significantly higher serum lysozyme activity (Fig. 1) compared to other experimental groups, which showed no significant difference among them. The same trend was followed in the post-challenge period. In the pre-challenge study, the control group exhibited the least NBT activity, which was statistically similar to the groups T3S and T4S (Fig. 2). The highest and significantly different activity was found in the T2F group. In the post-challenge study, the highest was found in the T2F group. All other groups showed no significant difference in the NBT activity. In both pre- and post-challenge periods the lowest A:G ratio was observed in the T2F group, whereas the highest A:G ratio was observed in the control group (Table 4). In the pre-challenge period, the highest phagocytic activity was found in the T2F group, whereas the least was found in the control group. The same trend was also followed in the post-challenge period (Fig. 3).

Sequence analysis partial β-defensin 1 and hepcidin genes

The sequence for β-defensin 1 received was 148 bp long and the sequence of hepcidin was 269 bp long. The nBLAST search in NCBI GenBank database with *L. rohita* partial β-defensin-1 showed sequence similarity of 100% with *Cyprinus carpio* beta defensin-like protein (JF343439.1) and *Megalobrama amblycephala* β-defensin 1 mRNA (KC351182.1) followed by 98% similarity with *Danio rerio* defensin, beta-like 1 (NM_001081553.1). *L. rohita* partial hepcidin mRNA sequence showed sequence similarity of 98% with *Cyprinus carpio* hepcidin (JX855261.1) and *Schizothorax richardsonii* hepcidin-like antimicrobial peptide mRNA (KC894741.1) followed by 96% similarity with *Puntius sarana* mRNA for hepcidin (FN429020.1). *L. rohita* partial hepcidin and defensin gene sequence was submitted to NCBI GenBank vide accession number KM 040782 and LC 259205, respectively.

Table 3 Growth parameters of different experimental groups fed different experimental diets at the end of the experiment

Treatments	Wt. gain %	SGR	FCR	PER
Control	43.68 ± 1.37	0.26 ± 0.01	2.75 ± 0.08	0.96 ± 0.02
T1F	46.39 ± 0.68	0.27 ± 0.03	2.58 ± 0.04	0.96 ± 0.01
T2F	41.89 ± 2.04	0.25 ± 0.01	2.87 ± 0.14	1.00 ± 0.05
T3S	45.10 ± 1.02	0.26 ± 0.001	2.66 ± 0.06	0.93 ± 0.02
T4S	45.15 ± 2.00	0.26 ± 0.01	2.66 ± 0.12	0.93 ± 0.04

Mean values (n = 3) in the same column with no superscript do not differ significantly ($P < 0.05$). Data were expressed as mean ± SE

Wt. gain % weight gain percentage, *SGR* specific growth rate, *FCR* feed conversion ratio, *PER* protein efficiency ratio

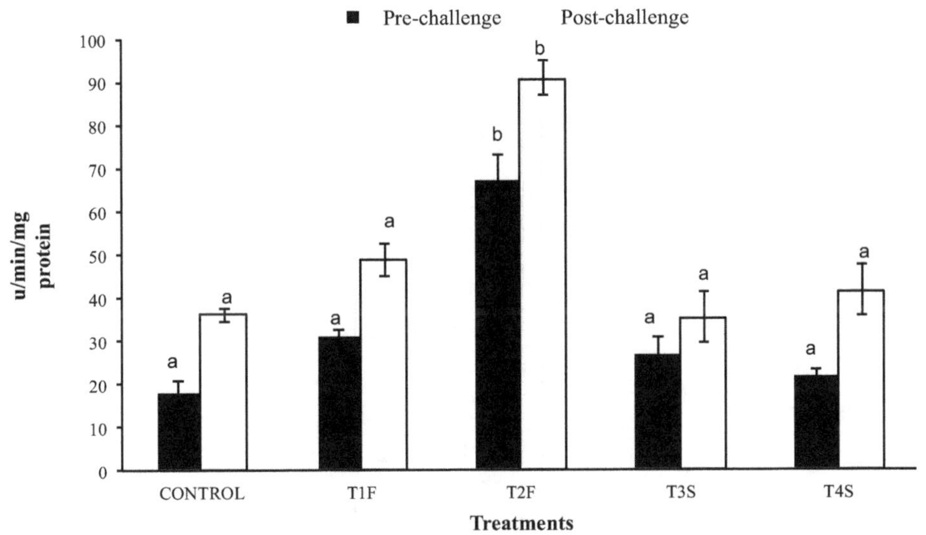

Fig. 1 Pre- and post-challenge serum lysozyme activities of different experimental groups. Mean values (n = 3) in the experimental group with different superscripts, (a, b) differ significantly ($P < 0.05$). Data were expressed as mean ± SE. Control—0% FRSE, T1F—1% FRSE, T2F—2% FRSE, T3S—3% seaweed powder, T4S—6% seaweed powder

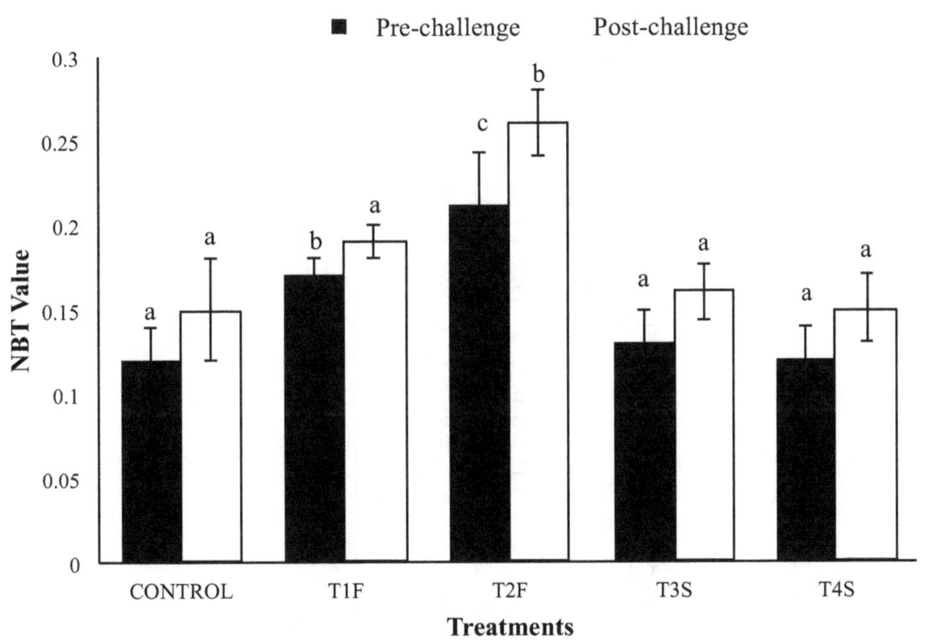

Fig. 2 Pre- and post-challenge NBT values of different experimental groups. Mean values (n = 3) in the experimental group with different superscripts (a, b, c) differ significantly ($P < 0.05$). Data were expressed as mean ± SE. Control—0% FRSE; T1F—1% FRSE; T2F—2% FRSE; T3S—3% seaweed powder; T4S—6% seaweed powder

Tissue distribution and real-time PCR analysis of the expression levels of β-defensin 1 and hepcidin

There was a variation in the tissue level expression of β-defensin 1 and hepcidin in *Labeo rohita*. Defensin mRNA expression was observed in gill, skin, liver and intestine whereas hepcidin was expressed in liver, kidney, spleen and intestine (Fig. 4). On the 60th day of feeding of experimental diets, the skin tissue of T2F group showed a 7.9-fold increase ($P > 0.05$) in β-defensin 1 expression compared to the control group (Fig. 5). The expression in the liver was increased 2.1-fold in the T2F group as compared to the control group. A slight but significant ($P > 0.05$) increase of 1.8-fold was found in the β-defensin 1 expression of the

Table 4 Serum total protein, albumin, globulin and A:G ratio of different experimental groups

	Pre-challenge				Post-challenge			
	Total protein	Albumin	Globulin	A:G Ratio	Total protein	Albumin	Globulin	A:G
Control	$1.82^{a} \pm 0.01$	$1.21^{b} \pm 0.01$	$0.61^{a} \pm 0.01$	$1.97^{d} \pm 0.03$	$1.63^{a} \pm 0.03$	$1.23^{a} \pm 0.01$	$0.40^{a} \pm 0.020$	$3.03^{a} \pm 0.12$
T1F	$2.38^{c} \pm 0.02$	$1.16^{ab} \pm 0.01$	$1.22^{c} \pm 0.03$	$0.95^{b} \pm 0.020$	$2.06^{b} \pm 0.71$	$1.18^{ab} \pm 0.01$	$0.88^{ab} \pm 0.71$	$1.33^{b} \pm 0.34$
T2F	$2.58^{d} \pm 0.04$	$0.99^{a} \pm 0.01$	$1.58^{d} \pm 0.03$	$0.62^{a} \pm 0.013$	$2.38^{ab} \pm 0.04$	$1.13^{ab} \pm 0.03$	$1.24^{ab} \pm 0.04$	$0.91^{b} \pm 0.051$
T3S	$1.66^{a} \pm 0.08$	$0.99^{a} \pm 0.06$	$0.67^{a} \pm 0.02$	$1.46^{c} \pm 0.05$	$1.36^{a} \pm 0.10$	$0.99^{b} \pm 0.07$	$0.36^{a} \pm 0.06$	$2.92^{a} \pm 0.57$
T4S	$2.13^{b} \pm 0.08$	$1.12^{ab} \pm 0.01$	$1.00^{b} \pm 0.05$	$1.13^{b} \pm 0.15$	$1.8^{ab} \pm 0.11$	$1.13^{ab} \pm 0.12$	$0.66^{ab} \pm 0.08$	$1.79^{b} \pm 0.40$

Data were expressed as mean ± SE

Mean values ($n = 3$) in the same column with different superscript (a, b, c, d, ab) differ significantly ($P < 0.05$)

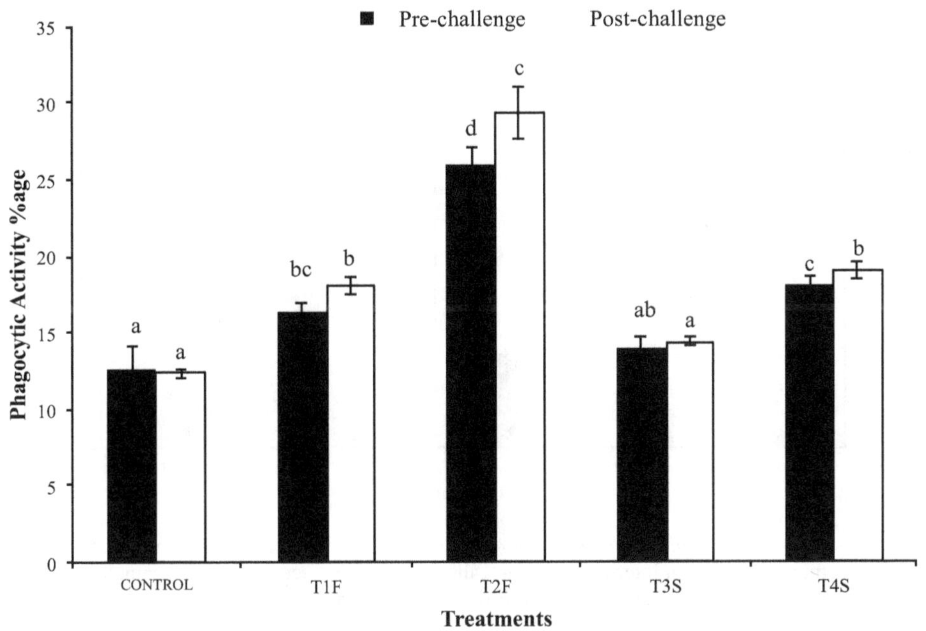

Fig. 3 Pre- and post-challenge phagocytic activity of different experimental groups. Mean values with different superscript (a, b, c, d, ab) differ significantly ($P < 0.05$). Data were expressed as mean ± SE, $n = 3$

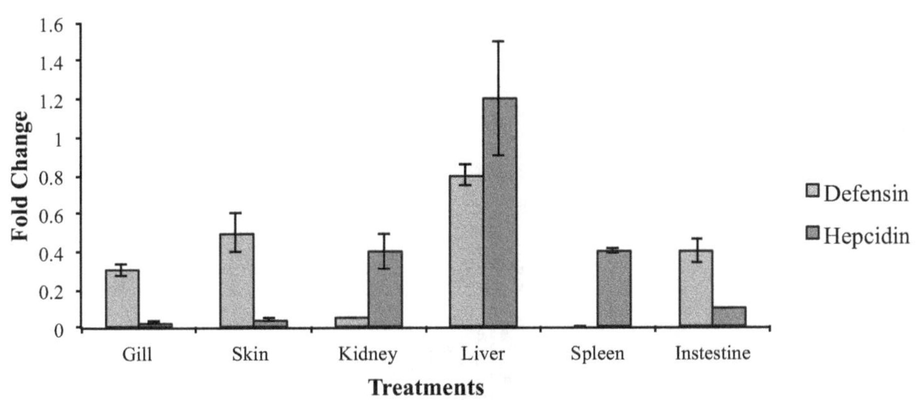

Fig. 4 Distribution of mRNA expression of antimicrobial peptides in different tissues of juvenile *Labeo rohita*. The relative expression of defensin and hepcidin normalized with beta-actin in samples was calculated using comparative C_t method. The fold change was compared with beta-actin of muscle tissue in which the copy number was considered as onefold. Data for real-time PCR were expressed as mean ± SE ($n = 3$)

intestine, whereas no significant increase was found in the gill. The crude fucoidan extract fed at dietary level of 1 and 2% showed a 3.4- and 6.4-fold higher expression of hepcidin in the liver (Fig. 6). Compared to the control group, the expression of hepcidin in the spleen was slightly, but significantly, higher in all the other groups.

LD_{50} value and challenge study

The LD_{50} value obtained after 5 days of observation period for the viable cells of *A. hydrophila* (ATCC 7986) injected to *Labeo rohita* fingerlings was found to be 1.79×10^4 cfu/mL. After injecting 0.1 mL/g of the same concentration bacteria in the experimental fish, the first mortality was observed at 11 h. At the end of the 14 days challenge period, the highest relative percentage of survival was found in the T2F group, fed 2%

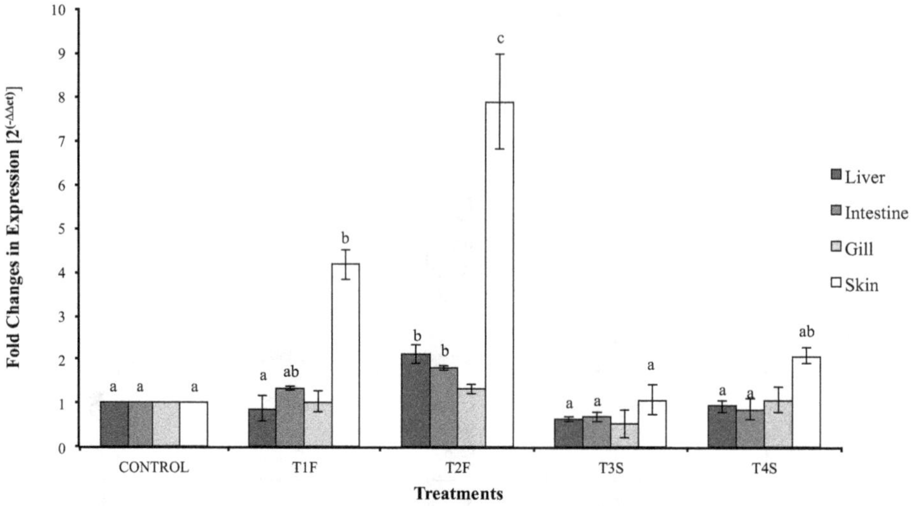

Fig. 5 Relative expression of *Labeo rohita* defensin mRNA in the liver, intestine, gill and skin after 60 days of experimental feeding regimen. Beta-actin was used as an internal control to calibrate the cDNA template for all samples. The data are presented as a fold increase of normalized mRNA copies in tissues of fish fed with different experimental diets to normalized mRNA copies in the tissues of fish fed with control diet. Data were expressed as mean ± SE, $n = 3$. Mean values of the same tissue with different superscripts (a, b, ab and c) differ significantly ($P < 0.05$). Mean values of the same tissue with no superscript do not differ significantly ($P < 0.05$)

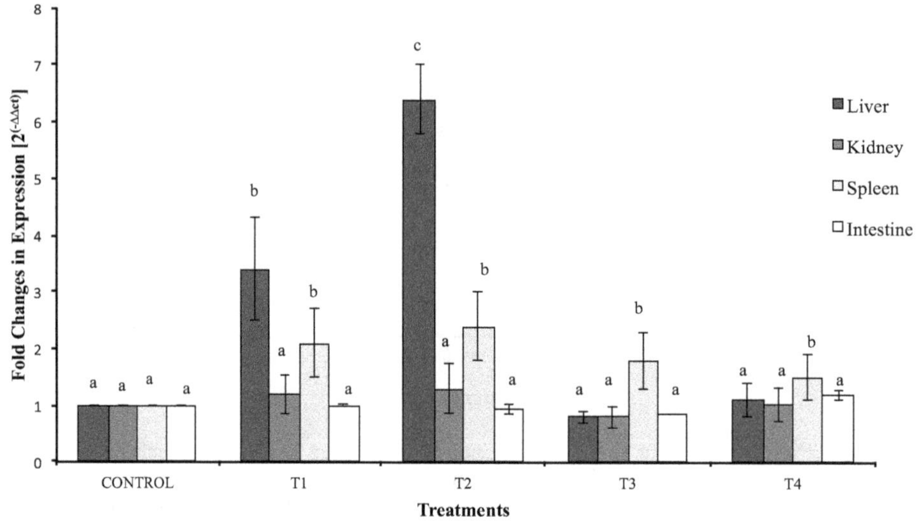

Fig. 6 Relative expression of *Labeo rohita* hepcidin mRNA in the different tissues after 60 days of experimental feeding regimen. Beta-actin was used as an internal control to calibrate the cDNA template for all samples. The data are presented as a fold increase of normalized mRNA copies in liver of fish fed with different experimental diets to normalized mRNA copies in the liver of fish fed with control diet. Data are expressed as mean ± SE, $n = 3$. Mean values with different superscripts (a, b and c) differ significantly ($P < 0.05$). Mean values of the same tissue with no superscript do not differ significantly

dietary fucoidan (Table 5). The groups fed 3 and 6% seaweed powder, T3S and T4S, exhibited the lowest relative percentage survival.

Discussion

The culture conditions, which deviate from the natural habitat, induce stress in fish (Barton and Iwama 1991). Stress increases susceptibility to diseases, possibly due to chronic elevated cortisol levels, causing

Table 5 Percentage mortality and relative percentage of survival of the different treatment groups after challenge with *Aeromonas hydrophila* at the end of the experimental trial of 60 days

Treatments	Percentage mortality	Relative percentage of survival
Control	$55.55^{a} \pm 6.4$	00.00 ± 00.00
T1F	$40.74^{b} \pm 3.7$	$26.66^{b} \pm 6.66$
T2F	$29.63^{c} \pm 3.7$	$46.66^{c} \pm 6.66$
T3S	$50.00^{a} \pm 3.2$	$10^{a} \pm 5.77$
T4S	$53.70^{a} \pm 1.8$	$3.33^{a} \pm 3.33$

Data of percentage mortality expressed as mean ± SE ($n = 3$). Relative percentage of survival compared against mean value of percentage mortality of control group (55.55%) which is assumed to be 100%. Mean values in the same column with different superscripts (a, b, c) differ significantly ($P < 0.05$)

immunosuppressive and catabolic actions (Refstie 1977). Although the use of prophylactic or metaphylactic antibiotics is a logical approach to avoid infections, it suffers from certain inherent drawbacks (Hektoen et al. 1995). Immune system of fish can be stimulated through the diet either by providing certain dietary nutrients in excess than the requirements (Pohlenz et al. 2012) or by administration of certain immunostimulant non-nutritive ingredients (Rieder and Samuelsen 2012). However, cost is the major deciding factor for including those ingredients in the feed. In the present study, emphasis was laid on the feasibility and cost effectiveness of the extraction process of fucoidan and the yield was calculated to be 5.6% in the crude extract. The yield of fucoidan varies widely depending upon the method of extraction employed and also the species of seaweeds (Ale et al. 2012). The present inexpensive method of alcohol water extraction does not involve the use of acid and can be employed for large-scale preparation of fucoidan-rich seaweed extract.

The effect of fucoidan-rich seaweed extract on growth performance of carps has not been studied yet. Fucoidan has shown significant increase in growth performance in aquatic animals (Traifalgar et al. 2010; Immanuel et al. 2012; Tuller et al. 2014; Prabu et al. 2016). We found no significant difference in growth performance of the experimental groups. There are number of reports which suggested that fucoidan is a potential inhibitor of cellular division (Religa et al. 2000; Yoshimoto et al. 2015) and cellular growth in various organisms (Aisa et al. 2004; Haneji et al. 2005; Maruyama et al. 2006; Alekseyenko et al. 2007). These studies indicate that fucoidan may not induce growth performance and hence support our findings. The growth promoting effect of fucoidan in some organisms may be an indirect effect through stress mitigation. However, it remains to be confirmed and needs further research.

The non-specific immune system of fish can respond to external factors (Tort et al. 2003) including dietary compounds (Magnadottir 2010). Respiratory burst (NBT reduction) is an important component of the non-specific immune system involved in the early eradication of pathogen (Bellavite 1988). Respiratory burst activity in fish may be modulated through production of reactive oxygen species in response to immunostimulants (Liu et al. 2014). Fucoidan-fed fish groups exhibited higher NBT reduction activity of the phagocytes compared to the control and the seaweed fed groups. This indicates that fucoidan boosts the immune system and health status of fish (Yang et al. 2014; Prabu et al. 2016). Phagocytic activity represents an immediate response carried out by the phagocytes to kill the pathogenic bacteria as a part of their defense mechanism (Neumann et al. 2001). It is a key element in host defenses against bacterial infections (Kantari et al. 2008). We found a dose-dependent increase in the phagocytic activity of the macrophages with the highest phagocytic activity observed in the 2% FRSE treatment, which has also been observed previously by El-Boshy et al. (2014). Lysozyme has a bactericidal effect by destroying cellular walls of bacteria. Lysozyme also stimulates phagocytic activity and participates in the regulation of immune cell differentiation and proliferation (Ghiasi et al. 2010). The dietary FRSE at 2% dietary level could increase the lysozyme activity of the fish compared to the control groups. Serum lysozyme activity has been found to be responsive to dietary fucoidan (El-Boshy et al. 2014; Prabu et al. 2016). At 2% dietary FRSE, the A:G ratio in the serum of fish reduced significantly compared to the control. The results indicate that the non-specific immune parameters are stimulated by dietary fucoidan-rich seaweed extract in a dose-dependent manner. It will be interesting to study the possible level of inclusion of fucoidan in the diet of carps providing immunostimulation at the least cost.

Antimicrobial compounds are important components of the innate immune system and are crucial to host defense. Out of the 2159 antimicrobial peptides identified from animals, 114 have been isolated from fish (The Antimicrobial Peptide Database 2017). Two antimicrobial peptides, β-defensin 1 and hepcidin were partially characterized from *Labeo rohita* with sequences of 148 and 269 bp, respectively. No study has been carried out to study the tissue level expression of antimicrobial peptides in response to feeding FRSE in fish. Antimicrobial peptides are the primitive components of the innate immune system, known to express in liver skin, gills and gut, (Marel et al. 2012) and act as the first line of defense against the bacterial pathogens. Therefore, expression study of the antimicrobial peptides in such tissues may provide significant information regarding the stimulation of peptide expression by immunostimulants and involvement in defense against bacterial pathogens. We found highest expression of β-defensin 1 in the skin, followed by the liver, whereas no significant upregulation was observed in the gill. The mechanism of immunostimulation by fucoidan in fish is not clearly known. However, the increased expression of β-defensin 1 in the skin may be explained on the basis of water-soluble nature of fucoidan and the possibility of interaction of fish with the pellets or with the leachate (Casadei et al. 2013). The expression of β-defensin 1 in the skin of fish has been induced by pathogenic challenge (Chen et al. 2013) as well as by immunostimulants in fish (Marel et al. 2012). Oral administration of FRSE had significant effect on the β-defensin 1 mRNA expression in the gut of the fish. The fish gut is a complex of endogenous probiotic bacteria, pathogenic bacteria, mucus, enzyme and bile secretions and the semidigested food. The endogenous probiotic bacteria and the gut mucosa prevent the colonization of the pathogenic bacteria and could also exert immunostimulatory effect on the gut-associated lymphoid system (Picchietti et al. 2007). Thus, the use of immunostimulant may alter the eubiosis of the gut or may stimulate the immune system. The effect of fucoidan on the gut microbiota of teleost fish has not been studied so far. Fucoidan is unique in the sense that it is an immunostimulant-possessing antibacterial property (Zapopozhets et al. 1995). Indeed, it will be interesting to study the interaction of fucoidan on the gut microbiota and its effect on immune system of the fish. Besides β-defensin 1, hepcidin is another most studied antimicrobial peptide (Shi and Camus 2006) that has been established as an iron-regulating hormone. For the expression study of hepcidin, the liver tissue was chosen, as it is the main site of expression of hepcidin (Krause et al. 2000). The expression of hepcidin gene in the liver was upregulated almost 6.5 times on the 60th day by feeding 2% FRSE. Hepcidin expression has been found responsive in different fish species when challenged with lipopolysaccharides (Hirono et al. 2005) and pathogenic bacteria (Wang et al. 2009). Hepcidin also shows inhibiting effect on *A. hydrophila* (Cai et al. 2012). However, only a limited number of studies have sought to investigate the effect of dietary immunostimulants on the gene expression of hepcidin in teleost fishes (Skov et al. 2012).

A. hydrophila is a Gram-negative, rod-shaped bacteria with single flagellum. Antimicrobial peptides have been shown to be active against Gram-negative bacteria (Hsieh et al. 2010). The mechanisms of multiple antimicrobial peptides have been analyzed and a model, termed the Shai–Matsuzaki–Huang model, has been proposed in which cationic antimicrobial peptides may use their positive charge to bind to the surface anionic membrane components of bacterial cells (Zasloff 2002). This binding is suggested to result in the leakage of bacterial cellular components, thereby killing the bacterial cells. Higher expression of antimicrobial peptides was recorded in the skin, liver and the intestine of *L. rohita*. These tissues are known to be the portal of entry and predilection sites of *A. hydrophila* in the fish. Therefore, the reduced post-challenge mortality of the fucoidan-fed fish may be attributed to the protection provided by increased expression of antimicrobial peptides and also stimulation of innate immune system.

The activation of antimicrobial peptides gene expression in particular, and the non-specific immune system in general, by feeding sulphated polysaccharides (FRSE) is thought to be mediated primarily through their recognition by specific receptors. The interaction of immunostimulatory polysaccharides with cell receptors may trigger signaling pathways, ultimately leading to induction of gene transcription (Xu et al. 2014). Therefore, it may be assumed that immunostimulatory action of fucoidan in the fish follows the same mechanism as other polysaccharides. However, at present, it remains to be confirmed and requires further research.

Seaweed powder is obtained from drying and pulverizing of the seaweed *Sargassum wightii*. The seaweed powder is the base material for the extraction of crude fucoidan. In the T3S and T4S group, seaweed powder was mixed with the feed at the rate 3 and 4%, respectively. Syad et al. (2013) found that *Sargassum wightii* has a fiber content of 17% dry weight and crude protein content of 1.4 mg/g dry weight. This means that the chance of any nutritional supplementation of the fish by feeding dry seaweed powder is very meager. Seaweed was included in T3S and T4S to evaluate any effect of dietary fucoidan from the seaweed powder on the fish. The growth performance and immune status and gene expression of antimicrobial peptides in these treatments had no significant difference from the control group. We observed that only fucoidan extracted from the seaweed powder has beneficial effect on fish immunity. Although there is considerable amount of fucoidan in T3S and T4S feeds coming from the seaweed powder, it is possible that seaweed bound fucoidan is in some inactive form and hence unable to incite any response in the fish.

Conclusion

Careful examination of the changes in genes expression and other immune parameters could reveal the processes that are involved in triggering the immune system of the fish by seaweed compounds. In the present study, we report characterization of two antimicrobial peptide genes β-defensin 1 and hepcidin with 148 and 269 bp, respectively, in *L. rohita*. Although there was no improvement in the growth of fish due to feeding fucoidan-rich seaweed extract, the expression of hepcidin was significantly upregulated in the liver. β-defensin 1 expression was also upregulated in the liver, skin and intestine. The non-specific immune parameters also showed response when the fish were fed with fucoidan rich seaweed extract. The post-challenge survival was highest in the FRSE fed groups. It concludes that fucoidan at the dietary level of 2% is a potential immunostimulant for *L. rohita*, although it does not have any growth promoting effect. The information generated through this study can be useful for designing studies on farmed fish to demonstrate the suitability of seaweed polysaccharides in preventive health care.

Acknowledgements The authors are grateful to Director, ICAR-Central Institute of Fisheries Education, Mumbai, India, for providing support and facilities to carry out this research. This work is dedicated to late Ms. Pooja Bhatt.

Conflict of interest All authors report no conflict of interest.

References

Aisa Y, Miyakawa Y, Nakazato T, Shibata H, Saito K, Ikeda Y, Kizaki M (2004) Fucoidan induces apoptosis of human HS-Sultan cells accompanied by activation of caspase-3 and down-regulation of ERK pathways. Am J Hematol 78:7–14. https://doi.org/10.1002/ajh.20182

Ale MT, Mikkelsen JD, Meyer AS (2012) Designed optimization of a single-step extraction of fucose-containing sulfated polysaccharides from *Sargassum* sp. J Appl Phycol 24:715–723. https://doi.org/10.1007/s10811-011-9690-3

Alekseyenko TV, Zhanayeva SY, Venediktova AA, Zvyagintseva TN, Kuznetsova TA, Besednova NN, Korolenko TA (2007) Antitumor and antimetastatic activity of fucoidan, a sulfated polysaccharide isolated from the Okhotsk sea *Fucus evanescens* brown alga. Bull Exp Biol Med 143:730–732. https://doi.org/10.1007/s10517-007-0226-4

Álvarez CA, Guzmán F, Cárdenas C, Marshall SH, Mercado L (2014) Antimicrobial activity of trout hepcidin. Fish Shellfish Immunol 41:93–101. https://doi.org/10.1016/j.fsi.2014.04.013

Amend DF (1981) Potency testing of fish vaccines. In: Anderson DP, Hennessen H (eds) Fish biologies: serodiagnostics and vaccines, vol 49. Development in biological standardization. Karger, Basel, pp 447–454

Anderson DP, Siwicki AK (1995) Basic haematology and serology for fish health programmes. In: Shharif M, Arthur JR, Subashinge RP (eds) Diseases in Asian aquaculture II. Fish Health Section, Asian Fisheries Society, Manilla, pp 185–202

Angulo FJ, Nargund VN, Chiller TC (2004) Evidence of an association between use of anti-microbial agents in food animals and anti-microbial resistance among bacteria isolated from humans and the human health consequences of such resistance. J Vet Med 51:374–379. https://doi.org/10.1111/j.1439-0450.2004.00789.x

Arunkumar K, Selvapalam N, Rengasamy R (2005) The antibacterial compound sulphoglycerolipid 1-0 palmitoyl-3-0(6′-sulpho-α-quinovopyranosyl)-glycerol from *Sargassum wightii* Greville (Phaeophyceae). Bot Mar 48:441–445. https://doi.org/10.1515/bot.2005.058

Barton BA, Iwama GK (1991) Physiological changes in fish from stress in aquaculture with emphasis on the response and effects of corticosteroids. Annu Rev Fish Dis 1:3–26. https://doi.org/10.1016/0959-8030(91)90019-G

Bellavite P (1988) The superoxide-forming enzymatic system of phagocytes. Free Radic Biol Med 4:225–261. https://doi.org/10.1016/0891-5849(88)90044-5

Cai L, Cai JJ, Liu HP, Fan DQ, Peng H, Wang KJ (2012) Recombinant medaka (*Oryzias melastigmus*) pro-hepcidin: multifunctional characterization. Comp Biochem Physiol B 161:140–147. https://doi.org/10.1016/j.cbpb.2011.10.006

Camus AC, Durborow RM, Hemstreet WG, Thune RL, Hawke JP (1998) *Aeromonas* bacterial infections-motile aeromonad septicemia. SRAC Publ 478:1–4

Casadei E, Bird S, Vecino JLG, Wadsworth S, Secombes CJ (2013) The effect of peptidoglycan enriched diets on antimicrobial peptide gene expression in rainbow trout (*Oncorhynchus mykiss*). Fish Shellfish Immunol 34:529–537. https://doi.org/10.1016/j.fsi.2012.11.027

Chen Y, Zhao H, Zhang X, Luo H, Xue X, Li Z, Yao B (2013) Identification, expression and bioactivity of *Paramisgurnus dabryanus* b-defensin that might be involved in immune defense against bacterial infection. Fish Shellfish Immunol 35:399–406. https://doi.org/10.1016/j.fsi.2013.04.049

Chotigeat W, Tongsupa S, Supamataya K, Phongdara A (2004) Effect of fucoidan on disease resistance of black tiger shrimp. Aquaculture 233:23–30. https://doi.org/10.1016/j.aquaculture.2003.09.025

Cuesta A, Meseguer J, Esteban MA (2011) Molecular and functional characterization of the gilthead seabream beta-defensin demonstrate its chemotactic and antimicrobial activity. Mol Immunol 48:1432–1438. https://doi.org/10.1016/j.molimm.2011.03.022

Defoirdt T, Sorgeloos P, Bossier P (2011) Alternatives to antibiotics for the control of bacterial disease in aquaculture. Curr Opin Microbiol 14:251–258. https://doi.org/10.1016/j.mib.2011.03.004

Deivasigamani B, Subramanian V (2016) Applications of immunostimulants in aquaculture: a review. Int J Curr Microbiol App Sci 5:447–453

Doumas BT, Watson W, Biggs HG (1971) Albumin standards and measurement of serum albumin with bromocresol green. Clin Chim Acta 31:87–96

Dubois M, Gilles KA, Hamilton JK, Rebers PA, Smith F (1956) Colorimetric method for determination of sugars and related substances. Anal Chem 28:350–356. https://doi.org/10.1021/ac60111a01

El-Boshy M, El-Ashram A, Risha E, Abdelhamid F, Zahran E, Gab-Alla A (2014) Dietary fucoidan enhance the non-specific immune response and disease resistance in African catfish, *Clarias gariepinus*, immunosuppressed by cadmium chloride. Vet Immunol Immunopathol 15:168–173. https://doi.org/10.1016/j.vetimm.2014.10.001

Falco A, Chico V, Marroqui L, Perez L, Coll JM, Estepa A (2008) Expression and antiviral activity of a beta-defensin-like peptide identified in the rainbow trout (*Oncorhynchus mykiss*) EST sequences. Mol Immunol 45:757–765. https://doi.org/10.1016/j.molimm.2007.06.358

Ghiasi F, Mirzargar SS, Badakhshan H, Shamsi S (2010) Effects of low concentration of cadmium on the level of lysozyme in serum, leukocyte count and phagocytic index in *Cyprinus carpio* under the wintering conditions. J Fish Aquat Sci 5:113–119. https://doi.org/10.3923/jfas.2010.113.119

Gong L, Wang H, Deng L (2014) Molecular characterization, phylogeny and expression of a hepcidin gene in the blotched snakehead *Channa maculate*. Dev Comp Immunol 44:1–11. https://doi.org/10.1016/j.dci.2013.11.007

Hamilton MA, Russo RC, Thurston RV (1977) Trimmed Spearman-Karber method for estimating median lethal concentrations in toxicity bioassays. Environ Sci Technol 11:714–719. https://doi.org/10.1021/es60130a004

Haneji K, Matsuda T, Tomita M, Kawakami H, Ohshiro K, Uchihara J, Masuda M, Takasu N, Tanaka Y, Ohta T, Mori N (2005) Fucoidan extracted from *Cladosiphon okamuranus* Tokida induces apoptosis of human T-cell leukemia virus type 1-infected T-cell lines and primary adult T-cell leukemia cells. Nutr Cancer 52:189–201. https://doi.org/10.1207/s15327914nc5202_9

Hektoen H, Berge JA, Hormazabal V, Yndestad M (1995) Persistence of antibacterial agents in marine sediments. Aquaculture 133:175–184. https://doi.org/10.1016/0044-8486(94)00310-K

Hirono I, Hwang JY, Ono Y, Kurobe T, Ohira T, Nozaki R, Aoki T (2005) Two different types of hepcidins from the Japanese flounder *Paralichthys olivaceus*. FEBS J 272:5257–5264. https://doi.org/10.1111/j.1742-4658.2005.04922.x

Hsieh JC, Pan CY, Chen JY (2010) Tilapia hepcidin (TH)2-3 as a transgene in transgenic fish enhances resistance to *Vibrio vulnificus* infection and causes variations in immune-related genes after infection by different bacterial species. Fish Shellfish Immunol 29:430–439. https://doi.org/10.1016/j.fsi.2010.05.001

Huys G, Rhodes G, McGann P, Denys R, Pickup R, Hiney M, Smith P, Swings J (2000) Characterization of oxytetracycline-resistant heterotrophic bacteria originating from hospital and freshwater fish farm environments in England and Ireland. Syst Appl Microbiol 23:599–606. https://doi.org/10.1016/S0723-2020(00)80036-3

Immanuel G, Sivagnanavelmurugan M, Marudhupandi T, Radhakrishnan S, Palavesama A (2012) The effect of fucoidan from brown seaweed *Sargassum wightii* on WSSV resistance and immune activity in shrimp *Penaeus monodon*. Fish Shellfish Immunol 32:551–564. https://doi.org/10.1016/j.fsi.2012.01.003

Kantari C, Pederzoli-Ribeil M, Witko-Sarsat V (2008) The role of neutrophils and monocytes in innate immunity. Contrib Microbiol 15:118–146

Kitikiew S, Chen J, Putra D, Lin Y, Yeh S, Liou C (2013) Fucoidan effectively provokes the innate immunity of white shrimp *Litopenaeus vannamei* and its resistance against experimental *Vibrio alginolyticus* infection. Fish Shellfish Immunol 34:280–290. https://doi.org/10.1016/j.fsi.2012.11.016

Krause A, Neitz S, Magert HJ, Schulz A, Forssmann WG, Schulz-Knappe P, Adermann K (2000) LEAP-1, a novel highly disulfide-bonded human peptide, exhibits antimicrobial activity. FEBS Lett 480:147–150. https://doi.org/10.1016/S0014-5793(00)01920-7

Liu L, Zhou Y, Zhao X, Wang H, Wang L, Yuan G, Asim M, Wang W, Zeng L, Liu X, Lin L (2014) Oligochitosan stimulated phagocytic activity of macrophages from blunt snout bream (*Megalobrama amblycephala*) associated with respiratory burst coupled with nitric oxide production. Dev Comp Immunol 47:17–24. https://doi.org/10.1016/j.dci.2014.06.005

Livak KJ, Schmittgen TD (2001) Analysis of relative gene expression data using real-time quantitative PCR and the $2^{-\Delta\Delta C_T}$ method. Methods 25:402–408. https://doi.org/10.1006/meth.2001.1262

Magnadottir B (2010) Immunological control of fish diseases. J Mar Biotechnol 12:361–379. https://doi.org/10.1007/s10126-010-9279-x

Marel M, Adamek M, Gonzalez S, Frost P, Rombout J, Wiegertjes G, Savelkoul H, Steinhagen D (2012) Molecular cloning and expression of two defensin and two mucin genes in common carp (*Cyprinus carpio* L.) and their up-regulation after β-glucan feeding. Fish Shellfish Immunol 32:494–501. https://doi.org/10.1016/j.fsi.2011.12.008

Maruyama H, Tamauchib H, Iizuka M, Nakano T (2006) The role of NK cells in antitumor activity of dietary fucoidan from *Undaria pinnatifida* Sporophylls (Mekabu). Planta Med 72:1415–1417. https://doi.org/10.1055/s-2006-951703

Neumann NF, Stafford JL, Barreda D, Ainsworth AJ, Belosevic M (2001) Antimicrobial mechanisms of fish phagocytes and their role in host defense. Dev Comp Immunol 25:807–825. https://doi.org/10.1016/S0145-305X(01)00037-4

Picchietti S, Mazzini M, Taddei AR, Renna R, Fausto AM, Mulero V, Carnevali O, Cresci A, Abelli L (2007) Effects of administration of probiotic strains on GALT of larval gilthead seabream: immunohistochemical and ultrastructural studies. Fish Shellfish Immunol 22:57–67. https://doi.org/10.1016/j.fsi.2006.03.009

Plumb JA (1999) Overview of warm water fish diseases. J Appl Aquac 2:1–10. https://doi.org/10.1300/J028v09n02_01

Pohlenz C, Buentello A, Mwangi W, Gatlin DM (2012) Arginine and glutamine supplementation to culture media improves the performance of various channel catfish immune cells. Fish Shellfish Immunol 32:762–768. https://doi.org/10.1016/j.fsi.2012.01.029

Poobalane S, Thompson KD, Diab A, Ardó L, Jeney G, Adams A (2008) Protein expression by *Aeromonas hydrophila* during growth in vitro and in vivo. Microb Pathog 45:60–69. https://doi.org/10.1016/j.micpath.2008.03.001

Prabu DL, Sahu NP, Pal AK, Dasgupta S, Narendra A (2016) Immunomodulation and interferon gamma gene expression in sutchi cat fish, *Pangasianodon hypophthalmus*: effect of dietary fucoidan rich seaweed extract (FRSE) on pre and post-challenge period. Aquac Res 47:199–218. https://doi.org/10.1111/are.12482

Refstie T (1977) Effect of density on growth and survival of rainbow trout. Aquaculture 11:329–334. https://doi.org/10.1016/0044-8486(77)90082-5

Reinhold JG (1953) Manual determination of serum total protein, albumin and globulin fraction by Biuret method. In: Reiner M (ed) Standard method of clinical chemistry. Academic Press, New York, p 88

Religa P, Kazi M, Thyberg J, Gaciong Z, Swedenborg J, Hedin U (2000) Fucoidan inhibits smooth muscle cell proliferation and reduces mitogen-activated protein kinase activity. Eur J Vasc Endovasc Surg 20:419–426. https://doi.org/10.1053/ejvs.2000.1220

Rieder A, Samuelsen AB (2012) Do cereal mixed-linked beta-glucans possess immune-modulating activities? Mol Nutr Food Res 56:536–547. https://doi.org/10.1002/mnfr.201100723

Robar N, Burness G, Murray DL (2010) Tropics, trophics and taxonomy: the determinants of parasite-associated host mortality. Oikos 119:1273–1280. https://doi.org/10.1111/j.1600-0706.2009.18292.x

Roberts RJ (1993) Motile aeromonad septicaemia. In: Inglis V, Roberts RJ, Bromage NR (eds) Bacterial diseases of fish. Blackwell Scientific Publications, London, pp 143–157

Sahu I, Das BK, Marhual N, Samanta M, Mishra BK, Eknath AE (2011) Toxicity of crude extracellular products of *Aeromonas hydrophila* on Rohu, *Labeo rohita* (Ham.). Indian J Microbiol 51:515–520

Sankaran K, Gurnani S (1972) On the variation in the catalytic activity of lysozyme in fishes. Indian J Biochem Biophys 9:162–165

Secombes CJ (1990) Isolation of salmonid macrophages and analysis of their killing activity. In: Stolen JS, Anderson DP, Robertson BS, van Muiswinkel WB (eds) Techniques in fish immunology. SOS Publications, Fair Haven, pp 137–154

Shi J, Camus AC (2006) Hepcidins in amphibians and fishes: antimicrobial peptides or iron-regulatory hormones? Dev Comp Immunol 30:746–755. https://doi.org/10.1016/j.dci.2005.10.009

Skov J, Kania PW, Holten-Andersen L, Fouz B, Buchmann K (2012) Immunomodulatory effects of dietary β-1,3-glucan from *Euglena gracilis* in rainbow trout (*Oncorhynchus mykiss*) immersion vaccinated against *Yersinia ruckeri*. Fish Shellfish Immunol 33:111–120. https://doi.org/10.1016/j.fsi.2012.04.009

Smith VJ, Desbois AP, Dyrynda EA (2010) Conventional and unconventional antimicrobials from fish, marine invertebrates and micro-algae. Mar Drugs 8:1213–1262. https://doi.org/10.3390/md8041213

Sørum H (2006) Antimicrobial drug resistance in fish pathogens. In: Aarestrup FM (ed) Antimicrobial resistance in bacteria of animal origin. American Society for Microbiology Press, Washington, DC, pp 213–238

Syad AN, Shunmugiah KP, Kasi PD (2013) Seaweeds as nutritional supplements: analysis of nutritional profile, physicochemical properties and proximate composition of *G. acerosa* and *S. wightii*. Biomed Prev Nutr 3:139–144

The Antimicrobial Peptide Database (2017). http://aps.unmc.edu/AP/main.php. Accessed 27 Dec 2017

Tort L, Balasch JC, Mackenzie S (2003) Fish immune system. A crossroads between innate and adaptive responses. Immunologia 22:277–286. http://www.inmunologia.org/Upload/Articles/6/0/602.pdf

Traifalgar RF, Kira H, Tung HT, Michael FR, Lainig A, Yokoyama S, Ishikawa M, Koshio S, Serrano AE, Corre V (2010) Influence of dietary fucoidan supplementation on growth and immunological response of juvenile *Marsupenaeus japonicas*. J World Aquac Soc 41:235–244. https://doi.org/10.1111/are.12029

Tuller J, Santis CD, Jerry DR (2014) Dietary influence of Fucoidan supplementation on growth of *Lates calcarifer* (Bloch). Aquac Res 45:749–754. https://doi.org/10.1111/are.12029

Wang KJ, Cai JJ, Cai L, Qu HD, Yang M, Zhang M (2009) Cloning and expression of a hepcidin gene from a marine fish (*Pseudosciaena crocea*) and the antimicrobial activity of its synthetic peptide. Peptides 30:638–646. https://doi.org/10.1016/j.peptides.2008.12.014

Xu W, Faisal M (2010) Defensin of the zebra mussel (*Dreissena polymorpha*): molecular structure, in vitro expression, antimicrobial activity, and potential functions. Mol Immunol 47:2138–2147. https://doi.org/10.1016/j.molimm.2010.01.025

Xu X, Yan H, Tang J, Chen J, Zhang X (2014) Polysaccharides in *Lentinus edodes*: isolation, structure, immunomodulating activity and future prospective. Crit Rev Food Sci 54:474–487. https://doi.org/10.1016/j.cbpb.2010.11.003

Yang Q, Yang R, Li M, Zhou Q, Liang X, Elmada ZC (2014) Effects of dietary fucoidan on the blood constituents, anti-oxidation and innate immunity of juvenile yellow catfish (*Pelteobagrus fulvidraco*). Fish Shellfish Immunol 41:264–270. https://doi.org/10.1016/j.fsi.2014.09.003

Yoshimoto M, Higaki K, Nanba E, Ikeguchi M (2015) Anti-proliferation activity of fucoidan in MKN45 gastric cancer cells and downregulation of phosphorylated ASK1, a cell cycle-regulated kinase. Yogana Acta Medica 58:1–7

Zapopozhets TS, Besednova NN, Loenko I (1995) Antibacterial and immunomodulating activity of fucoidan. Ministerstvo meditsinskoi i mikrobiologicheskoi promyshlennosti SSSR 40:9–13

Zasloff M (2002) Antimicrobial peptides of multicellular organisms. Nature 415:389–395. https://doi.org/10.1007/s10811-007-9303-3

Zubia M, Deslandes CP (2008) Alginate, mannitol, phenolic compounds and biological activities of two range- extending brown algae, *Sargassum mangarevense* and *Turbinaria ornata* (Phaeophyta: Fucales), from Tahiti (French Polynesia). J Appl Phycol 20:1033–1043. https://doi.org/10.1007/s10811-007-9303-3

4

Minerals content of extruded fish feeds containing cricket (*Acheta domesticus*) and black soldier fly larvae (*Hermetia illucens*) fractions

F. G. Irungu · C. M. Mutungi · A. K. Faraj · H. Affognon · C. Tanga · S. Ekesi · D. Nakimbugwe · K. K. M. Fiaboe

Abstract Animal food sources provide human beings with minerals considerably in adequate quantities. Fish is an indispensable reliable source of nutrients, as aquaculture is a sector that is fast growing and which provides 50% of the world's fish production. However, fish production is hampered by the increasing costs of feeds due to the ever rising cost of fish meal, an integral component of fish feeds. Substituting fish meal with cheap, yet highly nutritious ingredients in fish feeds is therefore paramount. This study investigated the effects of substituting fish meal with adult cricket meal (ACM) and black soldier fly meal (BSFM) on minerals content of extruded fish feeds, where four levels of substitution (0, 25, 50 and 75%) were used. The effect of feed moisture content on minerals was also studied where 20 and 30% feed moisture levels were used. Leaching effects of the pellets were studied as well. The results showed a significant increase ($P < 0.05$) in the levels of phosphorus and potassium as the level of fish meal substitution increased from 0 to 75%. On the other hand, iron and sodium levels reduced significantly ($P < 0.05$) as the level of fish meal substitution increased. Magnesium content increased with increasing level of substitution with BSFM, but decreased with increasing level of substitution with ACM. Copper, zinc and manganese were not greatly influenced by levels of fish meal substitution. Diets that had zero substitution showed higher leaching effect for most minerals than diets that were substituted with 75% ACM or BSFM. This study found that both ACM and BSFM can be used to substitute fish meal in fish feeds and obtain adequate mineral profile and low leaching effect.

Keywords Fish meal · Substitution · Cricket · Black soldier fly larvae · Minerals · Extrusion

---

F. G. Irungu (✉) · C. M. Mutungi · A. K. Faraj
Department of Dairy and Food Science and Technology, Egerton University, P. O. Box 536-20115, Egerton, Kenya
e-mail: gichuhofrancis@gmail.com

C. M. Mutungi · C. Tanga · S. Ekesi · K. K. M. Fiaboe
International Centre of Insect Physiology and Ecology, P. O. Box 30772-00100, Nairobi, Kenya

C. M. Mutungi
International Institute of Tropical Agriculture (IITA), Plot NO. 25, Mikocheni Light Industrial Area, Dar es Salaam, Tanzania

H. Affognon
International Crops Research Institute for the Semi-Arid Tropics, BP 320, Bamako, Mali

D. Nakimbugwe
Department of Food Technology and Nutrition, School of Food Technology, Nutrition and Bio-Engineering, Makerere University, P. O. Box 7062, Kampala, Uganda

Introduction

Food insecurity is a threat to many nations, especially the developing countries. In fact, 10.9% of the world's population (795 million people) is undernourished. This situation is worse in the sub-Saharan Africa where Middle Africa and East Africa have 41.3 and 31.5% of their population being undernourished, respectively (FAO 2015). Animal food sources provide protein of high value as well as a variety of micronutrients such as minerals in adequate quantities (FAO 2012; Murphy and Allen 2003). In addition, nutrients from animal food sources are more bio-available than most plant sources (Trumpo et al. 2001). Fish is one such source that provides 17% of animal protein to the world's population (Halden et al. 2014). Given the forecasted 9 billion in world population by the year 2050, fish becomes an indispensable reliable source of nutrients as aquaculture is also fast growing (Msangi et al. 2013; Halden et al. 2014; FAO 2014). Fishmeal, an integral ingredient in fish feed is becoming extremely expensive due to fluctuations in its production and this has contributed to high cost of feeds (FAO 2012, 2013, 2014; Njagi et al. 2013). There is thus the need to identify better alternatives to fishmeal. Insects such as crickets (*Acheta domesticus*) and black soldier fly (*Hermetia illucens*) have that potential (FAO 2013; Bondari and Sheppard 1981; Makkar et al. 2014), due to their high nutritional profile (Henry et al. 2015; FAO 2010; van Huis 2013), high bio-conversion ratio (Barroso et al. 2014) and their ability to produce less greenhouse gases and to be reared on less land area (Oonincx et al. 2010).

Traditionally, insects have been processed using different methods. For example, after harvest, mopane caterpillar is normally sun dried with the aim of achieving higher shelf life (Mutungi et al. 2017). In China, silkworm pupa, the most widely consumed insect, is normally fried in oil, boiled using water or in some cases milled into powder (Belluco et al. 2013). Other traditional processing methods of edible insects include sun drying, roasting, smoking, boiling as well as removal of unpalatable parts (Rumpold et al. 2014). Sun drying is cheap and easy to carry out. However, it is dependent on weather conditions and therefore is hampered by such climatic conditions as wet weather, strong winds as well as mist and fogs. The majority of these methods have far reaching negative impacts. Drying and toasting of termites and grasshoppers have been attributed to significant decrease in contents of individual vitamins, including pyridoxine, folic acid, ascorbic acid, niacin, riboflavin, retinol and α-tocopherol, depending on the insect species (Kinyuru et al. 2010). Additionally, the in vitro digestibility of proteins in tree locust has been shown to decline upon toasting and boiling, as the tannin and phytate content increased as a result of these treatments. According to Afiukwa and Okereke (2013), removal of wings as one of the unpalatable insects' parts reduces the amount of iron that can be obtained from *Trinervitermes germinates* (termite)—an edible insect from Nigeria. Thus, although these methods have been used to process edible insects in the past, they present a situation where their reliability is questioned (Chakravorty et al. 2014). Extrusion is a modern processing method that can be used to process insects and insects-based products such as fish feeds. It has the benefits of improving the nutritional value of end products, destroying undesirable enzymes and anti-nutritional factors (Nikmaram et al. 2015; Alam et al. 2016; Filipovic et al. 2010) and it has also been shown to enhance the apparent absorption of minerals (Razzaq et al. 2012; Alonso et al. 2001).

Minerals represent a minor proportion in the feeds' composition. However, they are required in almost every aspect of animal metabolism (Alonso et al. 2001; FAO 2017). They are classified either as macro elements, which are needed by the body in large quantities, or micro elements, needed by the body in small quantities (FAO 2017; Thompson et al. 2005). Macro elements include phosphorus, calcium, potassium, magnesium, sodium, chlorine and sulfur, while micro elements are composed of copper, zinc, manganese, iron, cobalt and iodine, among others (FAO 2017; Webster and Lim 2002). Fish requires these elements as they help in several functions such as transmission of nerve impulses, acid–base balance, as enzyme cofactors and activators, normal functioning of muscles, osmoregulation and in the formation of skeletal muscles (FAO 2017; Braga et al. 2016; Hassaan et al. 2013; NRC 1993). Fish may obtain some minerals such as calcium, sodium and potassium readily from the surrounding water environment (FAO 2017; NRC 1993), but cannot get adequate amounts of phosphorus, zinc, iron and copper from water (Oliva-Teles 2012; NRC 2011) and thus these must be supplied by the diet through nutritious feeds (Tang et al. 2012). The aim of this experiment was to study the effects of substituting fish meal with different levels of black soldier fly larvae and cricket meals on the mineral composition of extruded fish feeds. The effects of moisture content on the mineral profile of extrudates were also investigated.

Materials and methods

Research site

Extrusion of feeds was done using a small-scale single screw extruder at Bidii Fish Farmers, Luanda-Vihiga County, Kenya. Mineral analyses were carried out at Kenya Agricultural and Livestock Research Organization (KALRO), Njoro, Kenya.

Materials

Maize germ, sunflower cake, dried cassava chunks, wheat pollard and dried freshwater shrimps were purchased from a local vendor in Luanda town, Vihiga County, Kenya. Blanched and sun-dried adult house crickets (*Acheta domesticus* L.) were purchased from farmers in Homa Bay County, Kenya, while blanched and sun-dried black soldier fly larvae (*Hermetia illucens* L.) were obtained from Sanergy Limited, Nairobi, Kenya, a commercial insect rearing farm. Each ingredient was then ground separately into fine meal using a hammermill (Model 4, ARTHUR H. THOMAS, Philadelphia, PA, USA) and finally passed through a 1.0 mm aperture sieve.

Experimental design

A 2 × 4 × 2 factorial arrangement in a completely randomized design was used. The factors investigated were insect type (two levels; black soldier fly larvae and adult cricket), level of freshwater shrimp meal (FWSM) protein substitution with black soldier fly larvae meal (BSFM) or adult cricket meal (ACM) (four levels: 0, 25, 50 and 75%), and feed moisture content (two levels: 20 and 30% on wet weight basis). The experiment was replicated three times.

Feed formulation

The quantities of the ingredients required to formulate iso-proteinaceous diets (26% protein; 5 kg formulation) and the amount of water needed to achieve a required moisture content (on wet weight basis) were calculated using Microsoft Excel® function on Windows 2007 and weighed using an electronic digital weighing scale (Model no.: 7765, Ashton Meyers, China), into a 20-L bucket. These were mixed manually for 2 min and then transferred to a multi-vane paddle mixer (Model: MX-25, Unitech, New Delhi, India) for further mixing for 2 min. The control diet and the insect-based formulations are shown in Table 1.

Extrusion process

A single screw extruder (Model: DOLLY, Unitech, New Delhi, India), with a screw operating at a speed of 200 rpm was used. The extruder was equipped with a pre-conditioning chamber where blended ingredients were introduced through a manual hopper at a feed rate of 1 kg/min and preheated with steam at an inlet pressure of 4 bars for approximately 2 min. The mixture was then channeled into the extrusion jacket that housed a single screw with a barrel length and diameter of 55 and 6 cm, respectively. The extruder was set to operate at 120 °C and fitted with a die that had a diameter of 2 mm at the exit. Each extruded batch was collected into separate 20 L bucket and then dried in a solar tent for about 4 h to constant weight. About 500 g of the dried pellets were taken in duplicate and packed into zip-lock bags, which were placed in cool and dry wooden locker boxes for further analysis.

Minerals analyses

Freshwater shrimps, cricket, black soldier fly larvae and extruded sample diets were milled using an oscillating mill (Model: MM400; S/N: 129251116; Retsch, Germany) and weighed using an electronic weighing balance (Model: ABS 220-4; S/N: WB1210455; KERN & SOHN GmbH, Germany). Mineral contents were then

Table 1 Inclusion levels of insects' meals in freshwater shrimp meal substitution

Ingredient	Control	Inclusion level of insects (%)					
		BSFM25	BSFM50	BSFM75	ACM25	ACM50	ACM75
Maize germ	19	18.5	18.1	17.6	19.2	19.4	19.6
Sunflower cake	19	18.5	18.1	17.6	19.2	19.4	19.6
Cassava flour	5	4.9	4.8	4.6	5	5.1	5.2
Wheat pollard	28.5	27.8	27.1	26.5	28.8	29.1	29.4
FSWM	28.5	20.9	13.6	6.6	21.6	14.5	7.3
BSFM	–	9.5	18.4	27.1	–	–	–
ACM	–	–	–	–	6.3	12.5	19.1

FWSM freshwater shrimp meal, *BSFM* black soldier fly meal, *ACM* adult cricket meal, *BSFM25* black soldier fly meal substituting 25% of the protein supplied by FWSM in the control, *BSFM50* black soldier fly meal substituting 50% of the protein supplied by FWSM in the control, *BSFM75* black soldier fly meal substituting 75% of the protein supplied by FWSM in the control, *ACM25* adult cricket meal substituting 25% of the protein supplied by FWSM in the control, *ACM50* adult cricket meal substituting 50% of the protein supplied by FWSM in the control, *ACM75* adult cricket meal substituting 75% of the protein supplied by FWSM in the control

determined by atomic absorption spectrometry in an atomic absorption spectrophotometer (Model: AA-6300; S/N: A305243009165A; Shimadzu, Japan) after digestion in sulfuric acid and selenium powder. Calcium, potassium, magnesium, sodium, copper, zinc, manganese and iron were measured at wavelengths of 422.7, 766.5, 285.2, 589.0, 324.8, 213.9, 279.5 and 248.3 nm, respectively. Phosphorus was determined using a UV–visible spectrophotometer (Model; UV-1700; S/N: A11024302429LP; Shimadzu, Japan) and measured through a wavelength of 880 nm.

Leaching activity

For each measurement, three replicates were performed. About 5 g of sample was weighed and introduced into a 250 mL measuring cylinder containing 200 mL distilled water. Aliquots (5 mL) of the supernatant were drawn after 4, 8, 24 and 48 h and sodium azide added at a rate of 0.02% into the tubes containing the aliquots. Mineral analyses were then determined as outlined above.

Statistical analysis

Analysis was done using general linear model procedure (PROC GLM) of the Statistical Analysis System (SAS) software version 9.1.3 (SAS Institute Inc., USA) for analysis of variance (ANOVA). Means were separated using Tukey's HSD (honestly significant difference) test at $P < 0.05$ level of significance.

Results

The mineral profile of freshwater shrimp meal, adult cricket meal and black soldier fly meal

Phosphorus and potassium were significantly ($P < 0.05$) higher in BSFM than in FWSM, though the two elements did not show significant ($P < 0.05$) difference between BSFM and ACM (Table 2). Calcium was significantly ($P < 0.05$) higher in FWSM than both in ACM and BSFM. Magnesium was significantly ($P < 0.05$) higher in BSFM than in FWSM and ACM, while FWSM had significantly ($P < 0.05$) higher magnesium than ACM (Table 2). Sodium, copper and zinc were higher ($P < 0.05$) in both BSFM and ACM than in FWSM. Adult cricket meal showed significantly ($P < 0.05$) higher amounts of manganese than FWSM and BSFM. Iron was higher ($P < 0.05$) in ACM and FWSM than in BSFM (Table 2).

Table 2 Mineral composition of freshwater shrimp meal (FWSM), adult cricket meal (ACM) and black soldier fly larvae meal (BSFM)

Ingredient	P (ppm)	Ca (ppm)	K (ppm)	Mg (ppm)	Na (ppm)	Cu (ppm)	Zn (ppm)	Mn (ppm)	Fe (ppm)
FWSM	4385.8 ± 960.6^{b}	3932.4 ± 95.8^{a}	$10{,}065.2 \pm 962.9^{b}$	2191.5 ± 35.7^{b}	$21{,}246.1 \pm 23.5^{b}$	1512.5 ± 23.5^{b}	454.9 ± 14.7^{b}	1932.1 ± 95.6^{b}	922.2 ± 25.6^{a}
ACM	6542.5 ± 1246.7^{ab}	1600.7 ± 170.7^{b}	$11{,}981.6 \pm 957.6^{ab}$	1467.7 ± 26.8^{c}	$22{,}633.6 \pm 97.4^{a}$	1628.8 ± 30.6^{a}	566.4 ± 23.6^{a}	3126.1 ± 96.7^{a}	931.9 ± 24.8^{a}
BSFM	8579.7 ± 900.5^{a}	1971.4 ± 212.4^{b}	$13{,}853.1 \pm 978.7^{a}$	3542.2 ± 12.5^{a}	$22{,}546.9 \pm 92.4^{a}$	1609.4 ± 36.7^{a}	576.5 ± 14.7^{a}	1762.2 ± 86.7^{b}	845.1 ± 23.8^{b}

For each component, means followed by the same superscript letters along the same column are not significantly different at $P < 0.05$

FWSM freshwater shrimp meal, *BSFM* black soldier fly meal, *ACM* adult cricket meal, *P* phosphorus, *Ca* calcium, *K* potassium, *Mg* magnesium, *Na* sodium, *Cu* copper, *Zn* zinc, *Mn* manganese, *Fe* iron

Macro minerals as influenced by different diets extruded at various feed moisture contents

The level of phosphorus in control diets was significantly ($P < 0.05$) lower than in all the insect-based diets (Table 3). As the level of insect substitution increased, so did the level of phosphorus. However, for each level of insect substitution, BSFM-based diets recorded higher values of phosphorus than ACM diets. Moisture content had a significant ($P < 0.05$) effect on phosphorus only for the BSFM50 and BSFM75 diets. The highest level of phosphorus was recorded for the BSFM75 diet that had feed moisture content of 30%.

The control diet that was extruded with feed moisture content of 20% gave the highest amounts of calcium (Table 3). This level, however, did not show significant ($P < 0.05$) difference with the BSFM25 and ACM25 diets that were produced at 20% feed moisture content, or the control diet that had 30% feed moisture content. The increase in the level of insect substitution led to a corresponding decrease in calcium levels. However, BSFM-based diets had higher calcium levels than the corresponding ACM diets (Table 3). The moisture content was also found to have a significant ($P < 0.05$) influence on calcium, where for each formulation calcium was slightly higher for the extrudates that had 20% feed moisture content than those that had a feed moisture content of 30%.

The level of potassium appeared to increase with increasing level of fish meal substitution for each insect-based diet (Table 3). This increase was significant ($P < 0.05$), apart from the ACM25 diet that did not show significant ($P < 0.05$) difference with the control diet. Another peculiar observation was that for each diet, the moisture content showed a significant ($P < 0.05$) influence on potassium, where increasing the amount of feed moisture content from 20 to 30% resulted in a corresponding slight increase in the level of potassium (Table 3).

The amount of available magnesium seemed to increase as the level of fish meal substitution with BSFM increased (Table 3). On the contrary, increasing the level of fish meal substitution with ACM led to a slight decrease in the amount of magnesium. In addition, increasing the level of feed moisture content resulted in

Table 3 Effects of different levels of freshwater shrimp meal substitution with different insect species extruded at different feed moisture contents on phosphorus (P), calcium (Ca), potassium (K), magnesium (Mg) and sodium (Na) of the extrudates

Diet	MC (%)	P (ppm)	Ca (ppm)	K (ppm)	Mg (ppm)	Na (ppm)
Control	20	625.9 ± 19.9^{e}	$15{,}063.8 \pm 300.6^{a}$	6220.8 ± 0.7^{e}	3287.6 ± 4.5^{e}	1268.3 ± 9.2^{a}
	30	649.4 ± 13.5^{e}	$14{,}476.2 \pm 114.6^{ab}$	6229.4 ± 2.4^{e}	3318.9 ± 5.8^{de}	1157.8 ± 31.5^{bc}
BSFM25	20	712.6 ± 26.6^{d}	$14{,}696.9 \pm 420.8^{ab}$	6335.4 ± 0.59^{d}	3436.1 ± 31.4^{abcde}	1223.9 ± 7.0^{ab}
	30	735.9 ± 39.8^{d}	$10{,}752.5 \pm 37.8^{cd}$	6341.8 ± 3.8^{c}	3519.1 ± 4.4^{ab}	1107.6 ± 19.7^{c}
BSFM50	20	810.6 ± 10.1^{b}	$13{,}077.1 \pm 1412.7^{abc}$	6359.5 ± 8.55^{b}	3451.9 ± 13.4^{abcd}	991.9 ± 44.4^{d}
	30	789.3 ± 6.7^{c}	$10{,}386.2 \pm 370^{cd}$	6370.5 ± 10.4^{b}	3467.6 ± 10^{abcd}	804.7 ± 7.4^{ef}
BSFM75	20	804.2 ± 14.9^{b}	9646.2 ± 1053.9^{cde}	6362.5 ± 2.3^{b}	3491.6 ± 60.9^{abc}	492.1 ± 16.4^{g}
	30	840.9 ± 4.9^{a}	8855.8 ± 218.6^{de}	6385.9 ± 5.3^{a}	3597.5 ± 49.1^{a}	442.9 ± 5.3^{g}
ACM25	20	718.2 ± 2.3^{d}	$14{,}613.2 \pm 392.4^{ab}$	6226.3 ± 3.6^{e}	3450.3 ± 42.4^{abcd}	1145.7 ± 18.4^{bc}
	30	715.9 ± 23.3^{d}	$12{,}824.8 \pm 409.3^{bc}$	6243.1 ± 6.7^{e}	3383.8 ± 14.2^{bcde}	1049.6 ± 10.1^{d}
ACM50	20	749.4 ± 6.5^{cd}	8911.8 ± 1519.3^{de}	6359.7 ± 9.2^{b}	2956.6 ± 32.6^{f}	812.4 ± 33.6^{ef}
	30	769.9 ± 30.7^{bc}	4326.1 ± 39.2^{f}	6383.5 ± 8.4^{a}	2837.6 ± 38.4^{f}	717.9 ± 11.6^{f}
ACM75	20	782.1 ± 12.7^{bc}	6439.6 ± 392.4^{ef}	6385.3 ± 0.6^{a}	3334.8 ± 12.02^{cde}	878.2 ± 4.8^{e}
	30	767.8 ± 21.6^{bc}	4074.9 ± 357.7^{f}	6389.5 ± 3.3^{a}	2934.4 ± 17.8^{f}	766.09 ± 18.47^{f}

Extrusion was done at a barrel temperature of 120 °C and feed exited through a 2 mm die

For each component, means followed by the same superscript letters along the same column are not significantly different at $P < 0.05$

MC moisture content, *P* phosphorus, *Ca* calcium, *K* potassium, *Mg* magnesium, *Na* sodium, *Control* 0% insect substitution of FWSM, *BSFM25* black soldier fly meal substituting 25% of the protein supplied by FWSM in the control, *BSFM50* black soldier fly meal substituting 50% of the protein supplied by FWSM in the control, *BSFM75* black soldier fly meal substituting 75% of the protein supplied by FWSM in the control, *ACM25* adult cricket meal substituting 25% of the protein supplied by FWSM in the control, *ACM50* adult cricket meal substituting 50% of the protein supplied by FWSM in the control, *ACM75* adult cricket meal substituting 75% of the protein supplied by FWSM in control

higher amounts of magnesium for the control and BSFM diets, but resulted in lower levels of magnesium for the ACM diets (Table 3).

Increasing the level of fish meal substitution with insect meal beyond 25% resulted in lower levels of sodium, where BSFM50, BSFM75, ACM25, ACM50 and ACM75 diets gave significant ($P < 0.05$) lower amount of sodium than the control diet (Table 3). Moreover, increasing the amount of feed moisture content from 20 to 30% resulted in a significant ($P < 0.05$) reduction in the amount of sodium (Table 3) for all the diets, apart from BSFM75 whose reduction was not significant ($P < 0.05$).

Micro minerals as influenced by different diets extruded at various feed moisture contents

Copper levels in control diets did not show significant ($P < 0.05$) difference compared with BSFM75, ACM25 and ACM50 diets that were extruded with 30% feed moisture content (Table 4). The ACM75 diet extruded at both 20 and 30% feed moisture gave the highest values of copper that significantly ($P < 0.05$) differed from the other formulations. Moisture content showed significant ($P < 0.05$) influence on the level of copper, in that, for each diet, copper levels increased with increasing moisture content (Table 4).

The control diet that had 20% feed moisture content did not have significant ($P < 0.05$) difference compared with BSFM25 (20% MC), BSFM50 (20% MC), BSFM75 (30% MC), ACM25 (30% MC) and ACM50 (both 20 and 30% MC) in amount of zinc (Table 4). Similarly, the control diet that had 30% feed moisture content did not show significant ($P < 0.05$) difference in amount of zinc, compared with BSFM50 and ACM75 diets that were also extruded at 30% feed moisture content as well as BSFM75 that had 20% feed moisture content. ACM75 diet with 20% feed moisture content gave the highest level of zinc that had significant ($P < 0.05$) difference with all other formulations.

For BSFM diets, there was no definite trend on how different levels of substitution as well as moisture content influenced manganese (Table 4). However, ACM diets showed slight increase in the levels of manganese as feed moisture content was increased from 20 to 30%. In addition, ACM75 diets gave the highest

Table 4 Effects of different levels of freshwater shrimp meal substitution with different insect species extruded at different feed moisture contents on copper (Cu), zinc (Zn), manganese (Mn) and iron (Fe) of the extrudates

Diet	MC (%)	Cu (ppm)	Zn (ppm)	Mn (ppm)	Fe (ppm)
Control	20	83.7 ± 0.4^{b}	158.1 ± 1.3^{efg}	14.9 ± 0.1^{cd}	356.9 ± 1.6^{cd}
	30	89.9 ± 1.5^{b}	181.2 ± 4.3^{b}	10.1 ± 0.2^{ghi}	376.7 ± 2.2^{cd}
BSFM25	20	12.5 ± 1.2^{d}	159.5 ± 1.6^{efg}	12.1 ± 0.8^{efgh}	347.4 ± 30.45^{d}
	30	16.8 ± 2.7^{d}	145.8 ± 0.5^{hi}	13.8 ± 0.7^{cde}	355.5 ± 11.2^{d}
BSFM50	20	19.4 ± 0.1^{d}	168.3 ± 1^{cde}	18.3 ± 0.3^{a}	271.6 ± 6.4^{e}
	30	52.8 ± 0.5^{c}	177.4 ± 1.2^{bc}	15.6 ± 0.9^{bc}	309.7 ± 15.3^{d}
BSFM75	20	44.8 ± 0.0^{c}	171.4 ± 2.8^{bcd}	15.6 ± 0.3^{bc}	207.8 ± 11.9^{f}
	30	73.1 ± 5.8^{b}	161.9 ± 3.2^{def}	12.4 ± 0.5^{efg}	255.5 ± 11.2^{e}
ACM25	20	50.9 ± 8.8^{c}	143.1 ± 2.7^{i}	9.9 ± 0.3^{hi}	429.2 ± 5.3^{b}
	30	75.1 ± 3.4^{b}	155.3 ± 0.9^{fgh}	12.8 ± 0.2^{def}	469.6 ± 5.9^{ab}
ACM50	20	43.5 ± 4.2^{c}	153.5 ± 1.5^{fghi}	10.6 ± 0.0^{fghi}	467.2 ± 3.2^{ab}
	30	75.7 ± 0.0^{b}	148.7 ± 0.8^{ghi}	12.9 ± 0.0^{def}	492.2 ± 6.4^{a}
ACM75	20	152.3 ± 3.8^{a}	199.4 ± 1.2^{a}	17.8 ± 0.2^{ab}	502.4 ± 1.7^{a}
	30	170.8 ± 5.6^{a}	186.1 ± 2.5^{b}	18.3 ± 0.3^{a}	521.4 ± 8.2^{a}

Extrusion was done at a barrel temperature of 120 °C and feed exited through a 2 mm die

For each component, means followed by the same superscript letters along the same column are not significantly different at $P < 0.05$

MC moisture content, *Cu* copper, *Zn* zinc, *Mn* manganese, *Fe* iron, *Control* 0% insect substitution of FWSM, *BSFM25* black soldier fly meal substituting 25% of the protein supplied by FWSM in the control, *BSFM50* black soldier fly meal substituting 50% of the protein supplied by FWSM in the control, *BSFM75* black soldier fly meal substituting 75% of the protein supplied by FWSM in the control, *ACM25* adult cricket meal substituting 25% of the protein supplied by FWSM in the control, *ACM50* adult cricket meal substituting 50% of the protein supplied by FWSM in the control, *ACM75* adult cricket meal substituting 75% of the protein supplied by FWSM in the control

values of manganese that did not differ significantly ($P < 0.05$) with BSFM50 diets as well as the BSFM75 diet that were extruded at 20% feed moisture content (Table 4).

An increase in level of fish meal substitution with BSFM resulted in a decrease in final iron content, where BSFM75 diets had significantly ($P < 0.05$) lower levels of iron than the control (Table 4). On the other hand, substitution with ACM resulted in increased levels of iron as the level of substitution increased with all the ACM diets giving significantly ($P < 0.05$) higher levels of iron than the control and BSFM diets. It was also noted that increasing feed moisture content gave a corresponding increase in the amount of iron for all the diets, though this increase was only significant ($P < 0.05$) for the BSFM50 and BSFM75 diets (Table 4).

Leaching activity

Leaching activity was determined on pellets processed from the control formulation and insect meal containing blends at 75% substitution (BSFM75 and ACM75) and 30% moisture content, because the three exhibited relatively good mineral profile and have also shown better physico-chemical properties in comparison to other treatments (Irungu et al. 2018). Figure 1 shows the leaching activity of minerals that were studied over a period of 48 h. Phosphorus, calcium, potassium, sodium and manganese were highly leached in the control diet (as shown in Fig. 1 with asterisks), where, during the period studied, the control showed significantly ($P < 0.05$) higher leaching effect as compared to the ACM75 and BSFM75 diets. The leaching of magnesium and zinc was significantly ($P < 0.05$) higher with the ACM75 diets than with BSFM75 and control diets throughout the 48 h study period. At 4 h, BSFM75 had significantly ($P < 0.05$) lower leached

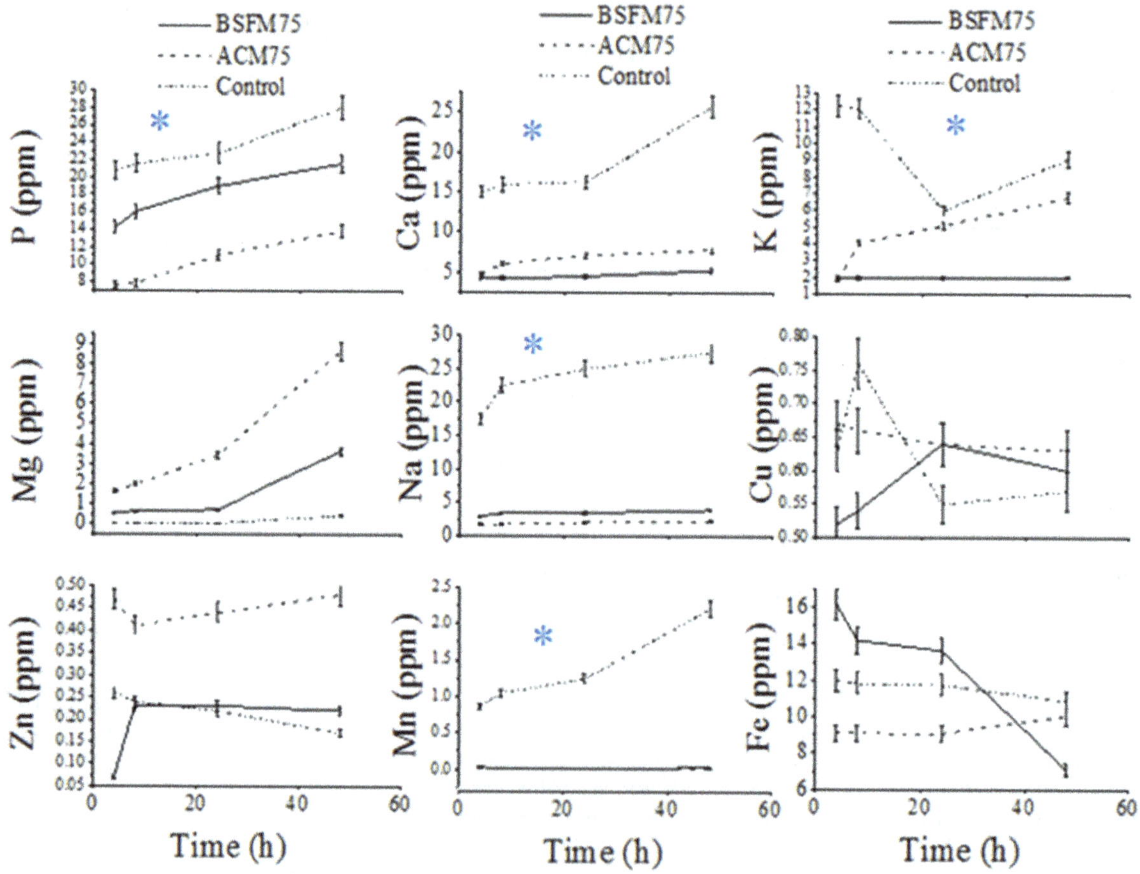

Fig. 1 Leaching activity of minerals (*P* phosphorus, *Ca* calcium, *K* potassium, *Mg* magnesium, *Na* sodium, *Cu* copper, *Zn* zinc, *Mn* manganese, *Fe* iron) in water introduced with pellets extruded from the control blend (round dotted line), black soldier fly larvae meal-containing blend (75% substitution; BSFM75) (solid line) and adult cricket meal-containing blend (75% substitution; ACM75) (square-dotted line) adjusted to 30% moisture content. Parameters with an asterisk show that the control diet gave significant ($P < 0.05$) higher leaching effect than ACM75 and BSFM75 diets

copper levels than ACM75 and the control, both of which were not significant at the time. However, at 48 h, all the three diets had no significant ($P < 0.05$) difference in the amount of leached copper. At 4, 8 and 24 h, BSFM75 had significantly ($P < 0.05$) higher amounts of leached iron levels than the control and ACM75 diets. This scenario changed at 48 h where BSFM75 showed significantly ($P < 0.05$) low amounts of leached iron than the control and ACM75 diets, both of which were not significant ($P < 0.05$). Another peculiar observation is that while it is expected for the level of leached minerals to increase as time progresses, BSFM75 showed a decreasing trend from 24 to 48 h with regard to copper, zinc and iron elements. The control also showed the same behavior with zinc and copper, while ACM75 only showed a decreased behavior with copper.

Discussion

The fact that ACM and BSFM were higher in the majority of minerals than FWSM indicates that the two insects are good sources of minerals and that the two are potential substitutes of FWSM in fish feeds.

Phosphorus and calcium are important minerals required by the fish for healthy bone formation as well as in enhancing the growth performance of fish (FAO 2016; Braga et al. 2016). The two have an antagonistic effect where an inadequate supply of one element affects the availability of the other due to the formation of the calcium–phosphorus complex (Chavez-Sanchez et al. 2000); therefore, both are discussed together. The high values of phosphorus recorded in BSFM-based diets can be attributed to the fact that BSFM had the highest amounts of phosphorus in comparison to ACM and FWSM (Table 2). Newton et al. (1977) and Barker et al. (1998) have also shown that BSFM has higher calcium and phosphorus than ACM. Likewise, the decrease in calcium amounts as the level of substitution increased is attributed to the high values of calcium that are found in FWSM in comparison to BSFM and ACM (Table 2). The low levels of calcium as the feed moisture content was increased to 30% could be due to the formation of solid bridge complexes (Tumuluru et al. 2011) that hindered the availability of the element. It could also be due to the fact that calcium is highly soluble in water (Nielsen 2010) and thus a lot of it could have been solubilized in water and leached during the extrusion process. On the other hand, phosphorus has poor solubility in water and this could explain why feed moisture content did not have an influence on the amount of available phosphorus. Unlike calcium, phosphorus must be supplied by the feed because fish cannot get it readily from the surrounding water environment (Tessenderlo 2005; Tang et al. 2012). As such, BSFM- and ACM-based diets are good sources of phosphorus, as both resulted in higher levels of phosphorus than the control (100% freshwater shrimps) diet. However, none of the study diets resulted in minimum requirements for phosphorus for tilapia (*Oreochromis Niloticus* [L]) and catfish (*Clarias gariepinus* [Burchell]), which is 5000 ppm and 4500 ppm, respectively (FDKS 2010; FAO 2017), and thus there is need to supplement the feeds with an inorganic source such as dicalcium phosphate so as to improve feed utilization and growth performance. Since calcium and phosphorus have an antagonist effect (Hassaan et al. 2013; FAO 2016), the Ca/P ratio of these diets on growth and development of fish need to be studied.

Potassium is essential in the growth and development of fish, as it regulates acid base balance, intracellular osmotic pressure and is also required for protein and glycogen synthesis (FAO 2017; NRC 1993). The increase in potassium as the level of substitution increased is due to the high potassium levels that are found in both ACM and BSFM in comparison to FWSM (Table 2). The fact that the moisture content influenced potassium levels shows that extrusion has an influence on this element, which could be due to chemical alteration (Singh et al. 2007) and destruction of polyphenols (Alonso et al. 2001) through extrusion at high feed moisture content, which allows phosphorus to become bio-available. Razzaq et al. (2012) also found an increase in potassium amounts of maize extrudates as moisture content was increased. The minimum potassium requirements for tilapia and catfish are 2100–3300 and 2600 ppm, respectively, and thus all our study diets meet these requirements.

Magnesium is required by fish, as it is an essential cartilage and bone component and also stimulates muscle contraction (Jahnen-Dechent and Ketteler 2012; FAO 2017). The decrease in the amount of magnesium as FWSM was substituted with BSFM is due to the higher magnesium content that is found in BSFM than in FWSM (Table 2). Likewise, ACM has lower magnesium content than FWSM (Table 2), the reason why diets substituted with ACM had lower magnesium content than the control diet. The fiber in ACM could have been more susceptible to restructuring due to the extrusion process to an extent of promoting chelating effects that could have rendered magnesium unextractable, and thus the low magnesium content in ACM diets (Cheftel 1986). According to FAO (2017), the minimum magnesium requirements for tilapia and catfish are 600–800 and 400 ppm, respectively, and all our study diets exceeded these levels. Thus, both BSFM and ACM can be used to substitute FWSM in fish feed and satisfy magnesium requirements for both tilapia and catfish, given that fish does not meet all its magnesium requirements from the surrounding water and feeds must supply the demand (FAO 2017).

Fish need sodium for maintenance of acid base balance and for regulation of osmotic pressure (FAO 2017; Webster and Lim 2002). The sodium in BSFM and ACM may not be readily available as in FWSM, probably due to differences in their structural formation, and this could explain why substituting FWSM with BSFM and ACM resulted in low amounts of sodium. Extrusion at high temperature and high moisture content may have led to the formation of phenolic complexes that changed the chemical structure of the compounds, thus rendering sodium unextractable (Alonso et al. 2001). In addition, high feed moisture content may have lowered the glass transition temperature, leading to the formation of solid bridges (Tumuluru 2014) that resulted in dense products that made sodium unavailable. There is little information on the minimum sodium requirements for both tilapia and catfish and thus it is impossible to conclude whether our study diets (especially insect-based diets) can successfully substitute fish meal and meet the sodium demand for these fishes.

Copper is directly involved in enzymes activity such as dopamine hydroxylase and cytochrome oxidase (Watanabe et al. 1997) and also acts as catalyst for some enzymes (Camire et al. 1990). The higher values of copper in the ACM75 diet than BSFM and control diets is due to the higher copper amounts that were found in ACM in comparison to BSFM and FWSM (Table 2). The increase in copper as feed moisture content increased could be due to the ability of water to increase the contact area of feed particles through van der Waals forces (Tumuluru 2014) and thus increasing the availability of copper. In addition, water used in extrusion could have had traces of copper, which led to more amount of the element as moisture content was increased to 30%. Wheat, one of the ingredients used, has also been shown to influence copper levels (Singh et al. 2007). All the study diets met the minimum requirements for copper in tilapia (6 ppm), as outlined by KEBS standards (FDKS 2010), as well as the minimum requirements for catfish (5 ppm), as specified by FAO (2017). Thus, the ACM75 diet can be used to substitute FWSM and still obtain the same or higher values of copper.

Zinc in fish is involved in many metabolic pathways, serves as a specific cofactor for various enzymes and has also been shown to assist in healing of wounds (Watanabe et al. 1997; FAO 2017). The KEBS minimum requirements for tilapia ranges between 50 and 100 ppm depending on the stage of fish that is being targeted (FDKS 2010), while the minimum requirement for catfish is 20 ppm (FAO 2017). The fact that all the study formulations were between 100 and 200 ppm and that the control diet did not differ from BSFM50 and ACM75 demonstrate that BSFM and ACM can be used to substitute FWSM up to 50 and 75%, respectively, with regard to zinc. The extrusion process has little effect on zinc composition, as the moisture content did not show significant influence on its content. Other studies have also documented the non-significant influence of extrusion on zinc and its bio-availability (Guy 2001; Kang and Chenoweth 2000).

Manganese is required in fish for normal brain functioning as well as for the metabolism of carbohydrates and lipids (Watanabe et al. 1997). The minimum requirements for manganese in feeds intended for tilapia and catfish are 12 and 2.4 ppm, respectively (FAO 2017). Findings from our study indicate that BSFM50 and ACM75 are best suited for feeding tilapia, as they contain more than 12 ppm of manganese. However, all the formulations met the minimum requirements for manganese that is required by catfish. Manganese is one of the elements that is least affected by the extrusion process and its parameters (Razzaq et al. 2012; Singh et al. 2007), partly due to its electronic configuration that lenders it less susceptible to the formation of chelates (Alonso et al. 2001). This explains why feed moisture content did not influence the manganese content of the final products.

Iron is necessary for hemoglobin and myoglobin pigments (FAO 2017) and is actively involved in oxidation/reduction reactions (Watanabe et al. 1997). The decrease in iron content as the level of FWSM substitution with BSFM increased is attributed to the low iron content that is found in BSFM in comparison to FWSM (Table 3). Similarly, the corresponding increase in iron content as FWSM is substituted with ACM is due to the higher iron content in ACM than in FWSM (Table 3). Other researchers have also shown that ACM has higher iron content than BSFM (Makkar et al. 2014; Finke 2002; Barker et al. 1998). It should be noted that all the formulations exceeded the minimum iron requirements for tilapia and catfish (60 and 30 ppm, respectively) (FAO 2017; FDKS 2010). The extrusion process influenced the amount of iron in the final products where, for each formulation, the 30% feed moisture content resulted in feeds having higher iron content. This could be attributed to the fact that moisture and high temperature of the extrusion (120 °C) increased the abrasion of the extruder barrel, screw and the die, thereby increasing the final iron content. Similar phenomena have been reported by others (Razzaq et al. 2012; Guy 2001; Alonso et al. 2001; Camire et al. 1990).

Leaching activity

Leaching activity is determined by water stability of feeds, where pellets with low water stability exhibit higher leaching effect (Saalah et al. 2010; Ayadi et al. 2011). Irungu et al. (2018) had reported that at 30% moisture content, ACM75- and BSFM75-based pellets had higher water stability than fish feeds that had 100% FWSM. This explains why the control diet had higher leaching effect for most of minerals than the ACM75 and BSFM75 diets. Our findings are in agreement with those of Haghbayan and Mehrgan (2015) and Hardy (2002), who reported that fish meal in fish feed increased the nutrient load in wastewater and thus contributed to higher leaching effect. Phosphorus and sodium leaching leads to eutrophication and algal blooms (Jia et al. 2015; Xiang and Zhou 2011; Saalah et al. 2010), while lack of adequate potassium affects nerve and muscle formation in fish (Wurts 1993) and thus there is need to substitute fish meal with either ACM75 or BSFM75 meals, as the two lead to low leaching of phosphorus, sodium and potassium as compared to the control. In addition, fish do not get phosphorus readily from the water environment (Tang et al. 2012), and thus feeds that have higher phosphorus leaching activity contribute to phosphorus deficiency in fish. Calcium and magnesium leaching do not pose a serious challenge, as both contribute to water hardness which is important for the growth of fish. For example, relatively high amounts of free calcium ions in water reduce the loss of sodium and potassium salts from the body of fish (Wurts 1993). However, high concentration of hard water may increase the susceptibility of fish to columnaris disease (Avant 2015). The decreasing copper, zinc and iron leached levels in the BSFM75 diets and copper and zinc in the ACM75 and control diets, respectively, could be due to formation of complexes by the macro minerals rendering these micro elements unavailable with time.

Conclusion

Both adult cricket meal and black soldier fly larvae meal can be used to substitute freshwater shrimp meal up to 75% in fish feeds and supply both tilapia and catfish with adequate quantities of minerals such as potassium. Further studies should investigate the bio-availability of minerals for fish from these diets.

Acknowledgements This work was supported by the 'INSFEED—Insect Feed for Poultry and Fish Production in Kenya and Uganda' project (Cultivate Africa Grant no.: 107839-001), co-funded by the International Development Research Centre, Canada (IDRC) and Australia Centre for International Agricultural Research (ACIAR).

Conflict of interest The authors declare that they have no conflict of interest.

References

Afiukwa J, Okereke C (2013) Evaluation of proximate and mineral contents of termite (*Trinervitermes germinatus*) from Abakaliki and Ndieze izzi, Ebonyi state, Nigeria. Am J Food Nutr 3(3):98–104. https://doi.org/10.5251/ajfn.2013.3.3.98.104

Alam MS, Kaur J, Khaira H, Gupta K (2016) Extrusion and extruded products: changes in quality attributes as affected by extrusion process parameters: a review. Crit Rev Food Sci Nutr 56(3):445–473

Alonso R, Rubio LA, Muzquiz M, Marzo F (2001) The effect of extrusion cooking on mineral bioavailability in pea and kidney bean seed meals. Anim Feed Sci Technol 94(1):1–13

Avant S (2015) Fish reared in hard water are more susceptible to columnaris disease. United States Department of Agriculture. https://www.ars.usda.gov/news-events/news/research-news/2015/fish-reared-in-hard-water-are-more-susceptible-to-columnaris-disease/. Accessed 2 June 2017

Ayadi FY, Muthukumarappan K, Rosentrater KA, Brown ML (2011) Twin-screw extrusion processing of rainbow trout (Oncorhynchus mykiss) feeds using various levels of corn-based distillers dried grains with solubles (DDGS). Cereal Chem 88(4):363–374

Barker D, Fitzpatrick MP, Dierenfeld ES (1998) Nutrient composition of selected whole invertebrates. Zoo Biol 17(2):123–134

Barroso FG, de Haro C, Sánchez-Muros MJ, Venegas E, Martínez-Sánchez A, Pérez-Bañón C (2014) The potential of various insect species for use as food for fish. Aquaculture 422–423:193–201. https://doi.org/10.1016/j.aquaculture.2013.12.024

Belluco S, Losasso C, Maggioletti M, Alonzi CC, Paoletti MG, Ricci A (2013) Edible insects in a food safety and nutritional perspective: a critical review. Compr Rev Food Sci Food Saf 12(3):296–313. https://doi.org/10.1111/1541-4337.12014

Bondari K, Sheppard DC (1981) Soldier fly larvae as feed in commercial fish production. Aquaculture 24:103–109

Braga WF, Araújo JG, Martins GP, Oliveira SL, Guimarães IG (2016) Dietary total phosphorus supplementation in goldfish diets. Latin Amer J Aquat Res 44(1):129–136

Camire ME, Camire A, Krumhar K (1990) Chemical and nutritional changes in foods during extrusion. Crit Rev Food Sci Nutr 29(1):35–57

Chakravorty J, Ghosh S, Jung C, Meyer-Rochow VB (2014) Nutritional composition of *Chondacris rosea* and *Brachytrupes orientalis*: two common insects used as food by tribes of Arunachal Pradesh, India. J Asia Pac Entomol 17(3):407–415. https://doi.org/10.1016/j.aspen.2014.03.007

Chavez-Sanchez C, Martinez-Palacios CA, Martinez-Perez G, Ross LG (2000) Phosphorus and calcium requirements in the diet of the American cichlid *Cichlasoma urophthalmus* (Gunther). Aquac Nutr 6(1):1–10

Cheftel JC (1986) Nutritional effects of extrusion-cooking. Food Chem 20(4):263–283

FAO (2010) The state of world fisheries and aquaculture. FAO, Rome

FAO (2012) The state of world fisheries and aquaculture. FAO, Rome, p 209

FAO (2013) Insects as animal feed. FAO, Rome

FAO (2014) The state of world fisheries and aquaculture. FAO, Rome, p 223

FAO (2015) The state of food insecurity in the world. Meeting the 2015 international hunger targets: Taking stock of uneven progress. http://www.fao.org/3/a-i4646e.pdf. Accessed 20 Mar 2017

FAO (2016) Minerals. FAO, Rome

FAO (2017) Aquaculture feed and fertilizer resources information system. FAO, Rome. http://www.fao.org/fileadmin/user_upload/affris/docs/North_African_Catfish/English/table_4.htm. Accessed 3 May 2017

Filipovic S, Kormanjos S, Sakac M, Filipovic J, Psodorov D, Okanovic D (2010) Effect of extrusion on nutritive value of animal feed. In 2nd workshop feed-to-food FP7 REGPOT-3. Extrusion technology in feed and food processing, Thematic Proceedings. Novi Sad, Serbia, 19–21 October, 2010. Institute for Food Technology, pp 97–116

Final Draft Kenya Standard (FDKS-2289-1) (2010) Compounded Tilapia feeds: specification. Kenya Bureau of Standards, Nairobi

Finke MD (2002) Complete nutrient composition of commercially raised invertebrates used as food for insectivores. Zoo Biol 21(3):269–285

Guy R (2001) Extrusion cooking: technologies and applications. Woodhead Publishing, Cambridge

Haghbayan S, Mehrgan MS (2015) The effect of replacing fish meal in the diet with enzyme-treated soybean meal (HP310) on growth and body composition of Rainbow trout Fry. Molecules 20(12):21058–21066

Halden AN, Lindberg JE, Masembe C (2014) Aquaculture—a fast growing food production sector. SLU Glob:42–45

Hardy RW (2002) Rainbow trout, *Oncorhynchus mykiss*. Nutrient requirements and feeding of finfish for aquaculture. In: Wibster CD (ed) Aquaculture research center. Aurburn, Alabama, ISD, pp 184–202

Hassaan MS, Soltan MA, Agouz HM, Badr AM (2013) Influences of calcium/phosphorus ratio on supplemental microbial phytase efficiency for Nile tilapia (*Oreochromis niloticus*). Egypt J Aquat Res 39(3):205–213

Henry M, Gasco L, Piccolo G, Fountoulaki E (2015) Review on the use of insects in the diet of farmed fish: past and future. Anim Feed Sci Technol 203:1–22

Irungu FG, Mutungi C, Faraj AK, Affognon H, Kibet N, Ekesi S, Nakimbugwe D, Fiaboe KKM (2018) Physicochemical properties of extruded fish feeds pellets containing black soldier fly (*Hermetia illucens*) larvae and adult cricket (*Acheta domesticus*) meals. J Insects Food Feed. https://doi.org/10.3920/JIFF2017.0008

Jahnen-Dechent W, Ketteler M (2012) Magnesium basics. Clin Kidney J 5(Suppl 1):i3–i14

Jia B, Tang Y, Tian L, Franz L, Alewell C, Huang JH (2015) Impact of fish farming on phosphorus in reservoir sediments. Sci Rep 5:16617

Kang SY, Chenoweth WL (2000) Zinc bioavailability in a semolina/soy protein mixture was not affected by extrusion processing. Nutr Sci 3(2):71–76

Kinyuru JN, Kenji GM, Njoroge SM, Ayieko M (2010) Effect of processing methods on the in vitro protein digestibility and vitamin content of edible winged termite (*Macrotermes subhylanus*) and grasshopper (*Ruspolia differens*). Food Bioprocess Technol 3(5):778–782. https://doi.org/10.1007/s11947-009-0264-1

Makkar HP, Tran G, Heuzé V, Ankers P (2014) State-of-the-art on use of insects as animal feed. Anim Feed Sci Technol 197:1–33

Msangi S, Kobayashi M, Batka M, Vannuccini S, Dey MM, Anderson JL (2013) Fish to 2030: prospects for fisheries and aquaculture. World Bank Report (83177-GLB)

Murphy SP, Allen LH (2003) Nutritional importance of animal source foods. J Nutr 133(11):3932S–3935S

Mutungi C, Irungu FG, Nduko J, Mutua F, Affognon H, Nakimbugwe D, Ekesi S, Fiaboe KKM (2017) Postharvest processes of edible insects in Africa: a review of processing methods, and the implications for nutrition, safety and new products development. Crit Rev Food Sci Nutr. https://doi.org/10.1080/10408398.2017.1365330

National Research Council (US)-Subcommittee on Fish Nutrition (1993) Nutrient requirements of fish. National Academy Press, Washington DC, p 114

National Research Council (US)-Subcommittee on Fish Nutrition (2011) Nutrient requirements of fish and shrimp. National Academy Press, Washington DC

Newton GL, Booram CV, Barker RW, Hale OM (1977) Dried larvae meal as a supplement for swine. J Anim Sci 44(3):395–400

Nielsen SS (2010) Food analysis. Springer, New York, p 550

Nikmaram N, Kamani MH, Ghalavand R (2015) The effects of extrusion cooking on anti-nutritional factors, chemical properties and contaminating micro-organisms in food. Int J Farm Allied Sci 4(4):352–354

Njagi KA, Njati IC, Huka GS (2013) Factors affecting profitability of fish farming under economic stimulus program in Tigania East District, Meru County, Kenya. IOSR J Bus Manag 15(3):25–36

Oliva-Teles A (2012) Nutrition and health of aquaculture fish. J Fish Dis 35(2):83–108

Oonincx DG, van Itterbeeck J, Heetkamp MJ, van den Brand H, van Loon JJ, van Huis A (2010) An exploration on greenhouse gas and ammonia production by insect species suitable for animal or human consumption. PLoS One 5:e14445

Razzaq MR, Anjum FM, Khan MI (2012) Effect of extruder variables on chemical characteristics of maize (*Zea mays* L.) extrudates. Pak J Food Sci 22(2):108–116

Rumpold BA, Fröhling A, Reineke K, Knorr D, Boguslawski S, Ehlbeck J, Schlüter O (2014) Comparison of volumetric and surface decontamination techniques for innovative processing of mealworm larvae (*Tenebrio molitor*). Innov Food Sci Emerg Technol 26:232–241

Saalah S, Shapawi R, Othman NA, Bono A (2010) Effect of formula variation in the properties of fish feed pellet. J Appl Sci 10(21):2537–2543

Singh S, Gamlath S, Wakeling L (2007) Nutritional aspects of food extrusion: a review. Int J Food Sci Technol 42(8):916–929

Tang Q, Wang C, Xie C, Jin J, Huang Y (2012) Dietary available phosphorus affected growth performance, body composition, and hepatic antioxidant property of juvenile yellow catfish (*Pelteobagrus fulvidraco*). Sci World J 2012:1–9

Tessenderlo G (2005) The role of phosphorus in the nutrition of Salmond fish. Tessenderlo Group, Brussels

Thompson KR, Muzinic LA, Engler LS, Webster CD (2005) Evaluation of practical diets containing different protein levels, with or without fish meal, for juvenile Australian red claw crayfish (*Cherax quadricarinatus*). Aquaculture 244(1):241–249

Trumpo P, Yates AA, Schlicker S, Poos M (2001) Dietary reference intakes: vitamin A, vitamin K, arsenic, boron, chromium, copper, iodine, iron, manganese, molybdenum, nickel, silicon, vanadium and zinc. J Am Diet Assoc 101(3):294–301

Tumuluru JS (2014) Effect of process variables on the density and durability of the pellets made from high moisture corn stover. Biosyst Eng 119:44–57

Tumuluru JS, Wright CT, Hess JR, Kenney KL (2011) A review of biomass densification systems to develop uniform feedstock commodities for bioenergy application. Biofuels Bioprod Biorefin 5(6):683–707

Van Huis A (2013) Potential of Insects as food and feed in assuring food security. Annu Rev Entomol 58(1):563–583

Watanabe T, Kiron V, Satoh S (1997) Trace minerals in fish nutrition. Aquaculture 151(1):185–207

Webster CD, Lim C (2002) Nutrient requirements and feeding of finfish for aquaculture. CABI Publishing, UK

Wurts WA (1993) Understanding water hardness. World aquaculture 24(1):18

Xiang SL, Zhou WB (2011) Phosphorus forms and distribution in the sediments of Poyang Lake, China. Int J Sedim Res 26(2):230–238

5

Efficacy of insect larval meal to replace fish meal in juvenile barramundi, *Lates calcarifer* reared in freshwater

Kumar Katya · M. Z. S. Borsra · Dev Ganesan · Giva Kuppusamy · Max Herriman · Andrew Salter · Sayed Azam Ali

Abstract The present experiment was conducted to evaluate the efficacy of dietary protein from black soldier fly, *Hermetia illucens,* larval meal (BSFL) to replace fish meal (FM) protein in juvenile barramundi, *Lates calcarifer*. Larvae of black soldier fly were fed with the underutilised crop, sesbania, *Sesbania grandiflora.* Five isonitrogenous (44% crude protein) and isocaloric (16.0 kJ available energy/g) experimental diets were formulated to replace FM using processed BSFL meal at 0 (control), 25% (BSFL25), 50% (BSFL50), 75% (BSFL75) and 100% (BSFL100). Data for proximate and amino acid analysis suggested BSFL meal as an inferior protein ingredient than FM, but parallel to soybean meal. At the end of 8 weeks of fish feeding trial, there were no significant differences in the average weight gain (WG) and specific growth rate among the group of fish-fed control, BSFL25 and BSFL50 diets ($P < 0.05$). Although numerical differences were recorded in the fish whole-body proximate composition, crude protein and moisture content were not much affected by the different dietary treatments. Essential amino acids including arginine, histidine, lysine and methionine were found to be higher in the whole body of fish-fed BSFL100 diet. Broken line regression analysis of average WG showed an optimum FM replacement level of 28.4% with BSFL meal. Therefore, the present experiment clearly demonstrates that the maximal dietary inclusion level of BSFL meal as FM protein replacer for the optimum growth of juvenile barramundi reared in freshwater could be greater than 28.4% but less than 50%, without any adverse effects on the fish whole-body proximate and amino acid composition.

Keywords Black soldier fly · Fish meal · Growth · Amino acids · Barramundi

Introduction

Aquaculture has become the fastest growing food-producing industry around the world; however, the production efficiency of aquaculture could be further increased through the development of nutritious and cost-effective alternatives to traditional and finite marine protein feedstuffs such as fish meal (FM) (Suarez et al. 2013; Katya et al. 2014). Numerous studies have been carried out to replace or reduce the inclusion level of FM and to identify promising alternative protein sources in aquafeeds. Some of the thoroughly investigated ingredients to partially replace FM in fish feed include plant protein sources such as soybean meal, corn gluten

K. Katya (✉) · M. Z. S. Borsra · D. Ganesan · G. Kuppusamy · M. Herriman · S. A. Ali
Crops for The Future, Jalan Broga, 43500 Kuala Lumpur, Malaysia
e-mail: kumar.katya@cffresearch.org

A. Salter
School of Biosciences, University of Nottingham, Nottingham, UK

meal and soy protein concentrate, by-products of mushroom (Carter and Hauler 2000; Refstie et al. 2001; Choi et al. 2004; Lim et al. 2004; Kim et al. 2008; Katya et al. 2014) and animal protein sources such as meat and bone meal, blood meal, feather meal, poultry by-product meal and lysine by-product (Bai et al. 1997, 1998; Lee and Bai 1997a, b). However, least attention has been given to access the use of unconventional but promising feedstuffs such as insect meal in commercial fish feed formulation.

In last few years, a considerable number of researches have shed the light on the efficacy of dietary insect meal in several aquaculture species including tilapia, *Oreochromis niloticus,* salmon, *Salmo salar*, African catfish, *Clarias gariepinus,* turbot, *Psetta maxima,* and rainbow trout, *Oncorhynchus mykiss* (Kroeckel et al. 2012; Fasakin et al. 2003; Lock et al. 2015; St-Hilaire et al. 2007a, b; Ogunji et al. 2007, 2008; Rana et al. 2015) feeds. Among a number of promising insect species, farming of black soldier fly, *Hermetia illucens* larvae/pre-pupae has received top priority for the commercial exploitation (Lock et al. 2015). The overall nutrient content viz. crude protein, lipid and essential amino acid content of insect meal closely resemble with FM, and the pre-pupae meal have been demonstrated to be suitably alternative for FM (Kroeckel et al. 2012; Lock et al. 2015; Rana et al. 2015) in several aquaculture species.

Barramundi or Asian sea bass, *Lates calcarifer* is one of the commercially important fish species widely cultured in several Asian countries. Nutrient requirements for this species has been well defined, considerable efforts have also been made in the three decades to develop nutritionally balanced and cost-effective diets (Glencross 2003). The commercial feed formulation for this species heavily depends upon the FM as the primary protein source. Although, a number of FM alternative ingredients of plant and animal origin have been investigated (Boonyaratpalin et al. 1998; Glencross 2003), use of locally available but nutritionally efficient feedstuff in barramundi feeds is warranted. Therefore, the present experiment was conducted to evaluate the replacement efficacy of protein from BSFL meal with FM protein in juvenile barramundi, reared in freshwater.

Materials and methods

Insects rearing and larval meal production

The insectarium and larval rearing units were set up at the field research center of Crops for The Future (CFF), Kuala Lumpur, Malaysia. Based upon the experience on behaviors of adult insects as well as using the knowledge on the life cycle of black soldier fly, a calm place was selected and natural environment were simulated to attract and induce the mating of adult insects. Six small plastic buckets filled with banana leaves and hay were hanged around the corners and center of the insectarium to collect insect eggs every morning. A series of round bottom 500 L capacity plastic tanks were filled one-third with the blend of underutilized crop sesbania, *Sesbania grandiflora.* Self-harvesting larvae collection chambers were prepared using flat wooden sheet fixed at 45° angle opening outside the plastic tanks. The BSFL production method were similar that is used by Ajani et al. (2004) and Adesulu and Mustapha (2000) for housefly maggot meal.

Experimental design and diets

Formulation and proximate composition of the experimental diets is shown in Table 1. Five isonitrogenous (44% crude protein) and isocaloric (16.0 kJ available energy/g) diets were formulated to replace fish meal using BSFL meal at 0 (control), 25% ($BSFL_{25}$), 50% ($BSFL_{50}$), 75% ($BSFL_{75}$) and 100% ($BSFL_{100}$). Fish-meal, soybean meal, corn gluten meal and squid liver powder served as the major protein sources in the experimental diets; fish oil was used as the lipid source while dextrin flour was the carbohydrate source. Fish meal was substituted at graded level and accordingly adjustments were done in other ingredients to balance the protein and energy content in all experimental diets. Feeds were designed and formulated to meet the optimum nutrient requirement of experimental fish, following the nutrient requirements extensively reviewed and recommended for juvenile size barramundi (Glencross 2003).

Procedures for diet preparation and storage were as previously described by Bai and Kim (1997). After thoroughly mixing the dry ingredients and fish oil with 30%-filtered tap water, experimental diets were pelleted with a laboratory pelleting machine without heating, using a 2-mm diameter module (MK-NG 1500

Table 1 Formulation and proximate composition of the experimental control diet (% dry matter base)

Ingredients	Control	$BSFL_{25}$	$BSFL_{50}$	$BSFL_{75}$	$BSFL_{100}$
Fish meal	40	30	20	10	0
BSF meal	0	15.4	30.8	46.2	61.6
Dextrin	16.4	12.3	8.2	4.1	0
Soybean meal	12	12	12	12	12
Corn meal	5.6	5.7	5.8	5.9	6.0
Squid liver powder	15	15	15	15	15
Fish oil	6.0	4.6	3.2	1.8	0.4
Vitamin premix[a]	3	3	3	3	3
Mineral premix[b]	2	2	2	2	2
Proximate composition[c]					
Crude protein	42.0	43.1	42.8	42.4	41.8
Crude lipid	12.0	11.8	12.1	12.0	12.4
Moisture	9.2	9.7	10.1	9.1	9.7
Crude ash	10.2	10.1	10.4	10.2	10.3

Diets represent; control (without black soldier fly larval meal inclusion), BSFL25 (25%), BSFL50 (50%), BSFL75 (75%) and BSFL100 (100% fish meal replacement using black soldier fly larval meal)

[a](as mg/kg in diets): ascorbic acid, 300; dl-calcium pantothenate, 150; choline bitatrate, 3000; inositol, 150; menadione, 6; niacin, 150; pyridoxine·HCl, 15; riboflavin, 30; thiamine mononitrate, 15; dl-α-tocopherol acetate, 201; retinyl acetate, 6; biotin, 1.5; folic acid, 5.4; B12, 0.06

[b]Contains (as mg/kg in diets): NaCl, 437.4; MgSO4·7H2O, 1379.8; NaH2P4·2H2O, 877.8; Ca(H2PO4)2·2H2O, 1366.7; KH2PO4, 2414; ZnSO4·7H2O, 226.4; Fe-citrate, 299; Ca-lactate, 3004; MnSO4, 0.016; FeSO4, 0.0378; CuSO4, 0.00033; Calcium iodate, 0.0006; MgO, 0.00135; NaSeO3, 0.00025

[c]Values are average from duplicate groups of samples

WSL food grinder). All the feeds were air-dried for 24 h and after processing, all diets were kept at −4 °C in the cold storage until use.

Experimental fish and feeding trial

Juvenile size barramundi, *Lates calcarifer* fish acclimated to freshwater were obtained from the hatchery Biodesaru, Gelang Patah Johor, Malaysia. Fish were transported to the experimental station (FishPLUS laboratory, FRC, Crops for The Future) and acclimated to the experimental conditions for 2 weeks before the feeding trial began. During the acclimation period, fish were fed control diet twice daily (10:00 and 1800 h) at approximately 3% of wet body weight/day. Twenty barramundi fish averaging 6.7 ± 0.1 g (mean ± standard deviation) were randomly distributed in fifteen individual fish groups to each of fifteen aquaria. Each aquarium was then randomly assigned to one of three replicates of five experimental diets. Triplicate groups of fish were fed one of the experimental diet twice daily (10:00 and 17:00 h) at approximately 2% of wet body weight/day at the beginning and 3% of wet body weight/day at the end of the feeding trial for 8 weeks. Total fish weight in each aquarium was determined every 2 weeks, and the amount of feeds fed to the fish was adjusted accordingly.

The feeding trial was conducted in an indoor semi-recirculation system with fifteen 300 L of aquaria receiving filtered freshwater from the center tank. Supplemental aeration was provided to maintain dissolved oxygen levels near 6.5 ± 0.5 (mean ± SD) ppm. Water temperature was 24 ± 1 °C (mean ± SD); pH was 7.5 ± 0.3 (mean ± SD) and the photoperiod of 12:12 (light: dark) was maintained throughout the experimental period.

Sample collection and analysis

At the end of the feeding trial, all of the fish were weighed and counted for the calculation of weight gain (WG), specific growth rate (SGR), feed conversion ratio (FCR) and survival rate. Growth performance parameters including WG, SGR and FCR were calculated using formulas.

$$\text{Weight gain }(\%) = [(\text{final weight} - \text{initial weight})/\text{initial weight}] \times 100.$$

$$\text{Specific growth rate }\left(\%\ \text{day}^{-1}\right) = [(\log_e \text{final wt.} - \log_e \text{initial wt.})/\text{days}] \times 100.$$

$$\text{Feed conversion ratio} = \text{feed}/\text{wet weight gain}.$$

Protocol for the feeding trial, fish anesthesia, sample collection and analysis were followed as prescribed and approved by the Animal ethics committee, University of Nottingham. After the final weighing, five fish were randomly collected from each aquaria and frozen at −20 °C for analysis. The proximate composition and amino acid analysis of the experimental diets, ingredients and fish whole body was performed by the standard methods of AOAC (1995). Samples were dried to a constant weight at 105 °C to determine moisture content. Ash was determined by incineration at 550 °C, crude lipid by soxhlet extraction using a Soxtec system 1046 (Tecator AB, Hoganas, Sweden), and crude protein by Kjeldahl method (N × 6.25) after acid digestion. Amino acid analysis was carried out using Ninhydrin method (Sykam Amino Acid Analyzer S433; Sykam, Eresing, Germany).

Statistical analysis

All data were analyzed by one-way ANOVA (Statistix 3.1, Analytical Software, St. Paul, MN, USA) to test the effects of the dietary treatments. When a significant treatment effect was observed, a least significant difference (LSD) test was used to compare means. Treatment effects were considered at $P < 0.05$ level of significance. Broken line model (Robbins et al. 1979) was used to evaluate the optimum dietary level of black soldier fly larval meal as fish meal replacer in barramundi.

Results

Nutrient profile of protein ingredients

Table 2 shows the proximate composition of FM, soybean meal and BSFL meal used as the major protein ingredients in this study. The crude protein content was found to be highest for FM followed by soybean meal and BSFL meal. On the other hand, crude lipid was recorded to be highest for BSFL meal followed by FM and soybean meal. Whereas, soybean meal exhibited higher fiber content compared to FM and BSFL meal. The amino acid analysis of FM and BSFL meal has been presented in Fig. 1.

Table 2 Analyzed nutrient composition (% DM base) of fish meal, soybean meal and black soldier fly, *Hermetia illucens* larval (BSFL) meal used as the major protein ingredients in the experimental diets formulation

Ingredients	Crude protein	Crude lipid	Crude ash	Moisture	Fiber
Fishmeal	69.9	11.5	9.0	9.3	<0.1
Soybean meal	46.3	1.0	6.2	11.9	29.9
BSFL meal	43.5	16.7	13.9	5.9	9.5

Values are average from duplicate groups of samples

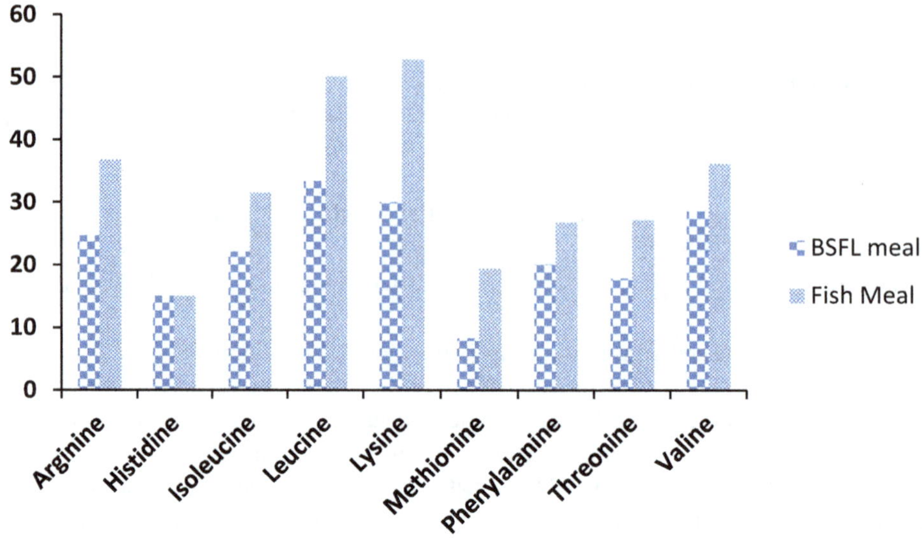

Fig. 1 Amino acid composition of black soldier fly, *Hermetia illucens* larval (BSFL) meal and fish meal. Values are average from duplicate groups of samples

Growth performances

The growth performances including weight gain (WG), specific growth rate (SGR), feed conversion ratio (FCR) and survival rate of barramundi-fed five experimental diets for 8 weeks have been summarized in Table 3. At the end of feeding trial, there were no significant differences in the average WG and SGR among the group of fish-fed control, $BSFL_{25}$ and $BSFL_{50}$ diets. However, WG and SGR of fish-fed $BSFL_{75}$ and $BSFL_{100}$ diets were significantly lower than that of fish-fed control diet. Whereas, data for FCR showed a significantly higher value for the group of fish-fed $BSFL_{100}$ diet than those of fish-fed all other diets. However, there was no significant difference in the FCR among the group of fish-fed control, $BSFL_{25}$, $BSFL_{50}$, and $BSFL_{75}$ diets. Average survival rate among different experimental groups numerically ranged between 80.5 and 100%. Although significant differences were recorded in the survival rate among the group of fish-fed different experimental diets, no clear trend in the effects of dietary treatments on fish survival rate could be drawn. Broken line regression analysis of the average WG indicated that BSFL meal could replace 28.4% of FM without any adverse effects on the growth performance of barramundi (Fig. 2).

Table 3 Growth performance of barramundi, *Lates calcarifer*-fed black soldier fly, *Hermetia illucens* larval (BSFL) meal as a fish meal replacer for 8 weeks

	Diets[1]				
	Control	$BSFL_{25}$	$BSFL_{50}$	$BSFL_{75}$	$BSFL_{100}$
WG (%)[2]	136.3 ± 6^{a}	119 ± 8^{ab}	110.3 ± 7^{ab}	92.3 ± 11^{b}	24.3 ± 6^{c}
SGR (%)/day[3]	2 ± 0.06^{a}	1.8 ± 0.08^{ab}	1.7 ± 0.08^{ab}	1.5 ± 0.13^{b}	0.5 ± 0.12^{c}
FCR[4]	2 ± 0.1^{b}	2.3 ± 0.1^{b}	2.4 ± 0.1^{b}	3.2 ± 0.1^{b}	11.3 ± 3.6^{a}
Survival (%)	94.4 ± 3.9^{ab}	100^{a}	94.4 ± 3.9^{ab}	80.5 ± 14.1^{b}	86.1 ± 10.3^{ab}

Means of triplicate groups of fish where values in the same row with different superscripts are significantly different ($P < 0.05$)

[1]Diets represent; control (without black soldier fly larval meal inclusion), BSFL25 (25%), BSFL50 (50%), BSFL75 (75%) and BSFL100 (100% fish meal replacement using black soldier fly larval meal)

[2]Weight gain (%) = [(final wt. − initial wt.)/initial wt.] × 100

[3]Specific growth rate (% day^{-1}) = [($\log_e$ final wt. − $\log_e$ initial wt.)/days] × 100

[4]Feed conversion ratio = feed given/wet weight gain

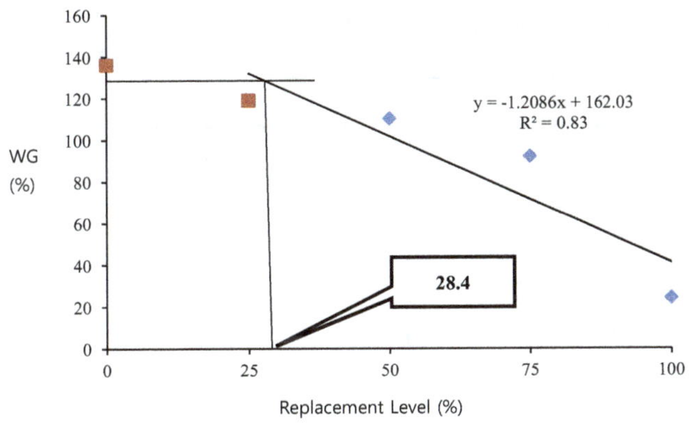

Fig. 2 Broken line regression analysis of average weight gain (WG) of barramundi, *Lates calcarifer*-fed black soldier fly, *Hermetia illucens* larval (BSFL) meal as a fish meal replacer for 8 weeks

Fish whole-body proximate composition

Whole-body proximate composition of barramundi-fed five experimental diets for 8 weeks has been shown in Table 4. Although numerical differences were recorded, crude protein and moisture content were found not to be much affected by the different dietary treatments. Whereas, fiber content was recorded highest for the group of fish-fed $BSFL_{100}$ diet.

Fish whole-body essential amino acid composition

Table 5 shows the whole body amino acid composition of barramundi-fed five different experimental diets for 8 weeks. Essential amino acids including arginine, histidine, lysine and methionine were found to be higher for the group of fish-fed $BSFL_{100}$ compared to fish-fed other experimental diets. Likewise, numerical differences were also observed for other essential amino acids, no clear trend among different dietary treatments could be drawn.

Discussion

Identifying promising alternative ingredients to minimize the inclusion level of dietary fish meal has been a much debated and extensively researched topic in aquaculture. Unfortunately, despite three decades of numerous replacement research, fish meal remains the prime protein ingredient in the majority of commercial fish feeds. Among plant protein, soybean meal is the extensively used ingredient in aquaculture diets due to its wide availability, comparatively low market price and promising nutrient profile (Akiyama et al. 1989; Gatlin et al. 2007; Trosvik et al. 2012; Cummins et al. 2013; Cummins et al. 2017). However, the price of soybean

Table 4 Whole body proximate composition (% dry weight basis) of barramundi, *Lates calcarifer*-fed black soldier fly *Hermetia illucens* larval (BSFL) meal as a fish meal replacer for 8 weeks

Diets[a]	Control	$BSFL_{25}$	$BSFL_{50}$	$BSFL_{75}$	$BSFL_{100}$
Moisture	74.3	71.9	79.2	74.4	78.0
Crude protein	62.2	63.1	55.7	69.4	66.6
Ash	15.6	15.1	17.4	19.8	19.0
Crude lipid	16.3	20.3	14.8	15.1	17.7
Fiber	6.0	6.1	4.5	3.6	8.1

Values are average from duplicate groups of samples

[a]Diets represent; control (without black soldier fly larval meal inclusion), BSFL25 (25%), BSFL50 (50%), BSFL75 (75%) and BSFL100 (100% fish meal replacement using black soldier fly larval meal)

Table 5 Whole body amino acids (ppm) composition of barramundi, *Lates calcarifer* fed black soldier fly, *Hermetia illucens* larval (BSFL) meal as a fish meal replacer for 8 weeks

Essential amino acids	Control	$BSFL_{25}$	$BSFL_{50}$	$BSFL_{75}$	$BSFL_{100}$
Arginine	6.6	6.6	6.6	6.5	9.4
Histidine	5.4	5.3	5.9	5.2	9.6
Isoleucine	1.0	1.0	1.0	1.0	0.9
Leucine	2.9	2.9	3.1	2.9	3.0
Lysine	2.8	2.8	2.8	2.7	3.2
Methionine	0.7	0.6	0.6	0.6	0.7
Phenylalanine	1.3	1.3	1.6	1.4	1.1
Threonine	2.4	2.2	2.2	2.2	2.1
Valine	0.5	0.5	0.5	0.5	0.5

Values are average from duplicate groups of samples

Diets represent; control (without black soldier fly larval meal inclusion), BSFL25 (25%), BSFL50 (50%), BSFL75 (75%) and BSFL100 (100% fish meal replacement using black soldier fly larval meal)

meal and other plant feedstuff has also increased dramatically due to their growing demand for the direct human consumption (FAO 2009). Therefore, attention could be directed towards the locally available and cheaper protein source, which may create flexibility in diet formulations and insect meal has been identified as one such ingredient (Rumpold and Schlüter 2013; Van der Spiegel et al. 2013; Van Huis 2013; Lock et al. 2015).

Black soldier fly was early been recognized as the potential candidate ingredient due to its rich nutritional profile comprised of 40% protein and 35% lipid (Bondari and Sheppard 1987). However, limited and scattered information on nutrient profile and culture techniques has restricted the use of insect larval meal in commercial fish feed formulation. Our observations for the protein content of BSFL meal closely resembled with those of previous reports, indicating 40–45% crude protein content in the larvae of black soldier fly (Yu and Chen 2009; Van Huis 2013). On the other hand, crude lipid content was found to be comparatively lower (16.7 vs. 30%) than those of values reported in these aforementioned studies. It is worthy to note that published reports on insect body composition show large differences between (Sanchez-Muros et al. 2015) and within species. It appears, the substrates used to feed the insect larvae might have profound effects on the final nutrient profile of processed larval meal. Whereas, data for amino acid comparison between fish meal and black soldier fly larval meal showed inferior but not limiting to be used in fish feed formulation, similar to the observation of Sanchez-Muros et al. (2015) with the insect, *Tenebrio molitor* meal. Overall performances for the the nutrient content (Table 2; Fig. 1) suggested, black soldier fly larvae meal as an inferior protein source than fish meal but closely similar with soybean meal.

The growth performance of barramundi, *Lates calcarifer* clearly demonstrated that larval meal of black soldier fly could be a promising alternative of fish meal in the feed of carnivorous fish species. Although WG and SGR decreased with fish meal replacement, no significant differences were recorded in these parameters among those of fish fed the BSFL-free diet and those fed-fish meal replaced with 25 and 50% of BSFL meal ($P < 0.05$). Whereas, published reports on BSFL meal as an alternative of fish meal are extremely few, but has been increasing recently (Cummins et al. 2017). Our results are in consistence with those of previous studies, suggesting an optimum fish meal replacement level of 20–50% using insect meal (Wing-Keong et al. 2001; St-Hilaire et al. 2007b; Alegbeleye et al. 2002; Cummins et al. 2017) in different aquaculture species. Similar to our observation, these aforementioned studies also reported the limitations in replacing fish meal beyond 50% of diet using insect meal. The possible reason for the reduction in fish growth performance with a corresponding increase in the dietary insect meal inclusion could be due to the substantial increase in chitin content (Sanchez-Muros et al. 2015) in addition to the deficiency in essential amino acid content (Cummins et al. 2017). Insect exoskeletons contain chitin and the b1, 4 bond in chitin has been reported to be indigestible for several fish species (Rust 2002). The ability of fish to degrade and digest chitin by the chitinase depends upon the fish species (Smith et al. 1989) and available gut bacteria. For instance, in red tilapia, *O. niloticus* × *O. hornorum* and nile tilapia, dietary fishmeal could be significantly replaced with shrimp meal, which is a rich source of chitin, without any adverse effects on the fish performance (El-Sayed 1998; Mansour 1998). Conversely, in the other species of tilapia, *O. niloticus* x *O. aureus*, the inclusion of chitin depressed growth (Shiau and Yu 1999). Our observations from the present study suggest, barramundi, *Lates calcarifer* might have limited efficiency to digest chitin thereby insect meal. Further investigation in this regard are recommended to study the chitin content in the larvae of black soldier fly and the ability of carnivorous fish species to digest chitin.

Further in the current experiment, whole-body proximate composition of fish was not much affected by the black soldier fly larvae inclusion as the fish meal replacer in the diets. Essential amino acids viz. arginine and histidine tend to be higher in the whole body of fish-fed fish meal-free diet, overall performance suggested no adverse effects of black soldier fly inclusion on the amino acid profile. Rainbow trout, *Onchrynchus mykiss* whole body showed a decreasing trend with a corresponding increase in the dietary black soldier fly inclusion (St-Hilaire et al. 2007a). Whereas, these authors found no difference in whole-body crude protein content in trout-fed different experimental diets. Likewise, in a recent study, the nutrient profile of shrimp whole body has been reported not much affected by fish meal replacement with dietary black soldier fly larval meal (Cummins et al. 2017). Our findings closely resembled with these two previous studies, suggesting negligible effects of dietary black soldier fly larval meal inclusion on the whole-body nutrient profile of aquaculture species. Overall observations from the present experiment are in agreement with those of previous reports, suggesting the potential benefit of using insect larval meal as fish meal alternative in fish feeds. Further studies are recommended to investigate the chitin content and efficiency of carnivorous fish species to digest it, with the view to further enhance the digestibility, and thereby use of insect meal in commercial fish feeds.

Conclusions

In conclusion, the current experiment opens a new avenue for the culture of insect larvae, using underutlised crop, Sesbania, *Sesbania grandiflora* as the substrate. Observations from the fish feeding trial, augment well for the use of black soldier fly larval meal as the dietary protein source in fish feeds. The maximal dietary inclusion level of protein from black soldier fly larval meal as fish meal protein replacer could be greater than 28.4 but less than 50%, characterized by the optimum growth performance, fish whole-body proximate and amino acid compositions in barramundi, *Lates calcarifer* reared in freshwater.

Acknowledgements This study was conducted under NEWTON Ungku Omar fund—Insect and Underutilised Crops Aquafeed Project (IUCAP), funded by British council and MIGHT, Malaysia. Authors also wish to thank, all the researchers, technical and field staffs of Crops for The Future, (CFF) for all their support during this project.

References

Adesulu EA, Mustapha AK (2000) Use of housefly maggots as a fishmeal replacer in tilapia culture: a recent vogue in Nigeria. In: Ftzimmons K, Filho JC (eds) Proceedings of the fifth international symposium on tilapia aquaculture, pp 138–143

Ajani EK, Nwanna LC, Musa BO (2004) Replacement of fishmeal with maggot meal in the diets of Nile tilapia, *Oreochromis niloticus*. J World Aquac Soc 35(52–277):54

Akiyama DM, Coelho SR, Lawrence AL, Robinson EH (1989) Apparent digestibility of feedstuffs by the marine shrimp *Penaeus vannamei* Boone. Nippon Suisan Gakkaishi 55:91–98

Alegbeleye WO, Obasa SO, Olude OO, Otubu K, Jimoh W (2002) Preliminary evaluation of the nutritive value of the variegated grasshopper (*Zonocerus variegatus L.*) for African catfish *Clarias gariepinus* (Burchell. 1822) fingerlings. Aquac Res 43:412–420

AOAC (1995) Official methods of analysis, 16th edn. Association of Official Analytical Chemists, Arlington

Bai SC, Kim KW (1997) Effects of dietary animal protein sources on growth and body composition in Korean rockfish, *Sebastes schlegelii*. J Aquac 10:77–85 **(in Korean with English Abstract)**

Bai SC, Jang HK, Kim KH (1997) Evaluation of feather meal and meat and bone meal as the fish meal replacer in Israeli carp diets. J Aquac 10:153–161

Bai SC, Jang HK, Cho ES (1998) Possible use of the animal by-product mixture as a dietary fish meal replacer in growing common carp (*Cyprinus carpio*). J Korean Fish Soc 31:380–385

Bondari K, Sheppard DC (1987) Soldier fly, *Hermetia illucens* L., larvae as feed for channel catfish, *Ictalurus punctatus* (Rafinesque), and blue tilapia, *Oreochromis aureus* (Steindachner). Aquac Res 18(3):209–220

Boonyaratpalin M, Suraneiranat P, Tunpibal T (1998) Replacement of fish meal with various types of soybean products in diets for the Asian sea bass, Lates calcarifer. Aquaculture 161:67–78

Carter CG, Hauler RC (2000) Fish meal replacement by plant meals in extruded feeds for Atlantic salmon, *Salmo salar* L. Aquaculture 185:299–311

Choi SM, Wang XJ, Lim SR, Kim KW, Bai SC, Shin IS (2004) Dietary dehulled soybean meal as a replacement for fish meal in fingerling and growing olive flounder, *Paralichthys olivaceus* (Temminck et Schlegel). Aquac Res 35:410–418

Cummins VC, Webster CD, Thompson KR, Velasquez A (2013) Replacement of fish meal with soybean meal, alone or in combination with distiller's dried grains with solubles, in practical diets for Pacific white shrimp, *Litopenaeus vannamei*, grown in a clear-water culture system. J World Aquac Soc 44:775–785

Cummins VC Jr, Rawles SD, Thompson KR, Velasquez A, Kobayashi Y, Hager J, Webster CD (2017) Evaluation of black soldier fly (*Hermetia illucens*) larvae meal as partial or total replacement of marine fish meal in practical diets for Pacific white shrimp (*Litopenaeus vannamei*). Aquaculture 473:337–344

El-Sayed A-FM (1998) Total replacement of fishmeal with animal protein sources in Nile tilapia (*Oreochromis niloticus*) feeds. Aquac Res 29:275–280

FAO (Food and Agricultural Organization of the United Nations) (2009) State of world fisheries and aquaculture 2008 (SOFIA). FAO Fisheries and Aquaculture Department, Rome

Fasakin EA, Balogun AM, Ajayi OO (2003) Evaluation of full-fat and defatted maggot meals in the feeding of clariid catfish *Clarias gariepinus* fingerlings. Aquac Res 34:733–738

Gatlin DM, Barrows FT, Brown P, Dabrowski K, Gaylord TG, Hardy RW, Herman E, Hu G, Krogdahl A, Nelson R, Overturf K, Rust M, Sealy W, Skonberg D, Souza EJ, Stone D, Wilson R, Wurtele E (2007) Expanding the utilization of sustainable plant products in aquafeeds: a review. Aquac Res 38:551–579

Glencross BD (2003) Can modelling assist with the determination of nutrient requirements in fish diet? In: van Barneveld RJ (ed) FRDC aquaculture nutrition subprogram annual workshop, Fremantle, WA, Australia. 29 May. pp 65–68

Katya K, Yun YH, Park G, Lee JY, Yoo G, Bai SC (2014) Evaluation of the efficacy of fermented by-product of mushroom, *Pleurotus ostreatus*, as a fish meal replacer in juvenile Amur catfish, *Silurus asotus*: effects on growth, serological characteristics and immune responses. Asian Australas J Anim Sci 27(10):1478–1486

Kim YC, Yoo GY, Wang X, Lee J, Shin IS, Bai SC (2008) Long term feeding effects of dietary dehulled soybean meal as a fish meal replacer in growing olive flounder, *Paralichthys olivaceus*. Asian Aust J Anim Sci 21:868–872

Kroeckel S, Harjes A-GE, Roth I, Katz H, Wuertz S, Susenbeth A, Schulz C (2012) When a turbot catches a fly: evaluation of a pre-pupae meal of the black soldier fly (*Hermetia illucens*) as fish meal substitute—growth performance and chitin degradation in juvenile turbot (*Psetta maxima*). Aquaculture 364:345–352

Lee KJ, Bai SC (1997a) Haemoglobin powder as a dietary fish meal replacer in juvenile Japanese eel, Anguilla japonica (Temminck et Schlegel). Aquac Res. 28:509–516

Lee KJ, Bai SC (1997b) Hemoglobin powder as a dietary animal protein source for juvenile Nile tilapia. Prog Fish-Cult. 59:266–271

Lim SR, Choi SM, Wang XJ, Kim KW, Shin IS, Min TS, Bai SC (2004) Effects of dehulled soybean meal as a fish meal replacer in diets for fingerling and growing Korean rockfish, *Sebastes schlegelii*. Aquaculture 231:457–468

Lock ER, ArisIwala T, Wagbo R (2015) Insect larval meal as an alternative source of nutrients in the diet of Atlantic salmon (*Salmo salar*) postmolt. Aquac Nutr 22(6):1–4

Mansour CR (1998) Nutrient requirements of red tilapia fingerlings. MSc. Thesis, Faculty of Science. University of Alexandria, Alexandria, Egypt, p 121

Muros MJS, De Haro C, Sanz A, Villareces S, Barroso FG (2015) Nutritional evaluation of *Tenebrio molitor* meal as fishmeal substitute for Tilapia (*Oreochromis niloticus*) diet. Aquac Nutr 22(5):943–955

Ogunji JO, Nimptsch J, Wiegand C, Schultz C (2007) Evaluation of the influence of housefly maggot meal (magmeal) diets on catalase, glutathione S-transferase and glycogen concentration in the liver of *Oreochromis niloticus* fingerling. Comp Biochem Physiol Part A: Mol Integr Physiol 147:942–947

Ogunji JO, Kloas W, Wirth M, Neumann N, Pietsch C (2008) Effect of housefly maggot meal (magmeal) diets on the performance, concentration of plasma glucose, cortisol and blood characteristics of *Oreochromis niloticus* fingerlings. J Anim Physiol Anim Nutr (Berlin) 92:511–518

Rana KS, Salam MA, Hashem S, Islam MA (2015) Development of black soldier fly larvae production technique as an alternate fish feed. Int J Res Fish Aquac 5(1):41–47

Refstie S, Storebakken T, Baeverfjord G, Roem A (2001) Long-term protein and lipid growth of Atlantic salmon (*Salmo salar*) fed diets with partial replacement of fish meal by soy protein products at medium or high lipid levels. Aquaculture 193:91–106

Robbins KR, Norton HW, Baker DH (1979) Estimation of nutrient requirements from growth data. J Nutr. 109(10):1710–1714

Rumpold BA, Schlüter OK (2013) Potential and challenges of insects as an innovative source for food and feed production. Innov Food Sci Emerg Technol 17:1–11

Rust MB (2002) Nutritional physiology. In: Halver JE, Hardy RW (eds) Fish nutrition. Academic Press, New York, pp 368–446

Shiau SY, Yu YP (1999) Dietary supplementation of chitin and chitosan depresses growth in tilapia, *Oreochromis niloticus* × *O. aureus*. Aquaculture 179:439–446

Smith JL, Opekun AR, Larkai E, Graham DY (1989) Sensitivity of the esophageal mucosa to pH in gastroesophageal reflux disease. Gastroenterology 96:683–689

St-Hilaire S, Cranfill K, McGuire MA, Mosley EE, Tomberlin JK, Newton L, Sealey W, Sheppard C, Irving S (2007a) Fish offal recycling by the black soldier fly produces a foodstuff high in omega-3 fatty acids. J World Aquac Soc 38:309–313

St-Hilaire S, Sheppard C, Tomberlin JK, Irving S, Newton L, McGuire MA, Mosley E, Hardy RW, Sealey W (2007b) Fly prepupae as a feedstuff for rainbow trout, *Oncorhynchus mykiss*. J World Aquac Soc 38:59–67

Suarez JA, Tudela C, Davis D, Daugherty Z, Taynor M, Glass L, Hoenig R, Buentello A, Beenetti DD (2013) Replacement of fish meal by a novel non-GM variety of soybean meal in cobia, *Rachycentron canadum*: ingredient nutrient digestibility and growth performance. Aquaculture 416–417:328–333

Trosvik KA, Rawles SD, Thompson KR, Metts LA, Gannam A, Twibell R, Webster CD (2012) Growth and body composition of Nile tilapia, *Oreochromis niloticus*, fry fed organic diets containing yeast extract and soybean meal as replacement for fish meal, with and without supplemental lysine and methionine. J World Aquac Soc 43:635–647

Van der Spiegel M, Noordam MY Van, der Fels-Klerx HJ (2013) Safety of novel protein sources (insects, microalgae, seaweed, duckweed, and rapeseed) and legislative aspects for their application in food and feed production. Comp Revis Food Sci Food Saf 12:662–678

Van Huis A (2013) Potential of insects as food and feed in assuring food security. Annu Revis Entomol 58:563–583

Wing-Keong N, Liew F-L, Ang LP, Wong KW (2001) Potential of mealworm (*Tenebrio molitor*) as an alternative protein source in practical diets for African catfish, *Clarias gariepinus*. Aquac Res 32:273–280

Yu GH, Chen YH (2009) Research progression on the larvae and prepupae of black soldier fly *Hermetia illucens* used as animal feedstuff. Chin Bull Entomol 46(1):41–45

6

Effect of dietary seaweed supplementation on growth performance, antioxidant and immune responses in European seabass (*Dicentrarchus labrax*) subjected to rearing temperature and salinity oscillations

Gaspar Lobo · Luís F. Pereira · José F. M. Gonçalves · Maria J. Peixoto · Rodrigo O. A. Ozório

Abstract The current study evaluated the effects of dietary seaweed supplementation in European seabass juveniles (*Dicentrarchus labrax*) subjected to rearing temperature and salinity oscillations, simulating natural variations in pond aquaculture conditions. Two experimental diets where formulated: a control diet (CTRL) with no supplementation and one supplemented with 7.5% seaweed mix (SW 2.5% *Fucus* sp., 2.5% *Gracilaria* sp. and 2.5% *Ulva* sp.). Seabass from both dietary groups (40.7 g initial body weight) was subjected to either combined salinity and temperature oscillations, or to fixed conditions. Growth performance, innate immune, and oxidative stress responses were evaluated. Results showed that seaweed supplementation had no significant effect on the analyzed parameters. However, environmental oscillations revealed significant effects on growth performance and oxidative stress response. Fish subjected to salinity and temperature oscillations had a significantly lower weight gain and daily growth index than those subjected to fixed conditions, regardless of dietary treatment. Total glutathione, oxidized glutathione, and catalase increased in fish subjected to oscillatory conditions. Lysozyme and peroxidase were not influenced by either diet or environmental conditions. In conclusion, this particular dietary seaweed mix supplementation did not mitigate the negative effects of environmental oscillations on growth performance and innate immune responses in European seabass.

Keywords Seaweeds · Dietary supplementation · Immunostimulants · European seabass · Environmental stressors · Immune and oxidative stress response

Introduction

In intensive aquaculture, fish are exposed to several stress factors of biological, chemical, and/or physical origin, that can affect fish well-being and immune status (Colombo et al. 1990). Sub-optimal biotic and abiotic

---

G. Lobo · L. F. Pereira · J. F. M. Gonçalves · M. J. Peixoto (✉) · R. O. A. Ozório
CIIMAR-Centro Interdisciplinar de Investigação Marinha e Ambiental, Universidade do Porto, Terminal de Cruzeiros do Porto de Leixões, Avenida General Norton de Matos s/n, 4450-208 Matosinhos, Portugal
e-mail: mjoaodiaspeixoto@gmail.com

G. Lobo · J. F. M. Gonçalves · M. J. Peixoto
ICBAS-Instituto de Ciências Biomédicas de Abel Salazar, Universidade do Porto, Rua de Jorge Viterbo Ferreira 228, 4050-313 Porto, Portugal

conditions are common causes for growth inhibition and mass mortalities (Sakai 1999). For instance, Vadstein et al. (1993) suggested that the majority of problems associated with fish production are caused by bacteria commonly present in hatcheries and emphasizes the importance of fish homeostasis. In addition, Björnsson and Ólafsdóttir (2006) showed a negative influence of ammonia and nitrogen high concentrations in cod rearing. It is important to highlight that fish have great adaptation strategies to counteract unfavorable environmental conditions. Hence, these constrictions will more often lead to decreases in their metabolic rates and reduction of feed intake than necessarily disease or death (Barnabe 2003).

In fish, feed intake and the kinetics of digestive processes are limited by temperature (Arana 1997). On a cellular level, temperature variations can affect protein and lipid conformation, as well as genetic modulation, altering transcription, translation, or path activation (Sengupta and Garrity 2013). Therefore, temperature influence over molecular kinetics modulates the occurrence of enzyme–subtract complexes, affecting numerous biochemical activities (Cornish-Bowden 1979). This kind of alterations induces an initial stress phase, increasing the formation of reactive oxygen species (ROS) and energy demands, potentially causing damage in proteins, lipids and DNA (Lushchak and Bagnyukova 2006). Similarly, salinity influences multiple aspects related to fish metabolism. Euryhaline fish have developed specific biochemical and physiological machinery that enables them to perceive and adapt to a wide range of salinity oscillations. When facing severe environmental conditions oscillations, compensation mechanisms are activated to relieve osmotic stress. Examples of some compensation mechanisms include alterations on fish standard metabolic rates (Lambert and Dutil 1997), feed intake, and feed conversion efficiency (Imsland et al. 2007). These adaptive strategies are a liability for aquaculture profitability, since they can cause immunosuppression, growth depression, and higher susceptibility to diseases. To solve this, alternative strategies of environmental-friendly techniques are constantly being explored. These alternatives include chemical therapy, vaccination (Bagni et al. 2000; Sakai 1999), or the application of dietary immunostimulants.

The use of dietary immunostimulants has been studied as a viable complement or alternative to traditional methods and has become widely accepted by fish farmers (Bricknell and Dalmo 2005). Immunostimulants, used as dietary supplements can improve fish immunity, increasing resistance during periods of great stress or more resilience during periods of deteriorating water conditions (Bagni et al. 2000). In vivo experiments with bacterial challenges recognized immunostimulants as antiparasitic, growth enhancers, and antibody production promoters (Bricknell and Dalmo 2005). Immunostimulatory substances, like glucans, have been reported to improve fish non-specific defense mechanisms (Jeney and Jeney 2002) and oral administration of yeast glucans increased protection against *Vibrio* spp. in Atlantic Salmon (Raa et al. 1992). Increased lysozyme activity was detected in seabass fed high levels of α-tocopherol (Obach et al. 1993) and enhanced phagocytic activity was reported in seabass fed diets supplemented with levamisole and glucans (Jeney et al. 1994).

Seaweeds have been recognized as an important resource for aquaculture. They are a source of bioactive compounds, such as laminarin, fucoidan, and β-glucans that act as immunostimulants in several fish species (Mustafa and Nakagawa 1995; Peixoto et al. 2016a, b, c).

The use of dietary seaweed supplementation has revealed significant effects on growth, feed utilization, stress response, physiological condition, and carcass quality of cultured fish. Supplementation with 5% *Ulva* spp. increased resistance to infection by *Pasteurela piscicida* in red seabream (Satoh 1987) and *Ulva* spp. and *Chondrus crispus* extracts have shown to increase respiratory burst and immune system stimulation in turbot and Atlantic salmon phagocytes (Castro et al. 2004). Besides immunocompetency, a positive correlation has been reported between seaweeds phenolic content and antioxidant capacity through lipid peroxidation inhibition (Heo et al. 2005). Extracts of red and brown seaweeds may also be effective therapeutic and prophylactic treatments against *Pseudomonas* spp. infection (Thanigaivel et al. 2015) and a 2.5% dietary supplementation of *Gracilaria* spp. led to improved antioxidant and innate system responses in European seabass (Peixoto et al. 2016a). In addition, according to Xu and Hirata (1990), the use of *Ulva* sp. as a feed additive for black seabream (*Acanthopagrus schlegeli*) and red seabream (*Pagrus major*) has been shown to be beneficial on growth and color. To date, studies on the dietary *Fucus* sp. supplementation in fish feed are very scarce. Most recently, however, Peixoto et al. (2016a, b, c) showed that *Fucus* sp. supplementation of up to 7.5% in practical diets for European seabass had no impact on growth performance.

To reduce economic losses caused by immunosuppression in intensive aquaculture, it is necessary to develop strategies that account for environmental stressors for farmed fish. Therefore, the aim of this study was to evaluate dietary seaweed supplementation on growth performance, antioxidant, and immune responses in seabass subjected to environmental stressors.

Materials and methods

All procedures were supervised by an accredited expert in laboratory animal science by the Portuguese Veterinary Authority (1005/92, DGV-Portugal, following FELASA category C recommendations), according to the guidelines on the protection of animals used for scientific purposes from the European directive 2010/63/UE.

Experimental diets

Two experimental diets were formulated according to the species requirements (isoproteic, 48%; isolipidic, 19%; isoenergetic, 23 kJ g^{-1}) (Table 1): a basal diet (CTRL) and a supplemented diet (SW), with an added seaweed mix at 7.5% level. The seaweed mix was supplied by ALGA+® and contained a mixture of dried and minced *Gracilaria* sp., *Ulva* sp., and *Fucus* sp. at a 1:1:1 ratio. Proximal composition of seaweeds is presented in Table 2.

Table 1 Feed formulation and proximate composition of the experimental diets

	Dietary treatments	
	CTRL	SW
Feed ingredients (% DM)		
Fishmeal standard	10.0	10.0
Fishmeal SOLOR	20.0	20.0
Soy protein concentrate (Soycomil)	11.8	10.3
Wheat gluten	4.0	4.0
Corn gluten	8.0	8.0
Soybean meal 48	12.0	12.0
Rapeseed meal	5.0	5.0
Wheat meal	9.0	3.0
Peas gelatinized (Aquatex 8071)	3.2	3.2
Fish oil: COPPENS	6.5	6.5
Soybean oil	4.0	4.0
Rapeseed oil	4.0	4.0
Vit & Min Premix PV01	1.0	1.0
Binder (Kieselghur)	0.5	0.5
Antioxidant powder (Paramega)	0.2	0.2
MCP	0.5	0.5
L-Lysine	0.2	0.2
DL-methionine	0.1	0.1
Gracilaria sp.	–	2.5
Ulva sp.	–	2.5
Fucus sp.	–	2.5
Proximate composition (%DM)		
Dry matter	94.7	94.8
Ash	8.6	10.6
Crude protein	47.8	47.9
Crude fat	19.1	19.3
Gross energy (kJ g^{-1} DM)	22.7	22.4

Table 2 Proximate composition (% DM) of the seaweeds used in this trial

	Gracilaria spp.	*Fucus* spp.	*Ulva* spp.
Dry matter	93.4	87.0	85.7
Crude protein	25.9	17.2	23.2
Ash	34.3	20.7	34.8
Crude fat	1.1	3.4	1.5
Gross energy (kJ g^{-1} DM)	12.8	15.1	12.1

Fish and experimental facilities

This trial was conducted in ICBAS (Instituto de Ciências Biomédicas Abel Salazar) at the Aquatic Engineering and Production Systems facilities. The experimental systems consisted of fiberglass tanks (80 L) connected to a single water recirculation system (TMC® System 5000P Marine). Six tanks were used for oscillatory conditions and in parallel, and six tanks were kept under constant conditions and used as control/fixed group. Both systems (fixed and oscillatory) were identical in design. The water flow rate was set to 4 L min^{-1}, with continuous aeration. Temperature and salinity were monitored twice a day. Oxygen, pH, ammonia, and nitrates were monitored and kept within optimal levels. Photoperiod was set to 12:12 h light:dark and water was sterilized by ultraviolet radiation.

The values for temperature and salinity oscillations were selected from a 3-year data log period to represent the most representative combination of temperature/salinity annually observed in a Portuguese seabass farm (Materaqua Lda, Ílhavo, Portugal) (Fig. 1). Variations in temperature were achieved using thermostat heaters (Trixie®—200 W) and water chillers (TECO® TR60). Variations in salinity were obtained by sea salt addition or replacing salt water with dechlorinated fresh water. Groups subjected to temperature and salinity oscillations were compared with groups subjected to fixed rearing conditions (temperature 25 °C, salinity 30 ppt).

European seabass juveniles were provided by IPMA (Olhão, Portugal) and kept in quarantine for 2 weeks. Then, fish with an average weight of 40.7 g were randomly distributed into the tanks (11 fish/tank). The experimental diets, CTRL and SW, were randomly attributed to each tank, assuring three replicates in the oscillatory group and three in the fixed group for each diet. Fish were hand fed, twice a day, until apparent visual satiation and feed consumption was recorded daily.

Sampling

After 63 days, fish were anesthetized with ethylene glycol monobutyl ether (0.25 mL L^{-1}), and sacrificed by decapitation. Weight was recorded for the entire lot. Blood and liver were sampled from two fish/tank. Blood was centrifuged for plasma collection and livers were immediately frozen in liquid nitrogen. All samples were then stored at − 80 °C until further analysis.

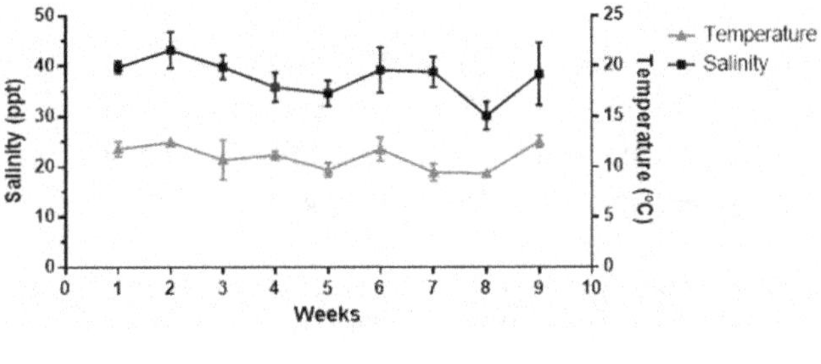

Fig. 1 Water salinity and temperature oscillations during the experimental period. Values chosen as the most representative of the natural variations in the aquaculture farm (Materaqua) over a 3-year period

Chemical analysis

Proximate composition of the diets was determined according to AOAC (2006) methods: dry matter by oven-drying at 103 °C for 16 h, ash by combustion in a muffle furnace (Nabertherm L9/11/B170, Bremen, Germany; 550 °C for 6 h), crude protein (N × 6.25) using a Leco nitrogen analyser (Model FP-528, Leco Corporation, St. Joseph, USA), lipid content by petroleum ether extraction using a Soxtherm Multistat/SX PC (Gerhardt, Königswinter, Germany; 40–60 °C), and gross energy in an adiabatic bomb calorimeter (Werke C2000, IKA, Staufen, Germany).

Humoral immune parameters

Peroxidase levels were measured according to the oxidation of TMB (3,3′,5,5′-tetramethylbenzidine) in the presence of H_2O_2 (Quade and Roth 1997). Final unit was presented as enzymatic units (EU), with one EU defined as producing an absorbance change of 1 optic density (OD). Lysozyme concentration in the plasma was determined by turbidimetric assay (Ellis 1990), measuring *Micrococcus lysodeikticus* lysis. Results are presented as enzymatic units, where 1 EU = 0.001 absorbance units per minute.

Oxidative stress

Livers were homogenized using K-phosphate buffer (pH 7.4, 0.1 M). Protein levels required for standardization of antioxidant parameters were quantified using Bradford method (Bradford 1976).

Lipid peroxidation (LPO) determination was based on malondialdehyde (MDA) level, measured as the amount of thiobarbituric acid reactive substances (TBARS) as a result of fatty acids oxidation (Ohkawa et al. 1979). Spectrophotometric readings were performed at 535 nm.

Catalase activity (CAT) was determined by reaction with H_2O_2 (Claiborne 1985). Glutathione-*S*-transferase (GST) activity was analyzed according to Habig et al. (1974) based on the quantification of the conjugate GSH-CDNB (1-chloro-2, 4-dinitrobenzene). Glutathione peroxidase (GPx) was quantified following the method described by Mohandas et al. (1984), measuring the formation of $NADP^+$. Glutathione reductase (GR) was determined based on NADPH oxidation measured at 340 nm (Cribb et al. 1989). Total and oxidized glutathione (TG and GSSG) were measured by the concomitant reaction of the GSH with 5,5′-dithiobis-(2-nitrobenzoic acid) (Baker et al. 1990), read at and absorbance of 412 nm. For GSSG evaluation, 2-vinyl-pyridine was used to trap the GSH present in the sample (Griffith 1980). Finally, GSH content was calculated by the difference between GSSG and TG levels.

Statistical analysis

Data were checked for normality (Shapiro/Wilk Test) and homogeneity of variances (Levene's test) and data transformation was applied when normality was not achieved. Two-way ANOVA was carried out using IBM SPSS Statistics 23. Tukey's HSD test was used for pairwise comparisons between treatments. A confidence level of 95% was considered in all statistical analysis.

Results

During the current trial, no mortalities were registered. Weight gain was significantly lower ($p < 0.05$) in fish subjected to oscillatory conditions (14.6 g and 22.6 g) when compared to fish subjected to the fixed condition (33.3 g and 40.1 g) (Table 3). The same pattern was observed for daily growth index (0.7 and 1.0 versus 1.47 and 1.6) (Table 3). Growth performance was not affected by dietary SW supplementation ($p > 0.05$) (Table 3).

Humoral immune responses (lysozyme and peroxidase activities) were not affected by dietary SW supplementation or rearing conditions ($p > 0.05$) (Fig. 2). The oxidative stress analyses showed that lipid peroxidation (LPO), glutathione peroxidase (GPx), glutathione *S*-transferase (GST), and reduced glutathione (GR) did not vary, regardless of the dietary and rearing conditions ($p > 0.05$; Figs. 3 and 4). Total glutathione

Table 3 Growth performance and feed utilization of seabass fed the experimental diets and subjected to oscillatory or fixed conditions

	Oscillatory		Fixed		p value		
	CTRL	SW	CTRL	SW	C	D	$C \times D$
Initial body weight (g)	26.5 ± 2.8	37.8 ± 1.7	42.9 ± 1.6	42.4 ± 0.4	0.193	0.546	0.459
Final body weight (g)	43.6 ± 1.2	59.3 ± 2.9	80.2 ± 5.7	85.7 ± 7.3	0.114	0.310	0.603
Weight gain (g)[A]	14.6 ± 4.1^b	22.6 ± 6.1^b	33.3 ± 12.4^a	40.1 ± 9.5^a	0.003	0.342	0.655
Daily growth index[B]	0.7 ± 0.2^b	1.0 ± 0.2^b	1.47 ± 0.2^a	1.6 ± 0.1^a	0.004	0.489	0.655
Feed conversion ratio[C]	1.18 ± 0.1	1.73 ± 0.04	1.14 ± 0.00	1.24 ± 0.24	0.050	0.409	0.942
Feed intake[D] (g kg ABW day^{-1})	118.6 ± 4.2	158.5 ± 28.1	157.5 ± 9.9	156.4 ± 11.9	0.184	0.342	0.202

Corresponding p values to rearing condition (C) and diet (D) factors, and interaction $C \times D$ are presented for each parameter. Values presented as mean ± standard deviation. Different superscript letters indicate significant differences ($p < 0.05$)

[A]Weight gain (g) = FBW − IBW, where FBW and IBW are the final and initial average body weights (g)

[B]Daily growth index (DGI) = $100 \times [(FBW)^{1/3} - (IBW)^{1/3}] \times$ trial duration in days

[C]Feed conversion ratio (FCR) = feed intake (g)/weight gain (g)

[D]ABW = (IBW + FBW)/2; day = feeding trial duration

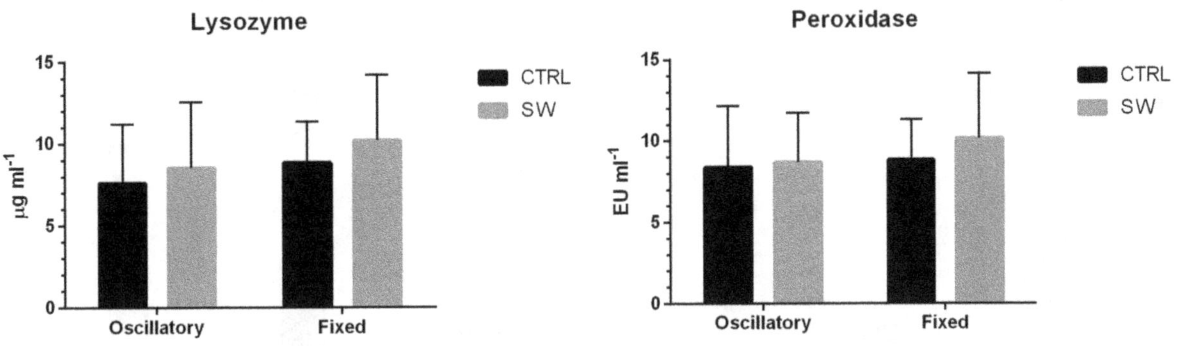

Fig. 2 Lysozyme and peroxidase activities determined in seabass fed the experimental diets and subjected to oscillatory or fixed conditions. Results presented as mean ± standard deviation

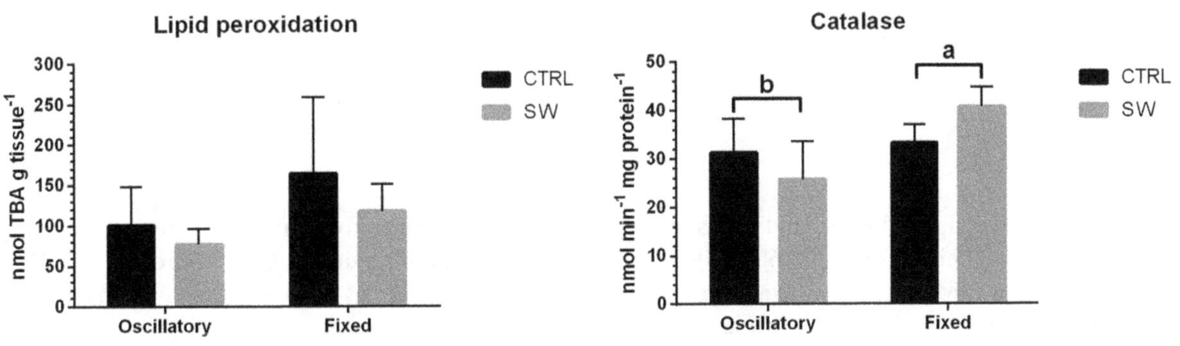

Fig. 3 Lipid peroxidation and catalase activity determined in the liver of seabass fed the experimental diets and subjected to oscillatory or fixed conditions. Results presented as mean ± standard deviation

(TG), oxidized glutathione (GSSG), and catalase (CAT) were significantly affected by rearing conditions; temperature and salinity oscillations led to a significant increase in TG and GSSG and to a significant decrease in CAT ($p < 0.05$, Fig. 4), with no influence of dietary SW supplementation ($p < 0.05$; Fig. 3).

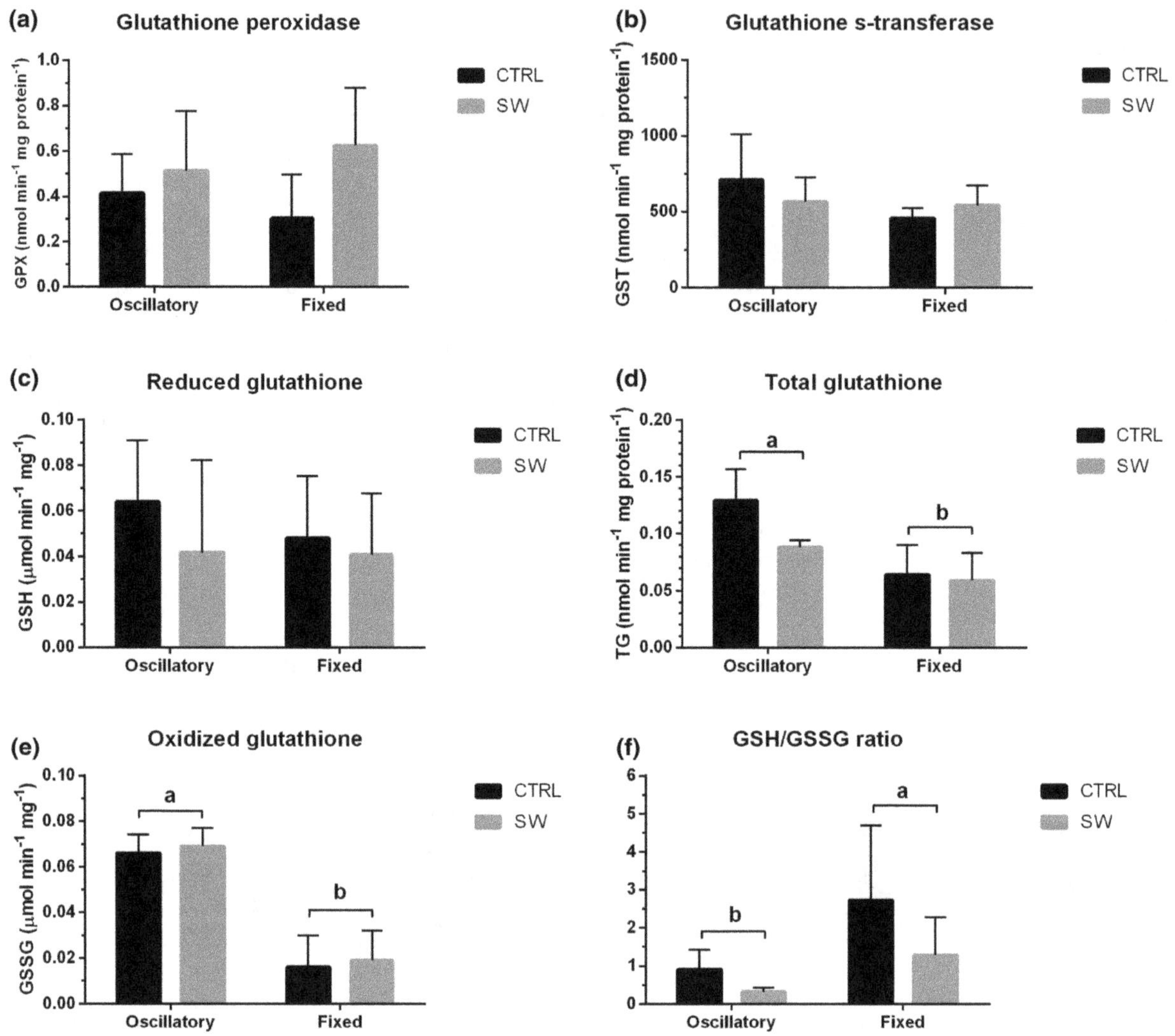

Fig. 4 Glutathione peroxidase (**a**), glutathione *S*-transferase (**b**), reduced glutathione (**c**), total glutathione (**d**) and oxidized glutathione (**e**) activities and GSH/GSSG ratio (**f**) determined in the liver of seabass fed the experimental diets and subjected to oscillatory or fixed conditions. In graphs **a–c**, absence of letters indicates non-significant differences and no interaction between conditions and diets. In graphs **d–f**, two-way ANOVA showed significant differences between conditions but not between diets, as indicated by letters

Discussion

Seaweeds have been previously studied as potential immunostimulants for many fish species including European seabass (Peixoto et al. 2016b), Nile tilapia (Güroy et al. 2007), rainbow trout (Güroy et al. 2013), and red sea bream (Mustafa and Nakagawa 1995). However, their effect as immunostimulants under environmental oscillations is yet to be accessed.

In this study, seabass growth was not influenced by dietary supplementation. This is consistent with results observed by Peixoto et al. (2016b) and Bagni et al. (2005), where no growth improvement was detected in European seabass fed diets supplemented with immunostimulants (algal extract with alginic acid and a yeast extract with β-glucans) at the same level of supplementation. Similarly, 5% *Ulva* sp. showed no effects on the growth performance of Nile tilapia (Güroy et al. 2007), while in European seabass, 10% inclusion of *U. rigida* or *Gracilaria bursa-pastoris* did not compromise growth (Valente et al. 2006). In addition, for striped mullet (Wassef et al. 2001) and Nile tilapia (Azaza et al. 2008), 10–25% inclusion with *Ulva* sp. had no effects on the growth performance of these species. On the other hand, Mustafa and Nakagawa (1995) showed that 5% inclusion of a seaweed mix (*Ascophyllum nodosum*, *Porphyra yezeoensis,* and *U. pertusa*) increased body weight and feed utilization of red seabream fingerlings.

Temperature and salinity combined effect has been reported as the main parameter affecting seabass growth (Conides and Glamuzina 2006). In this study, dietary seaweed supplementation had no effect on seabass growth performance, while the oscillatory environmental factors were the most accountable for the reduced growth performance. Both temperature and salinity played an important role in the outcome of the current study. Nevertheless, low temperatures possibly played a stronger role, as it is one of the most important environmental factor affecting the biochemical and physiological processes of aquatic organisms (Reynolds and Casterlin, 1979). Fish subjected to oscillatory conditions showed a lower growth performance in terms of daily growth index and weight gain when compared to those reared at fixed conditions. This is in agreement with Bagni et al. (2005) that associated growth suppression in seabass fed a combination of yeast β-glucan and alginic acid to sub-optimal water temperature.

The compounds associated with the immunostimulant properties of seaweeds still require more clarification. Nonetheless, polysaccharides, such as those present in seaweeds, have been linked to enhancements of fish innate immune system (Kim et al. 2012). Lysozyme and peroxidase are key components of fish innate immunity, acting against pathogens by disrupting their cell walls, either directly or through oxidative radicals action (Nayak 2010). In another study, the inclusion *G. lemaneiformis* showed to increased lysozyme activity in white-spotted spinefoot, associating it with the polysaccharides, vitamin C, and β-carotenes present in seaweeds (Xu et al. 2011). However, in the present trial, lysozyme and peroxidase activities of seabass were not affected by dietary seaweed supplementation nor by the environmental conditions. Similarly, previous studies report no effects in lysozyme and peroxidase activities for Nile tilapia fed diets supplemented with 5% and 10% *Ulva* spp. (Valente et al. 2006) and in European seabass fed diets with algal extract (Lauridsen and Buchmann 2010). On the other hand, an increase in lysozyme concentration was reported in European seabass when fed a seaweed supplemented diet (Peixoto et al. 2016c) and when fed a combination of dietary glucans and vitamins (Bagni et al. 2000), contradicting present results. Given this, we hypothesize that the beneficial effects of seaweeds on the innate immune response may be highly influenced by biotic and abiotic factors such as species, age, weight, rearing water conditions, and/or seaweed composition.

Environmental factors such as rearing temperature and salinity have been described to influence cellular antioxidant balance. Vinagre et al. (2012) showed that both LPO and CAT increased when European seabass was reared at temperatures outside the optimal range (24 °C). Similar results were presented by Madeira et al. (2013), where multiple estuarine fish species (including seabass) were collected and subjected to a temperature increase of 1 °C/h, starting at 24 °C. In this study, LPO, CAT, and GST simultaneously increased with temperature rise. In our work, no differences were found in these parameters. It is possible that temperature oscillation masks consistent reactions from these parameters. However, GSSG in this work was significantly higher in the oscillatory treatments. Eroglu et al. (2015) showed that fish exposed to heavy metals had reduced GSH/GSSG levels, due to the consumption of GSH and formation of GSSG. Leggatt et al. (2007) showed that besides metal toxicity, temperature alone could influence GSH levels, with fish being acclimatized within the range of comfort temperatures showed increased GSH transcription. Altogether, GSH appears to reduce in hazardous conditions, either by conception (promoting GSSG increase) or/and by reducing its formation.

Results from present study indicate that dietary seaweed supplementation had no influence on these oxidative stress biomarkers, which has also been observed by Peixoto et al. (2016b). However, the differences observed are tightly associated with temperature and salinity oscillations which is well described by several works (Loro et al. 2012; Madeira et al. 2013; Vinagre et al. 2012). In this work, TG levels were higher in fish subjected to the oscillatory condition. Similarly, Eroglu et al. (2015) observed an increase of oxidized glutathione (GSSG) leading to higher TG concentrations and lower GSH/GSSG ratio. Therefore, it appears plausible to consider that the oscillatory environmental conditions acted as ROS inducing event. TG levels behave as a potential antioxidant reserve that reduces in prolonged exposure to stress (Eroglu et al. 2015). In more detail, TG levels seem to oscillate as compensatory effect of GSH depletion, due to antioxidant activities (Guyonnet et al. 1999; Tan et al. 1998). According to Vinagre et al. (2012), CAT activities in juvenile seabass are highly sensitive to environmental temperature, particularly those outside the optimal range of the species, which is in accordance with our results.

This work showed that seaweed supplementation in European Seabass juveniles does not mitigate the effects of environmental oscillations. Therefore, dietary supplementation trials should focus on determining the bioactive compounds present in seaweeds and their association with specific responses. In addition, antioxidant responses in this work were altered solely in the glutathione's system suggesting a similarity to antioxidant responses in toxicity trials. However, antioxidant results were affected by the high deviation between trials suggesting a greater number of samples should be considered for these parameters.

Overall the differences observed in our study were mainly associated with the oscillating conditions and not the dietary seaweed supplementation, nor were interactions between factors detected for the analyzed parameters.

Acknowledgements The current study was supported by the Research Line INSEAFOOD of the project INNOVMAR-Innovation and Sustainability in the Management and Exploitation of Marine Resources (reference NORTE-01-0145-FEDER-000035), funded by the Northern Regional Operational Programme (NORTE 2020) through the ERDF.

References

Arana LV (1997) Princípios químicos de qualidade da água em aquicultura: uma revisáo para peixes e camaróes. UFSC, Florianopolis

Azaza MS, Mensi F, Ksouri J, Dhraief MN, Brini B, Abdelmouleh A, Kraïem MM (2008) Growth of Nile tilapia (*Oreochromis niloticus* L.) fed with diets containing graded levels of green algae ulva meal (*Ulva rigida*) reared in geothermal waters of southern Tunisia. J Appl Ichthyol 24:202–207. https://doi.org/10.1111/j.1439-0426.2007.01017.x

Bagni M, Archetti L, Amadori M, Marino G (2000) Effect of long-term oral administration of an immunostimulant diet on innate immunity in sea bass (*Dicentrarchus labrax*). J Vet Med Res 47:745–751. https://doi.org/10.1111/j.1439-0450.2000.00412.x

Bagni M et al (2005) Short-and long-term effects of a dietary yeast β-glucan (Macrogard) and alginic acid (Ergosan) preparation on immune response in sea bass (*Dicentrarchus labrax*). Fish Shellfish Immunol 18:311–325. https://doi.org/10.1016/j.fsi.2004.08.003

Baker MA, Cerniglia GJ, Zaman A (1990) Microtiter plate assay for the measurement of glutathione and glutathione disulfide in large numbers of biological samples. Anal Biochem 190:360–365. https://doi.org/10.1016/0003-2697(90)90208-Q

Barnabe G (2003) Aquaculture: biology and ecology of cultured species. Ellis Horwood series in aquaculture and fisheries support. CRC Press, Boca Raton

Björnsson B, Ólafsdóttir SR (2006) Effects of water quality and stocking density on growth performance of juvenile cod (*Gadus morhua* L.). ICES J Mar Sci 63:326–334. https://doi.org/10.1016/j.icesjms.2005.10.010

Bradford MM (1976) A rapid and sensitive method for the quantitation of microgram quantities of protein utilizing the principle of protein-dye binding. Anal Biochem 72:248–254. https://doi.org/10.1016/0003-2697(76)90527-3

Bricknell I, Dalmo RA (2005) The use of immunostimulants in fish larval aquaculture. Fish Shellfish Immunol 19:457–472. https://doi.org/10.1016/j.fsi.2005.03.008

Castro R, Zarra I, Lamas J (2004) Water-soluble seaweed extracts modulate the respiratory burst activity of turbot phagocytes. Aquaculture 229:67–78. https://doi.org/10.1016/S0044-8486(03)00401-0

Claiborne A (1985) Catalase activity. In: Greenwald RA (ed) CRC handbook of methods for oxygen radical research. CRC Press, Boca Raton, pp 283–284

Colombo L, Belvedere P, Pickering A, Schreck C (1990) Stress inducing factors and stress reaction in aquaculture. European Aquaculture Society Special Publication, Belgium

Conides AJ, Glamuzina B (2006) Laboratory simulation of the effects of environmental salinity on acclimation, feeding and growth of wild-caught juveniles of European sea bass *Dicentrarchus labrax* and gilthead sea bream, *Sparus aurata*. Aquaculture 256:235–245. https://doi.org/10.1016/j.aquaculture.2006.02.029

Cornish-Bowden A (1979) Chapter 8—Control of enzyme activity. In: Cornish-Bowden A (ed) Fundamentals of enzyme kinetics. Butterworth-Heinemann, pp 147–176. https://doi.org/10.1016/B978-0-408-10617-7.50013-4

Cribb AE, Leeder JS, Spielberg SP (1989) Use of a microplate reader in an assay of glutathione reductase using 5,5′-dithiobis (2-nitrobenzoic acid). Anal Biochem 183:195–196. https://doi.org/10.1016/0003-2697(89)90188-7

Ellis AE (1990) Lysozyme assays. In: Stolen JS, Fletcher TC, Anderson DP, Roberson BS, Van Muiswinkel WB (eds) Techniques in fish immunology. SOS Publications, Fair Haven, pp 101–103

Eroglu A, Dogan Z, Kanak E, Atli G, Canli M (2015) Effects of heavy metals (Cd, Cu, Cr, Pb, Zn) on fish glutathione metabolism. Environ Sci Pollut Res Int 22:3229–3237. https://doi.org/10.1007/s11356-014-2972-y

Griffith W (1980) Determination of glutathione and glutathione disulfide using glutathione reductase and 2-vinylpyridine. Anal Biochem 106:207–212. https://doi.org/10.1016/0003-2697(80)90139-6

Güroy BK, Cirik Ş, Güroy D, Sanver F, Tekinay AA (2007) Effects of *Ulva rigida* and *Cystoseira barbata* meals as a feed additive on growth performance, feed utilization, and body composition of Nile tilapia, *Oreochromis niloticus*. Turk J Vet Anim Sci 31:91–97

Güroy B, Ergün S, Merrifield DL, Güroy D (2013) Effect of autoclaved *Ulva* meal on growth performance, nutrient utilization and fatty acid profile of rainbow trout, *Oncorhynchus mykiss*. Aquacult Int 21:605–615. https://doi.org/10.1007/s10499-012-9592-7

Guyonnet D, Siess M-H, Le Bon A-M, Suschetet M (1999) Modulation of phase II enzymes by organosulfur compounds from allium vegetables in rat tissues. Toxicol Appl Pharmacol 154:50–58. https://doi.org/10.1006/taap.1998.8574

Habig WH, Pabst MJ, Jakoby WB (1974) Glutathione *S*-transferases: the first enzymatic step in mercapturic acid formation. J Biol Chem 249:7130–7139

Heo S-J, Park E-J, Lee K-W, Jeon Y-J (2005) Antioxidant activities of enzymatic extracts from brown seaweeds. Bioresour Technol 96:1613–1623. https://doi.org/10.1016/j.biortech.2004.07.013

Imsland AK, Björnsson BT, Gunnarsson S, Foss A, Stefansson SO (2007) Temperature and salinity effects on plasma insulin-like growth factor-I concentrations and growth in juvenile turbot (*Scophthalmus maximus*). Aquaculture 271:546–552. https://doi.org/10.1016/j.aquaculture.2007.07.007

Jeney G, Jeney Z (2002) Application of immunostimulants for modulation of the non-specific defense mechanisms in sturgeon hybrid: *Acipenser ruthenu* × *A. baerii*. J Appl Ichthyol 18:416–419. https://doi.org/10.1046/j.1439-0426.2002.00405.x

Jeney G, Galeotti M, Volpatti D (1994) Effect of immunostimulation on the non-specific immune response of sea bass *Dicentrarchus labrax*. In: International symposium on aquatic animal health, Seattle, Washington, USA, 4e8 September

Kim J-H, Lee J-S, Kang J-C (2012) Effect of inorganic mercury on hematological and antioxidant parameters on olive flounder *Paralichthys olivaceus*. Fish Aquat Sci 15:215–220. https://doi.org/10.5657/FAS.2012.0215

Lambert Y, Dutil J-D (1997) Can simple condition indices be used to monitor and quantify seasonal changes in the energy reserves of cod (*Gadus morhua*)? Can J Fish Aquat Sci 54:104–112. https://doi.org/10.1139/f96-149

Lauridsen JH, Buchmann K (2010) Effects of short-and long-term glucan feeding of rainbow trout (Salmonidae) on the susceptibility to *Ichthyophthirius multifiliis* infections. Acta Ichthyol Piscat 40:61–66. https://doi.org/10.3750/AIP2010.40.1.08

Leggatt R, Brauner C, Schulte P, Iwama G (2007) Effects of acclimation and incubation temperature on the glutathione antioxidant system in killifish and RTH-149 cells. Comp Biochem Physiol A Mol Integr Physiol 146:317–326

Loro V, Jorge M, Silva K, Wood C (2012) Oxidative stress parameters and antioxidant response to sublethal waterborne zinc in a euryhaline teleost *Fundulus heteroclitus*: protective effects of salinity. Aquat Toxicol 110:187–193. https://doi.org/10.1016/j.aquatox.2012.01.012

Lushchak VI, Bagnyukova TV (2006) Effects of different environmental oxygen levels on free radical processes in fish. Comp Biochem Physiol B Biochem Mol Biol 144:283–289. https://doi.org/10.1016/j.cbpb.2006.02.014

Madeira D, Narciso L, Cabral HN, Vinagre C, Diniz MS (2013) Influence of temperature in thermal and oxidative stress responses in estuarine fish. Comp Biochem Physiol A Mol Integr Physiol 166:237–243. https://doi.org/10.1016/j.cbpa.2013.06.008

Mohandas J, Marshall JJ, Duggin GG, Horvath JS, Tiller DJ (1984) Differential distribution of glutathione and glutathione-related enzymes in rabbit kidney: possible implications in analgesic nephropathy. Biochem Pharmacol 33:1801–1807. https://doi.org/10.1016/0006-2952(84)90353-8

Mustafa MG, Nakagawa H (1995) A review: dietary benefits of algae as an additive in fish feed. Isr J Aquacult (Bamid) 47:155–162. https://doi.org/10.1371/journal.pone.0124042

Nayak S (2010) Probiotics and immunity: a fish perspective. Fish Shellfish Immunol 29:2–14. https://doi.org/10.1016/j.fsi.2010.02.017

Obach A, Quentel C, Laurencin FB (1993) Effects of alpha-tocopherol and dietary oxidized fish oil on the immune response of sea bass *Dicentrarchus labrax*. Dis Aquat Organ 15:175–185. https://doi.org/10.3354/dao015175

Ohkawa H, Ohishi N, Yagi K (1979) Assay for lipid peroxides in animal tissues by thiobarbituric acid reaction. Anal Biochem 95:351–358. https://doi.org/10.1016/0003-2697(79)90738-3

Peixoto et al (2016a) Effects of dietary *Gracilaria* sp. and *Alaria* sp. supplementation on growth performance, metabolic rates and health in meagre (*Argyrosomus regius*) subjected to pathogen infection. J Appl Phycol 29:433–447. https://doi.org/10.1007/s10811-016-0917-1

Peixoto et al (2016b) Role of dietary seaweed supplementation on growth performance, digestive capacity and immune and stress responsiveness in European seabass (*Dicentrarchus labrax*). Aquacult Rep 3:189–197. https://doi.org/10.1016/j.aqrep.2016.03.005

Peixoto MJ et al (2016c) Diets supplemented with seaweed affect metabolic rate, innate immune, and antioxidant responses, but not individual growth rate in European seabass (*Dicentrarchus labrax*). J Appl Phycol 28:2061–2071. https://doi.org/10.1007/s10811-015-0736-9

Quade MJ, Roth JA (1997) A rapid, direct assay to measure degranulation of bovine neutrophil primary granules. Vet Immunol Immunopathol 58:239–248. https://doi.org/10.1016/S0165-2427(97)00048-2

Raa J, Rørstad G, Engstad R, Robertsen B (1992) The use of immunostimulants to increase resistance of aquatic organism to microbial infections. In: Sharif MI, Subasinghe RP, Arthur JR (eds) Diseases in Asian aquaculture. Fish health section. Asian Fisheries Society, Manila, pp 39–50

Reynolds WW, Casterlin ME (1979) Behavioral thermoregulation and the "final preferendum" paradigm. Am Zool 19(1):211–224

Sakai M (1999) Current research status of fish immunostimulants. Aquaculture 172:63–92. https://doi.org/10.1016/S0044-8486(98)00436-0

Satoh K (1987) Effect of Ulva meal supplementation on disease resistance of red sea bream. Nippon Suisan Nippon Suisan Gakk 53:1115–1120. https://doi.org/10.2331/suisan.53.1115

Sengupta P, Garrity P (2013) Sensing temperature. Curr Biol 23:304–307. https://doi.org/10.1016/j.cub.2013.03.009

Tan S, Sagara Y, Liu Y, Maher P, Schubert D (1998) The regulation of reactive oxygen species production during programmed cell death. J Cell Biol 141:1423–1432. https://doi.org/10.1083/jcb.141.6.1423

Thanigaivel S, Chandrasekaran N, Mukherjee A, Thomas J (2015) Investigation of seaweed extracts as a source of treatment against bacterial fish pathogen. Aquaculture 448:82–86. https://doi.org/10.1016/j.aquaculture.2015.05.039

Vadstein O, Øie G, Olsen Y, Salvesen I, Skjermo J, Skjåk-Bræk G (1993) A strategy to obtain microbial control during larval development of marine fish. Paper presented at the fish farming technology: proceedings of the first international conference of on fish farming technology, Rotterdam

Valente LMP, Gouveia A, Rema P, Matos J, Gomes EF, Pinto IS (2006) Evaluation of three seaweeds *Gracilaria bursa-pastoris*, *Ulva rigida* and *Gracilaria cornea* as dietary ingredients in European sea bass (*Dicentrarchus* labrax) juveniles. Aquaculture 252:85–91. https://doi.org/10.1016/j.aquaculture.2005.11.052

Vinagre C, Madeira D, Narciso L, Cabral HN, Diniz M (2012) Effect of temperature on oxidative stress in fish: lipid peroxidation and catalase activity in the muscle of juvenile seabass, *Dicentrarchus labrax*. Ecol Indic 23:274–279. https://doi.org/10.1016/j.ecolind.2012.04.009

Wassef E, El Masry M, Mikhail F (2001) Growth enhancement and muscle structure of striped mullet, *Mugil cephalus* L., fingerlings by feeding algal meal-based diets. Aquacult Res 32:315–322. https://doi.org/10.1046/j.1355-557x.2001.00043.x

Xu B, Hirata H (1990) Effects of feed additive Ulva reproduced in feedback culture system on the growth and color of red sea bream, *Pagrus major*. Aquacult Sci 38(2):177–182. https://doi.org/10.11233/aquaculturesci1953.38.177

Xu S, Zhang L, Wu Q, Liu X, Wang S, You C, Li Y (2011) Evaluation of dried seaweed *Gracilaria lemaneiformis* as an ingredient in diets for teleost fish *Siganus canaliculatus*. Aquacult Int 19:1007–1018. https://doi.org/10.1007/s10499-011-9418-z

7

Alteration in expression of atrogenes and IGF-1 induced by fasting in Nile tilapia *Oreochromis niloticus* juveniles

Caroline Nebo · Ken Overturf · Andreas Brezas · Maeli Dal-Pai-Silva · Maria Célia Portella

Abstract The growth rate of farmed fish is an important factor regarding aquaculture success. An understanding of the cellular events that occur in skeletal muscle when fish undergo periods of fasting and refeeding provides information useful in developing alternative feeding strategies for improving muscle growth in commercially cultivated species. To evaluate the effect of 1–3 weeks of fasting and 10 weeks of refeeding in Nile tilapia juveniles, we analyzed the growth performance and changes in muscle cellularity and the expression of the following growth and muscle related genes: MyoD, myogenin, IGF-1, IGF-1 receptor, MuRF-1, atrogin-1 and myostatin. Reduced body mass was observed in all three groups of fasted fish during their time off feed, and 10 weeks of refeeding resulted in partial compensatory growth of body mass. No differences in the frequency of white muscle fiber diameters were observed between fasted and fed control fish treatments. However, changes in gene expression induced by fasting and refeeding were found. IGF-1 receptor, ubiquitin ligases MuRF1 and atrogin-1 expression increased during the 1–3 weeks of fasting, while IGF-1 levels dropped significantly ($P < 0.001$) compared to the control treatment. Furthermore, myogenin mRNA level in fish submitted to 3 weeks of fasting was higher in comparison to the control treatment ($P < 0.05$). Overall, our results showed that 1–3 weeks of fasting can induce muscle atrophy activation in Nile tilapia juveniles, and 10 weeks of refeeding is enough to induce only partial compensatory growth.

Keywords Atrogin-1 · Fasting · IGF-1 · MuRF-1 · Muscle atrophy · Muscle regulatory factors

---

C. Nebo · M. C. Portella
Faculdade de Ciências Agrárias e Veterinárias, Universidade Estadual Paulista (Unesp), Jaboticabal, SP 14884-900, Brazil

K. Overturf (✉)
USDA-ARS, Hagerman Fish Culture Experiment Station, 3059F National Fish Hatchery Rd, Hagerman, ID 83332, USA
e-mail: Ken.Overturf@ARS.USDA.GOV

A. Brezas
University of Idaho, Hagerman Fish Culture Experiment Station, 3059F National Fish Hatchery Rd, Hagerman, ID 83332, USA

M. Dal-Pai-Silva
Universidade Estadual Paulista (Unesp), Instituto de Biociências, Botucatu, SP 18618-000, Brazil

Introduction

Fasting is a phenomenon that can occur naturally in the environment due to low food supply, during fish migration, overwintering or during reproduction, throughout which fish demonstrate the ability to survive without food for lengthy periods (Love 1970; Bower et al. 2009). After a period of growth depression and when favorable conditions are restored, there occurs a phase of accelerated growth, called compensatory growth (Ali et al. 2003). Partial, full and over-compensation growth have all been evoked in fish. However, over-compensation has only been observed when cycles of deprivation and satiation feedings have occurred. Compensatory growth has been partially linked to hyperphagia when rates of food consumption are significantly higher than those in fish that have not experienced a previous growth depression episode. The severity of the growth depression increases the duration of the hyperphagic phase rather than maximum daily feeding rate. In some studies, growth efficiencies were notably greater during the compensatory growth stage (Ali et al. 2003; Dobson and Holmes 1984). In teleosts, the maintenance of skeletal muscle development and growth is a complex process, and may be influenced by environmental factors and the nutritional and physiological states of the animal (Johnston et al. 2008, 2011). A significant amount of research effort has been expended in understanding the regulatory factors surrounding muscle accretion, specifically evaluation of the genes involved. Myogenic regulatory factors (MRFs) are a family of four transcription factors (MyoD, myogenin, myf5 and MRF4/myf6) that are highly conserved in vertebrates and are responsible for the control of myogenesis and muscle growth. Each gene has evolved to regulate a specific area of myogenesis. For instance, MyoD and myf5 are essential regulators of muscle cell proliferation determination, whereas myogenin and myf4 act as regulators of muscle cell differentiation (Rudnicki et al. 1993; Steinbacher et al. 2007; Johnston et al. 2008).

Muscle growth is also controlled by the regulator myostatin. Myostatin is a member of the transforming growth factor-β (TGF-β) superfamily that functions as a negative regulator of skeletal muscle development and growth in mammals; however, in fish, myostatin is also present in other tissues and in other forms (Maccatrozzo et al. 2001; Rios et al. 2002; Acosta et al. 2005; De Santis et al. 2012). Another regulator of muscle growth in vertebrates is the insulin-like growth factor-1 (IGF-1) which represents the main autocrine and paracrine regulatory mechanisms of skeletal muscle by activation of the PI3 k/Akt pathway which consequently increases protein synthesis, associated with muscle hypertrophy (Glass 2003; Sacheck et al. 2004; Kandarian and Jackman 2006; Fuentes et al. 2013).

Muscle protein degradation is regulated by several processes, one of which is the ubiquitin–proteasome pathway. Two members of these muscle-specific ubiquitin ligases are the muscle RING finger 1 (MuRF1) and muscle atrophy F-box (MAFbx/atrogin-1). Studies show that an increase in MuRF1 and atrogin-1 expression causes fiber reduction due to catabolic conditions (Glass 2005).

The Nile tilapia *Oreochromis niloticus* is globally one of the most important farmed finfish species and has been found to effectively adapt to a wide range of rearing conditions (Abdel-Hakim et al. 2009; Fox et al. 2009). However, tilapia culture is affected by water shortage, low food supply in floating cages, prolonged winter, fish reproductive period and other conditions. Understanding the molecular mechanisms that control muscle growth in tilapia would be beneficial for improving production under different conditions that occur such as short-term fasting.

Previously we showed that short periods of fasting promoted muscle cell fiber atrophy in small juvenile tilapia (0.6 g) and that refeeding caused compensatory mass gain correlated with increased expression of muscle growth related genes (MyoD, myogenin and myostatin) (Nebo et al. 2013). Thus, the aim of this study was to evaluate the effects in larger juveniles (25–30 g) under longer fasting and refeeding time intervals to test the hypothesis that the ubiquitin–proteasome pathway is activated in tilapia juveniles during periods of fasting, and that IGF-1 is overexpressed during the refeeding to induce compensatory growth.

Materials and methods

Ethical procedures

The study was approved by the Comitê de Ética no Uso de Animais (CEUA) of the Faculdade de Ciências Agrárias e Veterinárias, UNESP Univ Estadual Paulista (CEUA/FCAV/UNESP, 009436/11).

Experimental design

Fish were obtained from a commercial farm (Royal Fish, Itupeva—São Paulo, Brazil) and acclimated (15 days) to the experimental conditions at the Laboratório de Nutrição de Organismos Aquáticos, of the Centro de Aquicultura. A total of 1120 fish were stocked in thirty-two 150-L tanks (35 fish/tank) with constant water flow and aeration. Nile tilapia *Oreochromis niloticus* juveniles, Chitralada–Thai strain were weighed (initial weight 30.2 ± 0.9 g) and randomly distributed into four experimental treatments with eight replicates: (FC) control, fish were fed continuously during the 13 weeks; (F1) 1 week of fasting (1WF) and 10 weeks of refeeding (10WR), (F2) 2 weeks of fasting (2WF) and 10WR; (F3) 3 weeks of fasting (3WF) and 10WR. The experiment lasted 13 weeks, and a fixed period of 10 weeks of refeeding was maintained for all fasting treatments. When not fasted, fish were fed to apparent satiation with an extruded commercial diet (32% crude protein and 6.5% lipid) three times per day (9 am, 2 pm, and 5 pm).

White muscle samples ($n = 8$ per treatment) were collected to analyze morphology, morphometry and the expression of growth related and muscle atrophy genes, at the end of each period of fasting (1WF, 2WF and 3WF) and after 10WR. Water variables were monitored weekly and maintained at: temperature 30.1 ± 0.32 °C, pH 7.89 ± 0.14 and dissolved oxygen 4.74 ± 0.86 mg/L. For sampling, fish were anesthetized (0.1 g/L of benzocaine) and individually weighed and measured. The body mass (g) was measured at the beginning of the experiment (0 day) and after each period of fasting: one (1WF), two (2WF) and three (3WF) weeks, and after one (1WR), two (2WR), six (6WR) and 10 weeks of refeeding (10WR). A graphical representation of the experimental design is provided in Supplementary file 1.

Morphology and morphometry of muscle fibers

Fish were euthanized with benzocaine (0.4 g/L) and white muscle samples ($n = 8$ fish/treatment) were collected from the dorsal region, fixed in Karnovsky solution (8% paraformaldehyde and 2.5% glutaraldehyde in PBS—phosphate buffer saline) and embedded in Historesin® (Leica, Germany). Histological transverse sections (4 μm) were obtained and stained with hematoxylin-phloxine B to analyze muscle fiber diameter. To estimate the degree of muscle hypertrophy, hyperplasia or atrophy the diameter of 700 white muscle fibers for each treatment ($n = 8$ per treatment) were measured from between three to four fields per sample. The fibers were binned into classes according to their diameter (d, μm): class 10 = $d < 10$, class 30 = $10 > d < 30$, class 50 = $30 > d < 50$ and class 60 = $d \geq 50$, and white muscle fibers frequency in different diameter classes was determined. Images were captured with a Zeiss Axioplan 2 microscope and analyzed with Zeiss Axiovision 4.7 software (Carl Zeiss Microscope, LLC, USA). Morphology of muscle cells at different periods is provided in Fig. 1 and in color in Supplementary file 2.

Gene expression

White muscle samples ($n = 8$) were collected from the dorsal region of fish at the beginning of the experiment and at 1WF, 2WF, 3WF and after 10WR during the experiment and frozen in liquid nitrogen. Total RNA was extracted using TRIzol® Reagent according to the manufacturer's protocol (Life Technologies, Carlsbad, CA, USA). Extracted RNA integrity was confirmed by electrophoresis on a 1% agarose gel stained with GelRed (Biotium, Hayward, CA, USA). Quantity of extracted RNA was determined using a NanoVue™ Plus Spectrophotometer (GE Healthcare, Piscataway, NJ, USA). RNA purity was ensured by obtaining a 260/280 nm OD ratio ≥ 1.8. Total RNA were solubilized in RNase-free water and incubated with Dnase I (Life Technologies, Carlsbad, CA, USA) to remove any DNA present in the samples. For expression analysis using SYBR, total RNA (2 μg) was reverse transcribed using the High Capacity cDNA archive kit with RNase inhibitor (Life Technologies, Carlsbad, CA, USA), according to the manufacturer's instructions. At the end of the reaction, the samples were stored at − 20 °C.

Gene expression analysis was evaluated by reverse transcription quantitative real-time polymerase chain reaction (RT-qPCR). For each gene, specific primers were tested at different concentration levels to determine optimal amplification. Samples of cDNA were amplified with specific primers for *O. niloticus MyoD, myogenin, myostatin* (Nebo et al., 2013), IGF-1, IGF-1 receptor, and the ubiquitin regulating gene, MAFbx/atrogin-1 designed from cDNA nucleotide sequences, available in GenBank (http://www.ncbi.nlm.nih.gov/

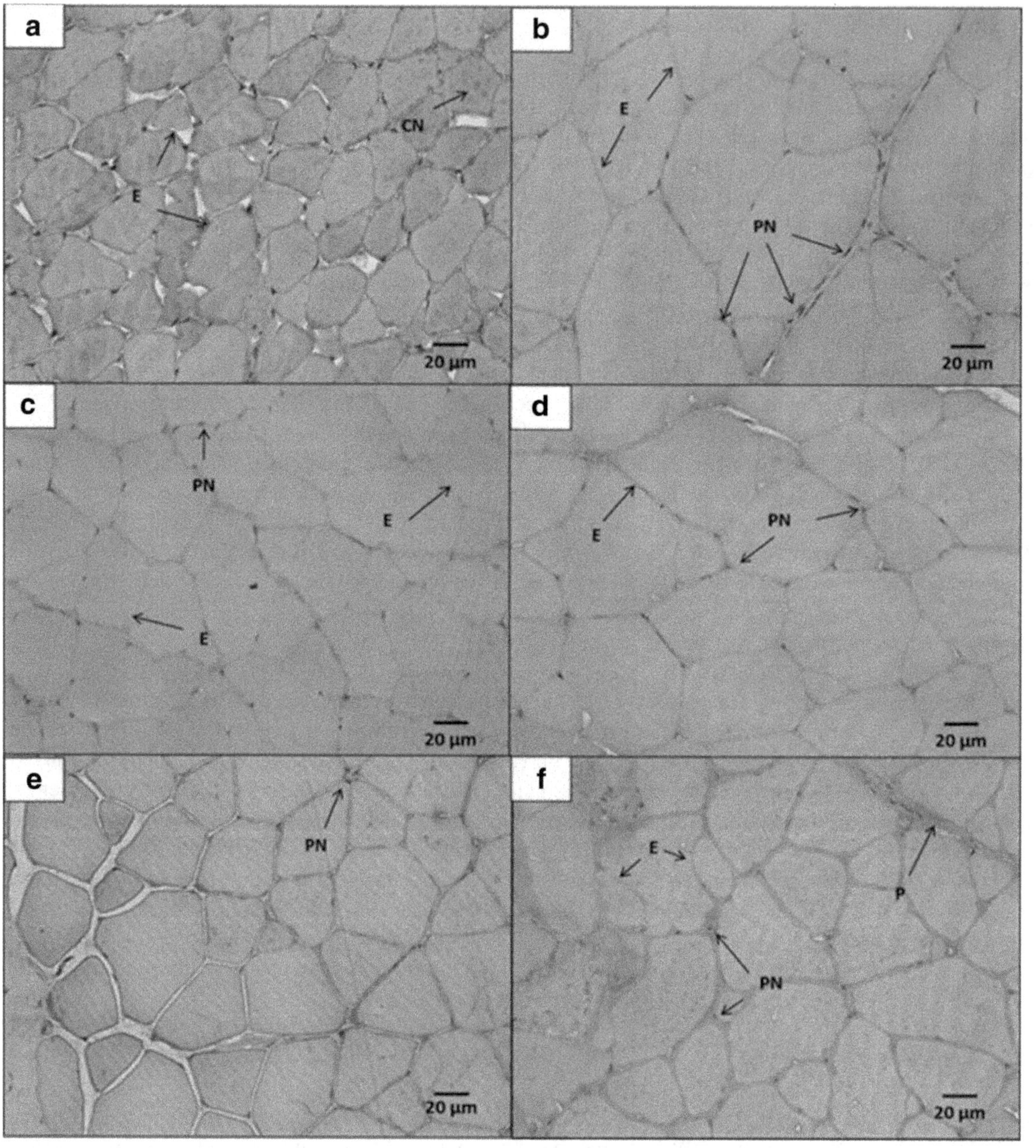

Fig. 1 Morphology of muscle cells after different periods of fasting and refeeding. Transverse section of white muscle fiber of Nile tilapia *Oreochromis niloticus* juveniles after 1 (1WF), 2 (2WF) and 3 (3WF) weeks of fasting and after 10 weeks of refeeding (10WR). **a** F1 at 1WF; **b** F1 at 10WR; **c** F2 at 2WF; **d** F2 at 10WR; **e** F3 at 3WF; **f** F3 at 10WR. Small fibers around big fibers, forming a growth mosaic pattern. Central nucleus (CN), peripheral nucleus (PN), perimysium (P) and endomysium (E). (FC) control, fed continuously during the 13 weeks; (F1) 1 week of fasting and 10WR, (F2) 2 weeks of fasting and 10WR; (F3) 3 weeks of fasting and 10WR. Hematoxylin–phloxine stain. Scale bars: 20 µm, 40 X

pubmed/nucleotide). The gene expression analyses were performed with an ABI 7500 Fast Real-Time PCR System (Life Technologies, Carlsbad, CA, USA). The reactions were run in duplicate using Fast SYBR Green Master Mix (Life Technologies, Carlsbad, CA, USA), according to the manufacturer's instructions. Target gene expression was normalized to ubiquitin-like protein 3 (ubl3) gene expression and melting dissociation curves were performed to confirm that only a single product was amplified. Control reactions lacking cDNA template were also included to check for reagent contamination. The normalization genes were previously tested and the gene with the least variation between samples was chosen.

For the analysis of MuRF1, atrogin-1 and beta-actin genes expression primers and probes were utilized. Reactions were run on an ABI Prism 7900HT Sequence Detection System (Life Technologies, Carlsbad, CA, USA) using Verso 1-Step RT-qPCR ROX Mix (Thermo Fisher Scientific, Waltham, MA, USA). The reactions were run in duplicate using a TaqMan Assay kit (Life Technologies, Carlsbad, CA, USA), according to the manufacturer's instructions. Samples were normalized to beta-actin gene expression. Relative gene expression was calculated using the ($2^{-\Delta\Delta CT}$) method using the FC group as the untreated sample (Livak and Schmittgen 2001). For IGF-1, IGF-1 receptor, beta-actin, MuRF1, and atrogin-1 genes primers and probes were designed using PrimerQuest (Integrated DNA Technologies, Coralville, IA) (Table 1).

Statistical analysis

Morphometric and gene expression data were analyzed by one-way ANOVA for differences between means, and where significant ($P < 0.05$) post hoc analyses tested with Tukey's, using the Statistical Analysis System SAS V9 software (SAS Institute Inc., Cary, NC, USA). Two comparisons were made: fed control (FC) between fasted treatments (F1, F2, and F3); and also fasted treatments were compared between groups after fasting (1WF, 2WF, and 3WF) and 10 weeks of refeeding. All data were tested for homogeneity of variance using Brown–Forsythe test ($P < 0.05$) and normality using Cramer–von Mises test ($P < 0.05$). Logarithmic

Table 1 Oligonucleotide primers and probes used for RT-qPCR amplification

Primer–Probe sequence (5′–3′)	GenBank accession no.
Atrogin-1	
Forward: GTGGGCTCGTTCAGGACAT	XM0054505921
Reverse: CGGTGTACCCAGATGTTGATGTTC	
Probe: CAGCACAGACTTGCCC	
B-actin	
Forward: GCCCCACGCCATCCT	XM0034431272
Reverse: GAGTAGCCACGCTCTGTCA	
Probe: CTGGCCGTGACCTCAC	
MuRF1	
Forward: AGGGTGAACTGAGTAATGCTATCGA	XM005456465
Reverse: GCAGGCATCTTCCATCTG	
Probe: CAGGCGACCATTGCTG	
IGF1	
Forward: CAAGAGCACCCAAGGTTAGTAG	EU2721491
Reverse: CAGGTGAAGGTCTCTTCTTGTT	
IGF1 receptor	
Forward: GTGGTCAAAGACGAACCAGA	KC5067771
Reverse: CTCATGACGGAGGCTTCATT	
Myogenin	
Forward: GCAGCCACACTGAGGGAGAA	Nebo et al. 2013GU246717
Reverse: AAGCATCGAAGGCCTCGTT	
Myostatin	
Forward: TGTGGACTTCGAGGACTTTGG	Nebo et al. 2013KT987208
Reverse: TGGCCTTGTAGCGTTTTGGT	
MyoD	
Forward: TCAGACAACCAGAAGAGGAAGCT	Nebo et al. 2013FJ907953
Reverse: CCGTTTGGAGTCTCGGAGAA	
Ubl3	
Forward: CCTGCGGCTGATTTTGGT	XM005463259
Reverse: CAGAGTCGTTGGGAGAGAAGA	

transformations (natural logarithm) were used, in analysis of body mass, IGF-1, IGF-1 receptor, myogenin, atrogin-1 and MuRF1, as relevant correction of original data.

Results

Body mass response to fasting and refeeding

Survival rate of Nile tilapia juveniles was greater than 83% in the FC (83.9%) F1 (86.1%), and F2 (87.5%) treatments, whereas in the F3 it was much lower (79%). However, these differences were not significant.

As expected, no body mass (BM) increase was observed during the fasting period (week 1–week 3); On the other hand, fasted fish lost significant weight during fasting in relation to FC ($P < 0.001$) (Table 2). BM was increased during the refeeding periods (1WR, 2WR, 6WR and 10WR), but it was lower in all fasted fish compared to FC group ($P < 0.001$). Regarding the 1WR treatment, the BM in fish from F1 and F2 was significantly higher compared to the F3 treatment. Similar results were observed for the treatment group 2WR, the BM was higher in fish from F1 and F2 treatments compared to F3. However, at 6WR, the BM increased in fish from F3 compared to F1 and F2 treatments. After 10WR, the BM in fish from F1 was significantly higher compared to the F2 and F3 treatments (Table 2).

Morphological and morphometric analyses

Muscle morphology analysis revealed that at 1WF the F1 treatments showed a high number of muscle fibers with a central nucleus, exhibiting a pattern of mosaic hyperplasia (Table 3). Nonetheless, the muscle fiber morphology at 2WF or 3WF in groups of fish from F2 and F3 treatments was different from what was observed the week before, showing an increased number of mature muscle fibers with multinucleated cells

Table 2 Body mass in response to fasting in Nile tilapia (*Oreochromis niloticus*) juveniles

	Body mass (g)				
	FC	F1	F2	F3	*P* value
Fasting					
1WF	40.20 ± 0.34[a]	30.20 ± 0.36A			< 0.001
2WF	48.75 ± 0.22[a]		29.44 ± 0.24AB		< 0.001
3WF	52.89 ± 0.35[a]			28.50 ± 0.31B	< 0.001
Refeeding					
1WR	48.68 ± 0.14[a]	36.47 ± 0.38A			< 0.001
	52.89 ± 0.35[a]		35.72 ± 0.24A		< 0.001
	64.52 ± 0.37[a]			32.50 ± 0.12B	< 0.001
2WR	52.89 ± 0.35[a]	40.21 ± 0.35A			< 0.001
	64.86 ± 0.37[a]		41.12 ± 0.35A		< 0.001
	74.76 ± 0.36[a]			36.86 ± 0.69B	< 0.001
6WR	88.45 ± 0.38[a]	73.38 ± 0.49B			< 0.001
	104.56 ± 1.08[a]		69.25 ± 0.83C		< 0.001
	133.42 ± 1.61[a]			79.12 ± 0.76A	< 0.001
10WR	152.31 ± 2.02[a]	119.03 ± 0.69A			< 0.001
	177.6 ± 3.08[a]		98.97 ± 1.15C		< 0.001
	182.67 ± 2.12[a]			106.8 ± 0.80B	< 0.001

Fish body mass (g, BM) after 1 (1WF), 2 (2WF) and 3 (3WF) weeks of fasting and after 1 (1WR), 2 (2WR), 6 (6WR) and 10 (10WR) weeks of refeeding. FC: control, continuously fed during the 13 weeks of treatment; F1: 1 week of fasting and 10WR; F2: 2 weeks of fasting and 10WR; F3: 3 weeks of fasting and 10WR. Uppercase letters compare between fasting treatments after fasting and refeeding

[a] Significant difference ($P < 0.05$) between FC and fasting treatments. Data are mean ± SD ($n = 8$)

with peripheral nuclei (Table 3). After 10WR, in the fasted treatments (F1, F2 and F3), muscle fiber morphology exhibited a typical mosaic pattern with polygonal shaped fibers, small fibers adjacent to big fibers (Table 3).

After 1WF, 2WF and 3WF, morphometric analyses showed a higher quantity (> 95%) of smaller muscle fibers in between classes 10 and 30, and a lower quantity (< 0.6%) in class 50. The frequency of muscle fibers in diameter classes did not differ significantly between fasted fish and the respective FC group (Table 3). However, after 10WR the distribution of white muscle fiber frequency in diameter classes changed in fish from F1, F2, and F3 treatments, showing higher numbers of big fibers in classes 50 and 60 compared to classes 10 and 30. Similar to the fasting period, no differences were found among FC and the fasted fish, after 10WR (Table 3).

Increased expression of atrogenes and IGF-1 reduction during fasting

The IGF-1 gene expression, at 1WF, 2WF and 3WF, was downregulated, exhibiting significant differences ($P < 0.028$) in comparison to the FC group; and after 10WR, the IGF-1 levels were increased, being similar to the IGF-1 mRNA level in the FC group, showing a substantial response to refeeding. Furthermore, after 10WR, the F1 group was found to express IGF-1 significantly higher compared to the F2 group ($P < 0.017$) (Fig. 2a). Conversely, the IGF-1 receptor was upregulated in the fish from the F1, F2 and F3 groups during the fasting periods, but after 10WR, only the expression of IGF-1 receptor in the F3 group was found to be significantly reduced ($P < 0.03$) in comparison to the FC group (Fig. 2b).

Table 3 Muscle fibers frequency in Nile tilapia (*Oreochromis niloticus*) juveniles

Frequency of fibers (%)				
Diameter class	10	30	50	60
1WF				
FC	32.1 ± 11.6	67.6 ± 11.4	0.4 ± 0.2	
F1	42.5 ± 10.8	57.5 ± 10.7		
P value	0.522	1.000		
2WF				
FC	24.8 ± 4.6	74.8 ± 4.5	0.5 ± 0.4	
F2	29.2 ± 7.3	70.6 ± 7.3	0.2 ± 0.1	
P value	0.960	1.000	0.763	
3WF				
FC	29.5 ± 8.9	70.5 ± 9.5		
F3	36.9 ± 7.7	63.1 ± 7.8		
P value	0.732	1.000		
10WR-F1				
FC	0.3 ± 0.2	8.5 ± 4.3	19.5 ± 7.5	71.8 ± 11.0
F1	0.4 ± 0.1	13.0 ± 6.6	22.2 ± 12.1	65.5 ± 15.3
P value	0.943	0.954	1.000	0.853
10WR-F2				
FC	0.3 ± 0.1	8.3 ± 5.2	16.6 ± 10.5	74.9 ± 13.3
F2	–	14.0 ± 6.4	26.7 ± 9.1	59.4 ± 14.0
P value		0.552	0.413	0.390
10WR-F3				
FC	–	8.0 ± 6.9	20.8 ± 11.2	71.2 ± 15.3
F3	0.4 ± 0.4	18.3 ± 11.3	26.0 ± 10.5	55.4 ± 20.0
P value		0.444	0.970	0.265

Frequency of white muscle fibers in certain diameter class groups after 1 (1WF), 2 (2WF) and 3 (3WF) weeks of fasting and 10 weeks of refeeding (10WR). Muscle diameters are grouped into four diameter classes (d, μm): class 10 = $d < 10$, class 30 = $10 \leq d < 30$, class 50 = $30 \leq d < 50$ and class 60 = $d \geq 50$. (FC) control, fed continuously during the 13 weeks; (F1) 1 week of fasting and 10WR; (F2) 2 weeks of fasting and 10WR; (F3) 3 weeks of fasting and 10WR. Data are mean ± SD ($n = 8$)

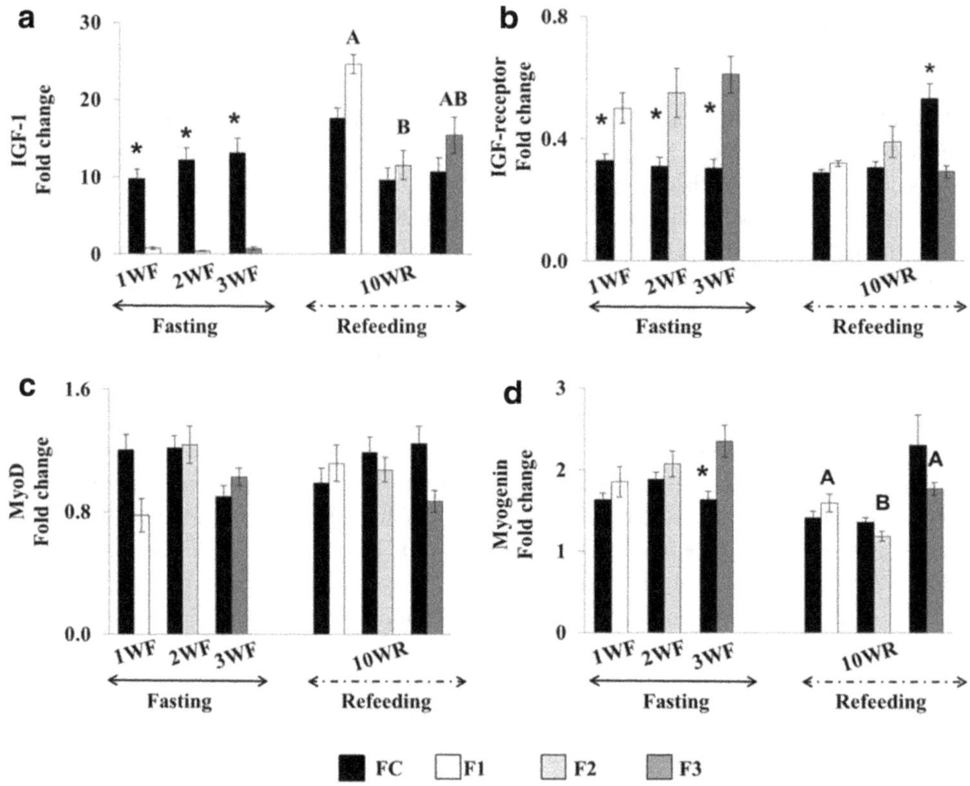

Fig. 2 Relative gene expression of IGF-1, IGF-1 receptor, MyoD and myogenin in white skeletal muscle of Nile tilapia *Oreochromis niloticus* juveniles subjected to fasting and refeeding. Gene expression was evaluated after 1 (1WF), 2 (2WF) and 3 (3WF) weeks of fasting and after 10 weeks of refeeding (10WR). **a** Relative expression of IGF-1; **b** relative expression of IGF-1 receptor; **c** relative expression of MyoD; **d** relative expression of myogenin. (FC) control, feeding continuously during the 13 weeks; (F1) 1 week of fasting and 10WR; (F2) 2 weeks of fasting and 10WR; (F3) 3 weeks of fasting and 10WR. Uppercase letters denote significant differences between F1, F2, and F3 treatments after fasting and refeeding, and *significant difference ($P < 0.05$) between control and fasted treatments. Data are mean ± SEM ($n = 8$)

MyoD gene expression was found to not differ among all the groups, either in the fasting or the refeeding group (Fig. 2c). Regarding myogenin expression, its mRNA levels were not found to differ significantly except at 3WF where they were higher compared to the FC group (Fig. 2d).

The myostatin mRNA levels, after fasting and refeeding, did not differ between control and fasted groups (Fig. 3a). The atrogin-1 and MuRF-1 gene expression levels, at 1WF, 2WF and 3WF, increased significantly ($P < 0.004$) compared to the FC group. However, after 10WR, the atrogin-1 and MuRF-1 levels in all fasted groups were similar to the FC (Fig. 3b, c). Atrogin-1 expression was serially higher after 10WR and significantly different between the F1, F2 and F3 groups, but not significantly different compared to the control (Fig. 3b).

Discussion

In the present work, we investigated if fasting for a certain interval followed by 10 weeks of refeeding would change muscle gene expression. Concurrently, we also wanted to evaluate if this starve/refeed regime would cause alterations in white muscle fiber plasticity and morphology in Nile tilapia *Oreochromis niloticus* juveniles.

The periods of fasting evaluated, 1, 2 and 3 weeks (1WF, 2WF, and 3WF, respectively) were not sufficient to cause any significant decrease in weight in fasted tilapia juveniles during the fasting period. However, significant differences were detected between the groups during the refeeding periods. The inability of the fasted groups to achieve body mass as the FC fish after refeeding, as previously reported, could be that the fasting periods were not of sufficient to induce a strong compensatory growth response in these animals at this phase of development (Wang et al. 2000).

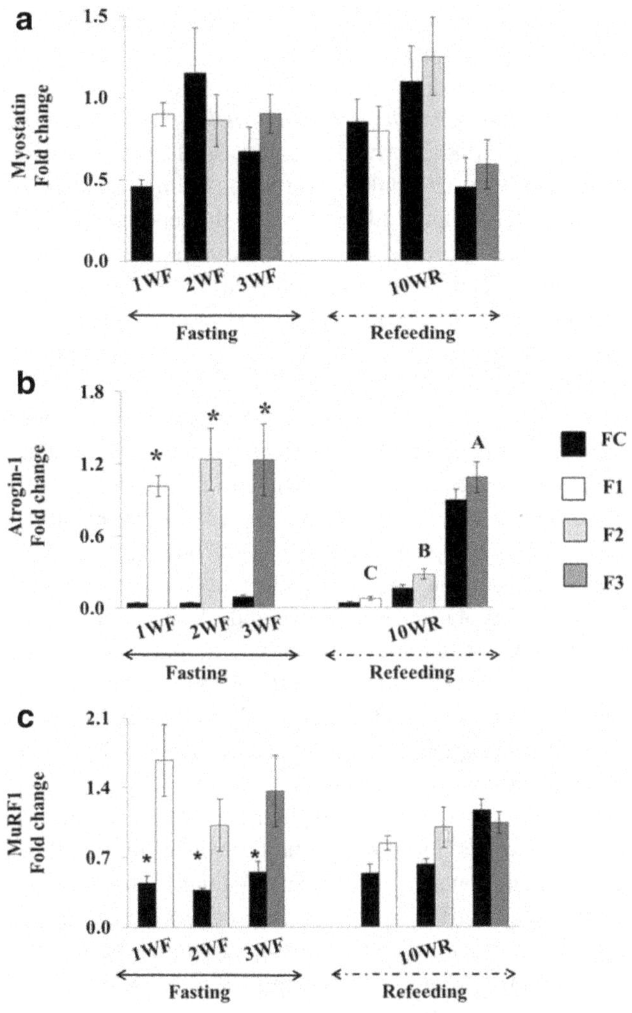

Fig. 3 Relative gene expression of myostatin, atrogin-1 and MuRF-1 in white skeletal muscle of Nile tilapia *Oreochromis niloticus* juveniles subjected to fasting and refeeding. The gene expression were calculated after 1 (WF), 2 (2WF) and 3 (3WF) weeks of fasting and after 10 weeks of refeeding (10WR). **a** Relative expression of myostatin; **b** relative expression of atrogin 1; **c** relative expression of MuRF1. (FC) control, fed continuously during the 13 weeks; (F1) 1 week of fasting and 10 weeks of refeeding; (F2) 2 weeks of fasting and 10WR; (F3) 3 weeks of fasting and 10WR. Uppercase letters denote significant differences between F1, F2, and F3 treatments after fasting and refeeding, and *significant difference ($P < 0.05$) between control and fasted treatments. Data are mean ± SEM ($n = 8$)

To understand if fasting periods were detrimental to fish skeletal muscle growth and to determine the response of fish refeeding after 10 weeks, muscle tissue was analyzed by evaluating the morphology and the diameters of muscle fibers to determine the relative degree of hypertrophy, hyperplasia or atrophy of fibers. High frequency (over 90%) of small white muscle fibers in classes 10 and 30 during one 1WF, two 2WF and three 3WF weeks, in all treatments, showed that muscle hyperplasia was occurring in the juvenile tilapia within this study. Hyperplasia was evident within the first week of the experiment by the observation of several small fibers with central nuclei. This morphological feature characterizes skeletal muscle development in mosaic hyperplasia, with small fibers containing central nuclei surrounding big fibers throughout the myotome (Johnston et al. 2003; Johnston and Hall 2004; Johnston et al. 2008; Steinbacher et al. 2007; Valente et al. 2013). However, after 10 weeks of refeeding the growth pattern changed, with increasing the frequencies of muscle fibers in the > 50 μm diameter classes and decreasing the frequencies in the 10 and 30 diameter classes. This change in fiber distribution was demonstrated by the fish in all the fasted and fed groups and can be considered as evidence of hypertrophic muscle growth (Veggeti et al. 1993; Valente et al. 1999; Zimmerman and Lowery 1999; Johnston 2006).

IGF-1 mRNA levels dropped significantly throughout periods of fasting (1WF, 2WF, and 3WF), but after subsequent refeeding, the IGF-1 muscle expression in previously fasted fish was identical to the FC group.

Among fasting treatments, the mRNA levels were higher in the F1 group. These findings are in agreement with other studies in which reductions in muscle IGF-1 levels as a result of fasting, and subsequent recovery following 2–4 weeks of refeeding have also been seen in studies with tilapia juveniles *Oreochromis mossambicus* (Fox et al. 2009), rainbow trout *Oncorhynchus mykiss* fry (Montserrat et al. 2007), and rainbow trout adults submitted to 10 weeks of food deprivation and 4–34 days of refeeding (Chauvigne et al. 2003).

In channel catfish, *Ictalurus punctatus* the IGF-1 expression decreased after 30 days of fasting, although after refeeding, its expression level was found to be similar to the fed control group (Peterson et al. 2009). In rainbow trout treated with IGF-1 during feed deprivation, inhibition of atrogin-1 expression was observed, along with a reduction of proteolytic activity (Cleveland et al. 2009). In the present study, IGF-1 receptor expression was upregulated significantly during 1–3 weeks of fasting in F1, F2, and F3, compared to FC. The high level of IGF-1 receptor in juvenile tilapias, during fasting periods, appears to correlate with a down-regulation of IGF-1 expression.

Other studies reported diverse results with different fish species and experimental conditions, and in general, fasting causes significant changes in MRF gene expression (Chauvigne et al. 2003; Valente et al. 2012). However, our results showed that there was no change in the expression of MyoD between FC and the fasted groups F1, F2 and F3 during the entire experimental period. Similar results were found in rainbow trout, where no significant change in MyoD expression levels was found in fish in response to starvation and refeeding (Johansen and Overturf 2006). The same results were also noted in another study with rainbow trout fasted for 1, 2 and 4 weeks, where MyoD levels in white muscle were not significantly affected by fasting (Montserrat et al. 2007). Furthermore, myogenin levels in Nile tilapia juveniles measured at the first 2 weeks of fasting (2WF) and after 10 weeks of refeeding (10WR) did not differ in relation to FC. In gilthead sea bream, myogenin gene expression was also found to not be affected after 30 days of food deprivation (Garcia et al. 2014). MyoD expression can be expected to be reduced in fasting fish as the lack of available nutrient energy would preclude muscle development and growth.

If the expression of myostatin was high, it would explain the relatively low expression of MyoD in fasted treatments. However, in our study, skeletal muscle myostatin levels were not different in Nile tilapia juveniles between fasted and control fish throughout the experiment. A long-term study of fasting in adult tilapia also showed that myostatin expression was unaffected by fasting (Rodgers et al. 2003) Our findings suggest that the level of starvation in these juveniles was enough to activate an alternative mechanism for regulating muscle growth rather than through myostatin. In rainbow trout deprived of food for 10 weeks and refeed for 34 days, the myostatin expression, similar to our findings, was not significantly changed during the experiment (Chauvigne et al. 2003). However, in tilapia fingerlings (0.6 g), myostatin expression was upregulated after 5 days of fasting in comparison to the fed control; nonetheless, after refeeding no differences were detected (Nebo et al. 2013). Likewise in another study performed with tilapia in larval phase, myostatin mRNA levels were upregulated by fasting and then reduced during refeeding (Rodgers et al. 2003). Taken into consideration the above, these results suggest that the degree of catabolism induced by fasting in tilapia is greater in larvae and fingerlings, likely due to lower body reserves, than in juvenile and adult fish that have more energy storage, and thus a higher capacity to tolerate fasting conditions.

The levels of atrogin-1 and MuRF1 expression in white skeletal muscle after 1–3 weeks of fasting increased rapidly, suggesting an increase in muscle turnover. However, the expression of atrogin-1 and MuRF1 after 10 weeks refeeding dropped to basal levels, and no significant differences were found among all groups ($P > 0.05$). Our results corroborate other reported findings with different fish species submitted to different fasting protocols. For instance, in rainbow trout juveniles, after 14 days of food deprivation the atrogin-1 levels increased and 12 h after refeeding decreased significantly (Seiliez et al. 2008); in salmon white muscle, the expression of MuRF1 increased in fasted fish compared to fed control (Tacchi and Bicherdike 2012); and in Atlantic salmon and zebrafish, atrogin-1 and MuRF1 were also found to be upregulated during fasting and downregulated after refeeding (Bower et al. 2009; Bower and Johnston 2010; Amaral and Johnston 2011).

It is well established in vertebrates that insulin-like growth factor-1 stimulates activation of the PI3 K/Akt/mTOR cascade, increasing protein synthesis and hypertrophic muscle growth (Engert et al. 1996; Zhang et al. 2008; Duan et al. 2010), and inactivates FOXO transcription factors, decreasing the expression of atrogin-1/MAFbx and MuRF1 (Sacheck et al. 2004; Edstron et al. 2006; Cleveland et al. 2009; Fuentes et al. 2013). In this context, the lower IGF-1 expression observed in fasted tilapia juveniles likely plays a role in the higher expression of MuRF1 and atrogin-1, consequently inducing muscle atrophy, over extended periods of time, and blocking somatic growth. On the other hand, after refeeding, the expression of IGF-1 increased to the basal level, equal to the control, resulting in growth recovery.

In conclusion, the results from the present study demonstrated that the corresponding upregulation of atrogin-1 and MuRF1 and the drop in IGF-1 expression during fasting periods correlate with an initial stage of muscle atrophy. However, no signs of atrophy were observed in muscle morphology, but the expression of atrophy genes changed during fasting. Nevertheless, the increased IGF-1 levels after 10 weeks of refeeding presumably activated the Akt pathway, which correspondingly decreased the expression of genes involved in the ubiquitin–proteasome pathway, thus reducing rates of protein degradation and contributing to growth recovery. Furthermore, we observed partial compensatory growth (fish grew, but did not reach the same body mass than the control fish), after refeeding in Nile tilapia juveniles. These results contribute to a better understanding of molecular control related to skeletal muscle growth during fasting and after refeeding. This information will be beneficial in developing best management plans for feeding optimized to economically maximize growth during production specifically if phase feeding or fasting can be used.

Acknowledgements The authors are grateful to the Fundação de Amparo à Pesquisa do Estado de São Paulo/FAPESP for the financial support by means of scholarships (2011/08426-3 and 2013/13100-3) and a research grant (2011/22326-0) and to the facilities and team of the University of Idaho—Hagerman Fish Culture Experiment Station. We thank technician Karen Frank at Hagerman Fish Culture Experiment Station for helping in gene expression analyses.

Author contribution Caroline Nebo was involved in planning and carrying out the experiment, data analysis and writing the manuscript. Ken Overturf was involved in planning, data analysis and writing the manuscript. Andrezas Brezas was involved with data analysis and writing. Maeli Dal-Pai-Silva was involved with experimental design, data analysis and writing of the manuscript. Maria Celia Portell was involved with experimental design, data analysis and manuscript writing.

References

Abdel-Hakim NF, Abo-State HA, Al-Azab AA, El-Kholy KHF (2009) Effect of feeding regimes on growth performance of juvenile hybrid tilapia (*Oreochromis niloticus* × *Oreochromis aureus*). World J Agric Sci 5:49–54

Acosta J, Carpio Y, Borroto I, González O, Estrada MP (2005) Myostatin gene silenced by RNAi show a zebrafish giant phenotype. J Biotechnol 119:324–331

Ali M, Nicieza A, Wootton RJ (2003) Compensatory growth in fishes: a response to growth depression. Fish Fish 4:147–190

Amaral IPG, Johnston IA (2011) Insulin-like growth factor (IGF) signaling and genome-wide transcriptional regulation in fast muscle of zebrafish following a single-satiating meal. J Exp Biol 214:2125–2139

Bower NI, Johnston IA (2010) Transcriptional regulation of the IGF signaling pathway by amino acids and insulin-like growth factors during myogenesis in Atlantic salmon. PLoS One 5:1–14

Bower NI, Taylor RG, Johnston IA (2009) Phasing of muscle gene expression with fasting-induced recovery growth in Atlantic salmon. Front Zool 6:1–13

Chauvigne F, Gabillard JC, Weil C, Rescan PY (2003) Effect of refeeding on IGF-1, IGF-II, IGF receptors, FGF2, FGF6, and myostatin mRNA expression in rainbow trout myotomal muscle. Gen Comp Endocrinol 132:209–215

Cleveland BM, Weber GM, Bleming KP, Silverstein JT (2009) Insulin-like growth factor-I and genetic effects on indexes of protein degradation in response to feed deprivation in rainbow trout (*Oncorhynchus mykiss*). Am J Physiol Regul Integr Comp Physiol 297:1332–1342

De Santis C, Gomes GB, Jerry DR (2012) Abundance of myostatin gene transcripts and their correlation with muscle hypertrophy during the development of barramundi, *Lates calcarifer*. Comp Biochem Physiol B 163:101–107

Dobson S, Holmes R (1984) Compensatory growth in the rainbow trout, *Salmo gairdneri* Richardson. J Fish Bio 25:649–656

Duan C, Ren H, Gao S (2010) Insulin-like growth factors (IGFs), IGF receptors and IGF-binding proteins: roles in skeletal muscle growth and differentiation. Gen Comp Physiol 167:344–351

Edstron E, Altun M, Hagglund M, Ulfnake B (2006) Atrogin-1/MAFbx and MuRF1 are downregulated in aging-related loss of skeletal muscle. J Gerontol Ser Biol Sci Med Sci 61:663–674

Engert C, Berglund EB, Rosenthal N (1996) Proliferation precedes differentiation in IGF-I stimulated myogenesis. J Cell Biol 135:431–440

Fox BK, Breves JP, Hirano T, Grau EG (2009) Effects of short- and long-term fasting on plasma and stomach ghrelin, and the growth hormone/insulin-like growth factor I axis in the tilapia, *Oreochromis mossambicus*. Domest Anim Endocrinol 37:1–11

Fuentes EN, Valdes JA, Molina A, Bjornsson BT (2013) Regulation of skeletal muscle growth in fish by the growth hormone—insulin-like growth factor system. Gen Comp Endocrinol 192:136–148

Garcia de la Serrana D, Codina M, Capilla E, Jimenez-Amilburu V, Navarro I, Du SJ, Johnston IA, Gutierrez J (2014) Characterization and expression of myogenesis regulatory factors during in vitro myoblast development and in vivo fasting in the gilthead sea bream (*Sparus aurata*). Comp Biochem Physiol B 167:90–99

Glass DJ (2003) Molecular mechanisms modulating muscle mass. Trends Mol Med 9:344–350

Glass DJ (2005) Skeletal muscle hypertrophy and atrophy signaling pathways. Int J Biochem Cell Biol 37:1974–1984

Johansen KA, Overturf K (2006) Alterations in expression of genes associated with muscle metabolism and growth during nutritional restriction and refeeding in rainbow trout. Comp Biochem Physiol 144:119–127

Johnston IA (2006) Environment and plasticity of myogenesis in teleost fish. J Exp Biol 209:2249–2264

Johnston IA, Hall TE (2004) Mechanisms of muscle development and responses to temperature change in fish larvae. Am Fish Soc Symp 40:85–116

Johnston IA, Fernandez DA, Calvo J, Vieira VLA, North AW, Abercromby M, Garland T (2003) Reduction in muscle fibre number during the adaptive radiation of notothenioid fishes: a phylogenetic perspective. J Exp Biol 206:2595–2609

Johnston IA, Macqueen DJ, Watabe S (2008) Molecular biotechnology of development and growth in fish muscle Fisheries for global welfare and environment. 5th World Fish Congr., pp 241–262

Johnston IA, Bower NI, Macqueen DJ (2011) Growth and regulation of myotomal muscle mass in teleost fish. J Exp Biol 214:1617–1628

Kandarian SC, Jackman RW (2006) Intracellular signaling during skeletal muscle atrophy. Muscle Nerve 33:155–165

Livak KJ, Schmittgen TD (2001) Analysis of relative gene expression data using real-time quantitative PCR and the 2^{-DDCT} method. Methods 25:402–408

Love RM (1970) The chemical biology of fishes, vol I. Academic Press, New York

Maccatrozzo L, Bargelloni L, Rodaelli G, Mascarello F, Patarnelo T (2001) Characterization of the Myostatin gene in the gilthead sea bream (*Sparus aurata*): sequence, genomic structure and expression pattern. Mar Biotechnol 3:224–230

Montserrat N, Gabillard JC, Capilla E, Navarro MI, Gutiérrez J (2007) Role of insulin, insulin-like growth factors, and muscle regulatory factors in the compensatory growth of the trout (*Oncorhynchus mykiss*). Gen Comp Endocrinol 150:462–472

Nebo C, Portella MC, Carani FR, Almeida FLA, Padovani CR, Carvalho RF, Dal-Pai-Silva M (2013) Short periods of fasting followed by refeeding change the expression of muscle growth-related genes in juvenile Nile tilapia (*Oreochromis niloticus*). Comp Biochem Physiol B 164:268–274

Peterson BC, Small BC, Waldbieser GC, Bosworth BG (2009) Endocrine response of fast and slow-growing families of channel catfish. N Am J Aquac 70:240–250

Rios R, Carneiro I, Arce VM, Devesa J (2002) Myostatin is an inhibitor of myogenic differentiation. Am J Physiol Cell Physiol 282:993–999

Rodgers BD, Weber GM, Kelley KM, Levine MA (2003) Prolonged fasting and cortisol reduce myostatin mRNA levels in tilapia larvae; short-term fasting elevates. Am J Physiol Regul Integr Comp Physiol 284:1277–1286

Rudnicki MA, Schnegeisberg PNJ, Stead RH, Braum T, Arnold HH, Jaenish R (1993) MyoD or myf5 is required for the formation of skeletal muscle. Cell 75:1351–1359

Sacheck JM, Ohtsuka A, Mclary SC, Goldberg AL (2004) IGF-I stimulates muscle growth by suppressing protein breakdown and expression of atrophy-related ubiquitin ligases, atrogin-1 and MuRF1. Am J Physiol Endocrinol Metab 287:591–601

Seiliez I, Panserat S, Skiba-cassy S, Fricot A, Vachot C, Kaushik S, Tesseraud S (2008) Feeding status regulates the polyubiquitination step of the ubiquitin-proteasome-dependent proteolysis in rainbow trout (*Oncorhynchus mykiss*) muscle. J Nutr 138:487–491

Steinbacher P, Haslett JR, Obermayer A, Marschanllinger J, Bauer HC, Sanger AM, Stoiber W (2007) MyoD and Myogenin expression during myogenic phases in brown trout: a precocious onset of mosaic hyperplasia is a prerequisite for fast somatic growth. Dev Dyn 236:1106–1114

Tacchi L, Bicherdike R (2012) Muscle-specific RING finger (MuRF) cDNAs in Atlantic salmon (*Salmo salar*) and their role as regulators of muscle protein degradation. Mar Biotechnol 14:35–45

Valente LMP, Rocha E, Gomes EFS, Silva MW, Oliveira MH, Monteiro RAF, Fauconneau B (1999) Growth dynamics of White and red muscle fibers in fast and slow growing strains of rainbow trout. J Fish Biol 55:675–691

Valente LM, Bower NI, Johnston IA (2012) Postprandial expression of growth-related genes in Atlantic salmon (*Salmo salar* L.) juveniles fasted for 1 week and fed a single meal to satiation. Br J Nutr 108:2148–2157

Valente LM, Moutou KA, Conceição LEC, Engrola S, Fernandes JMO, Johnston IA (2013) What determines growth potential and juvenile quality of farmed fish species? Rev Aquac 5:5168–5193

Veggeti A, Mascarello F, Scapolo PA, Rowlerson MD, Carnevali CMD (1993) Muscle growth and myosin isoform transitions during development of a small teleost fish (*Poecilia reticulate)* (Atheriniformes, Poeciliidae): a histochemical, immunohistochemical, ultrastructural and morphometric study. Anat Embryol 187:353–361

Wang Y, Cui Y, Yang Y, Cai F (2000) Compensatory growth in hybrid tilapia, *Oreochromis mossambicus* × *O niloticus*, reared in seawater. Aquaculture 189:101–108

Zhang XD, Zhu YF, Cai LS, Wu TX (2008) Effects of fasting on the meat quality and antioxidant defenses of market-size farmed large yellow croaker (*Pseudosciaena crocea*). Aquaculture 280:136–139

Zimmerman AM, Lowery MS (1999) Hyperpastic development and hypertrophic growth of muscle fibers in the white seabass (*Atractoscion nobilis*). J Exp Zool 284:299–308

8

Dietary short-chain organic acids enhanced resistance to bacterial infection and hepatopancreatic structural integrity of the giant freshwater prawn, *Macrobrachium rosenbergii*

Wing-Keong Ng · Chia-Ling Lim · Nicholas Romano · Beng-Chu Kua

Abstract The use of short-chain organic acids in the diets of aquacultured animals as a prophylactic to bacterial pathogens is receiving increasing research and commercial interest. After giant freshwater prawns, *Macrobrachium rosenbergii,* were fed diets supplemented with formic (FA), citric (CA), lactic (LA), propionic acid (PA), an organic acid blend (OAB) at 1% or a control diet (no additives) for 6 weeks, the prawns were subjected to *Vibrio harveyi* challenge for 2 weeks. From all remaining prawns, the hepatopancreatic histopathology was examined. Results showed that the survival of prawns to *V. harveyi* challenge was significantly higher ($P < 0.05$) when fed the LA diet, followed by the CA diet, than the other treatments. This was likely due to less bacterial-induced hepatopancreatic damage and higher energy reserves compared to those fed the other diets. This is the first report to compare the efficacy of different organic acids to a crustacean as well as showing hepatopancreatic protective properties to freshwater prawns when challenged with pathogenic bacteria. The present study indicated that dietary organic acids can be a viable alternative to the use of harmful antibiotics in the prawn farming industry.

Keywords Organic acids · Lactic acid · *Macrobrachium rosenbergii* · *Vibrio* challenge · Hepatopancreas

Introduction

An increasing trend in aquaculture is to reduce the reliance on antibiotics since these can be an effective prophylactic against disease as well as growth promoter when used at low doses, the long-term consequences can be harmful to the host animal, environment and potentially the human consumer (reviewed by Defoirdt et al. 2009; Marshall and Levy 2011). A promising alternative to antibiotics are organic acids, which include short-chain fatty acids (SCFA), which are "generally regarded as safe" compounds with one or more carboxyl groups and less than six carbons in their molecular formula. These have been used for many decades in the terrestrial animal feed industry as an antimicrobial and growth promoter (Thompson and Hinton 1997; Kluge et al. 2006). In more recent years, organic acids such as lactic, formic, citric and propionic acids and their

W.-K. Ng (✉) · C.-L. Lim · N. Romano
Fish Nutrition Laboratory, School of Biological Sciences, Universiti Sains Malaysia, 11800 USM, Penang, Malaysia
e-mail: wkng@usm.my

N. Romano
Department of Aquaculture, Faculty of Agriculture, Universiti Putra Malaysia, Serdang, Selangor, Malaysia

B.-C. Kua
National Fish Health Research Centre, Batu Maung, Penang, Malaysia

associated salts have been the focus of research for use in aquatic animal feeds (Ng and Koh 2011, 2016; Sukor et al. 2016).

The beneficial effects of dietary organic acids to the gastrointestinal tract of aquatic animals can include reducing potentially harmful bacteria in favor of more beneficial bacteria (e.g., lactic acid bacteria); however, the effectiveness of organic acids to disease resistance can be highly dependent on the host species and the type of organic acid used, often leading to contrasting findings (reviewed by Ng and Koh 2016). It is known that bacterial pathogens such as *Vibrio* spp. can be highly pathogenic to various crustacean species during the larviculture or post-larvae stages under stressful conditions (Tonguthai 1992; Lavilla Pitogo et al. 1998; Jayaprakash et al. 2006). Although *Vibrio* spp. is predominately found in brackish and marine waters, these bacteria have been detected in diseased *Macrobrachium rosenbergii* within some freshwater ponds in Thailand (Tonguthai 1992) even though the main route of infections is during the larval stages, which require brackish water (Jayaprakash et al. 2006). Moreover, when injected, *V. harveyi* was shown to be highly virulent to *M. rosenbergii* juveniles, which was suggested to become an increasing disease problem to freshwater hosts (Siripornadulsil et al. 2014). Apart from anti-bacterial properties, organic acids including SCFA can have a beneficial impact on farmed aquatic animals by enhancing nutrient availability, feed palatability, digestive enzyme activity as well as gut morphology (as reviewed in Ng and Koh, 2016). Dietary organic acids have also been reported to enhance the non-specific immune response of shrimp and have shown substantial promise in the control of vibriosis (Su et al. 2014; Romano et al. 2015; Ng et al. 2015).

The global farming of *M. rosenbergii* has been steadily growing over the past decade from 196,848 metric tons in 2004 to 230,333 metric tons in 2014 (FAO 2016). This industry has been limited, however, due to the threat of bacterial diseases leading to the widespread use of antibiotics in some areas to control bacterial diseases in *M. rosenbergii* farming (Paul et al. 2011). The exact extent of this practice is difficult to determine; nevertheless, finding eco-friendly prophylactics to bacterial disease would clearly benefit this industry. We recently reported that *Litopenaeus vannamei* and *Penaeus monodon* fed organic acid-treated diets had significantly enhanced resistance to *V. harveyi* (Romano et al. 2015; Ng et al. 2015); to date and to the best of our knowledge, no information exists that compares the efficacy of different dietary organic acids on the disease resistance and subsequent histopathology of a freshwater crustacean. The aim of the current study was to compare different dietary additions of organic acids at 1% that included formic acid (FA), citric acid (CA), lactic acid (LA), propionic acid (PA) or an organic acid blend (OAB), on susceptibility and hepatopancreatic histopathology of *M. rosenbergii* to *V. harveyi* challenge.

Materials and methods

Experimental diets

A total of six isonitrogenous and isolipidic practical diets were formulated to contain a 1% level of different organic acids which included formic acid (FA; HCOOH), citric acid (CA; $C_6H_8O_7$), lactic acid (LA; $C_3H_6O_3$), propionic acid (PA; CH_3CH_2COOH) and an organic acid blend (OAB). The proprietary OAB (Universiti Sains Malaysia) consisted of five organic acids (formic acid, lactic acid, malic acid, tartaric acid and citric acid; 75% w/w) adsorbed onto a silicon dioxide-based inert carrier (25% w/w) as previously described (Koh et al. 2016). Since LA, FA and PA were in a solution form, these were made into a powder by mixing with a silicon dioxide-based inert carrier at a 1:1 ratio. Therefore, when these were added in the experimental diets at 2%, this yielded 1% LA, FA and PA, respectively. The CA and the OAB were already in powder form and were added at 1% (Table 1). Since all the organic acids were made into powder form before use, they were added to the mixture of ingredients and mixed homogenously.

The main source of dietary protein was fishmeal (9.78%) and soybean meal (64.16%), while fish oil and soybean oil were included as the main lipid sources. All major dietary ingredients were purchased from local stores except otherwise indicated. Fish and soybean oils, corn starch and α-cellulose were purchased from Liang Traco Ltd. (Malaysia). Cholesterol (Solvay Ltd., The Netherlands) was added at 0.5% in all diets to satisfy the reported requirements for *M. rosenbergii* (Briggs et al. 1988). The proximate composition of the experimental diets was measured according to standard AOAC (1997) methods and values were generally similar among all diets (Table 1). Crude dietary lipid and protein were relatively constant at 7.7–8.3 and

Table 1 The ingredient and proximate composition (% dry matter) of the experimental diets for *Macrobrachium rosenbergii* with various added organic acids

	Experimental diets					
	Control	CA	LA	FA	PA	OAB
Fishmeal[a]	9.78	9.78	9.78	9.78	9.78	9.78
Soybean meal[b]	64.16	64.16	64.16	64.16	64.16	64.16
Fish oil	3.35	3.35	3.35	3.35	3.35	3.35
Soybean oil	2.70	2.70	2.70	2.70	2.70	2.70
Cholesterol	0.50	0.50	0.50	0.50	0.50	0.50
Corn starch	5.97	5.97	5.97	5.97	5.97	5.97
Vitamin premix[c]	3.00	3.00	3.00	3.00	3.00	3.00
Mineral premix[c]	4.00	4.00	4.00	4.00	4.00	4.00
CMC[d]	2.00	2.00	2.00	2.00	2.00	2.00
Citric acid[e]	0.00	1.00	0.00	0.00	0.00	0.00
Lactic acid[e]	0.00	0.00	2.00	0.00	0.00	0.00
Formic acid[e]	0.00	0.00	0.00	2.00	0.00	0.00
Propionic acid[e]	0.00	0.00	0.00	0.00	2.00	0.00
Organic acid blend[f]	0.00	0.00	0.00	0.00	0.00	1.00
α-Cellulose	4.53	3.53	2.53	2.53	2.53	3.53
Proximate composition						
Dry matter	97.70	93.40	94.00	94.3	91.80	91.70
Crude protein	35.22	35.50	35.07	35.18	35.18	35.42
Crude lipid	8.33	8.04	7.88	7.91	7.78	7.71
Ash	8.92	9.37	9.40	9.68	9.33	9.20
Crude fiber	1.94	2.22	1.33	1.81	2.38	2.71
pH	6.20	5.56	5.63	5.20	5.76	5.51

[a]Danish fish meal (Sri Putra Ltd., Malaysia) contained (% dry weight): 72.8 crude protein and 9.5 crude lipid

[b]Solvent-extracted soybean meal (Soon Soon Ltd., Malaysia) contained (% dry weight): 46.2 crude protein and 2.1 crude lipid

[c]Vitamin and mineral premix according to Kim et al. (2013)

[d]Carboxymethyl cellulose (Liang Traco Ltd., Malaysia)

[e]An inert carrier, silicone dioxide (SiO_2), was mixed with liquid organic acids (*LA* lactic acid, *FA* formic acid, *PA* propionic acid) at a 1:1 ratio, while citric acid (CA) was added in powder form. All organic acids were purchased from BioBasic Inc., Malaysia

[f]OAB, organic acid blend, consisted of five organic acids (formic acid, lactic acid, malic acid, tartaric acid and citric acid; Koh et al. 2016)

34.6–35.3%, respectively. This feed formulation was based on the known nutrient requirements for *M. rosenbergii* (D'Abramo and Sheen 1994). Dietary pH was measured according to Romano et al. (2015).

The experimental diets were prepared by thoroughly mixing the dry ingredients with oil and water in a Hobart mixer. The moist dough was then screw-pressed through a 2-mm die in a meat mincer and the feed pellets formed were fan-dried. After forced air-drying the pellets for at least 6 h to improve water stability, the diets were then crumbled into appropriate sizes (0.5–1.5 mm) and stored at −20 °C in airtight polyethylene bags until required.

Source of animals and acclimation

Giant freshwater prawn, *M. rosenbergii,* post-larvae were obtained from a commercial hatchery (Sitiawan, Perak) and acclimated in a one ton stocking tank that ran on a freshwater flow-through system for 3 weeks upon arrival to our laboratory. The water temperature was ambient (28 ± 2 °C) and the prawns were fed twice per day to apparent satiation with a commercial prawn feed (45% crude protein) (Gold Coin Ltd., Malaysia).

The water was sourced from the city water supply which was carbon and sand filtered prior to entering the tank and several air stones provided continuous aeration.

Experimental set-up and design

After the acclimation period, 360 prawns (initial weight 0.17 ± 0.01 g, mean ± SD) were randomly distributed into 18 glass aquariums [90-L capacity; 30 cm (W) × 70 cm (L) × 46 cm (H)]. Treatments were randomly assigned for a total of three replicates for each dietary treatment, and in each replicate there were 20 prawns. All aquaria ran on a freshwater flow-through system after being carbon and sand filtered and the set-up and conditions of the culture system were previously described by Kim et al. (2013). In each aquarium, one air stone provided gentle aeration and three groups of polyvinyl chloride (PVC) pipes (each pipe were 21 × 6.5 × 10.5 cm, radius × diameter × length), arranged as a pyramid, were allocated to minimize cannibalism by acting as shelters. The water temperature range was 29 ± 2 °C throughout the feeding period.

The prawns were fed with the experimental diets three times daily (08:30, 12:30 and 16:30 h) to apparent satiation (approximately 15% of wet body weight per day). To reduce nutrient leaching, the amount of feed given at each feeding time was dependent on feeding activity of the prawns and the introduction of feeds was stopped when it was observed that active prawn feeding activity had ceased. Each day, the aquaria were siphoned to remove feces and any molted exoskeletons and the feeding duration lasted 6 weeks. Growth and survival data were not monitored during the 6-week feeding period since it was known from an immediate prior 8-week feeding trial that prawns fed the same diets in the same experimental set-up elicited a percentage weight gain of 630–812% and a survival of 68–93% (unpublished data). The prawns were fed 6 weeks in the present experiment for the prawns to achieve sizes (1.2–1.5 g) suitable for the bacterial challenge test and subsequent hepatopancreatic histopathological analysis, which is the focus of the present study.

Bacterial challenge and histology

After 6 weeks, all prawns were transferred to the nearby National Fish Health Research Centre (Department of Fisheries, Malaysia) for hepatopancreas histology preparations and bacterial challenge test. For the bacterial challenge test, ten apparently healthy prawns of similar size (1.2–1.5 g) were distributed into each triplicate tank and fed their respective experimental diet twice daily (8:30 and 16:30 h) for 1 week. Prawns used for the bacterial challenge test were considered healthy prawns as they showed good growth and were actively feeding. All tanks had one air stone providing gentle aeration and PVC pipes acted as shelters. Each day, any uneaten feed or feces were siphoned out in the morning and approximately 30% of the water was exchanged. Remaining prawns from each dietary treatment not used for the disease challenge test were dissected and the hepatopancreas was immersion fixed in Davidson's formalin solution for 24 h. Samples were then transferred to 70% ethanol until processing using a Leica TP 1020 automatic tissue processor (Leica Microsystems, Nussloch, Germany). The samples were then embedded in paraffin wax and sectioned (5–6 μm thickness) using a Leica RM 2145 rotary microtome. Sections were mounted on poly-L-lysine (Sigma) frosted slides, stained with hematoxylin and eosin and examined under a light microscope.

In preparation for the bacterial disease challenge, *V. harveyi* (strain no. ATCC 14126), which was stored in a Marine broth 2216 containing beads and 20% (v/v) glycerol (Qrec, New Zealand) (pH 7.4) at −80 °C was grown on a TSA plate (Trypticase Soy Agar) at 28 °C for 24-h. A single colony was selected and cultured on a TSA plate for another 24-h at 28 °C before being transferred to a Tryptic Soy Broth for 24-h at 28 °C. The broth cultures were then stored in an incubator shaker for 12 h. The broth culture was then centrifuged at 10,000 rpm for 15 min. The supernatant was removed and the bacterial pellet was re-suspended in a saline solution (0.85% NaCl) and then diluted to a standard concentration equal to an optical density (OD) of 1.0 at 545 nm. As determined by standard dilutions and plating methods, this standard suspension of bacteria contained 1×10^9 CFU/ml.

Using this bacterial suspension, the prawns were injected intramuscularly with 50 μl between the third and fourth abdominal segments (Sahoo et al. 2006) at a concentration of 1×10^5 CFU/ml. This concentration was based on a pilot study to determine the relationship between prawn mortality and the challenge dose (10^7, 10^6, 10^5, 10^4 and 10^3 CFU/ml) and the most appropriate dose was chosen based on the occurrence of 50% mortality of prawns over a 2-week period. Each day the prawns were fed their respective diet twice to apparent satiation along with a 15% water exchange. Mortalities were recorded daily and the challenge period lasted 2 weeks. After the 2-week bacterial challenge, the hepatopancreas from the remaining prawns were immersion fixed for histological examination. Three replicate prawns in each treatment were dissected for the hepatopancreas and examined for the histopathology. The prevalence of various hepatopancreatic cells (R cells, B cells and E cells) and tubule diameter (using a micrometer, 100 μm) was quantified within 20 randomly selected tubules from triplicates in each treatment.

Statistical analysis

All data were subjected to one-way ANOVA using SPSS 11.5 (SPSS, Chicago, IL, USA) after confirmation of homogeneity of variance. When a significant treatment effect was observed, a Duncan's multiple range test was used to compare means. Treatment effects were considered significant when $P < 0.05$.

Results

Histology of the hepatopancreas without pathogen challenge

The hepatopancreas of *M. rosenbergii* fed the control diet without pathogen challenge is shown in Fig. 1 and consisted of "star-like" tubules of varying sizes that contained different epithelial cells including E cells ("embryonalzellen" or embryonic cells), B cells ("blasenzellen" cells), R cells ("restzellen" cells) and F cells ("fibrenzellen" cells). The E cells are undifferentiated containing a relatively large nucleus, while the R cells are known to contain lipid droplets and these cells were generally located medially and distally from the lumen in each tubule. The B cells were the largest among the cells, but with high size variability, and often located medially and proximally from the lumen. The F cells were basophilic and were observed to be much less prevalent than the other cells. There was no significant difference ($P > 0.05$) to the prevalence of the R cells, E cells or B cells among the dietary treatments and ranged from 4.25 to 5.61, 16.88–19.42 and 7.42–8.41 (mean number/tubule), respectively. No significant differences in tubule diameter were observed.

Bacterial challenge

The cumulative mortalities of prawns fed different organic acid diets after the *V. harveyi* challenge test are shown in Fig. 2. Results showed that prawns fed the LA diet had significantly lower ($P < 0.05$) mortalities when challenged with *V. harveyi* than all other dietary treatments after 2 weeks. This was followed by the prawns fed the CA diet, which was significantly lower than those fed the control, FA, PA or OAB diet. The

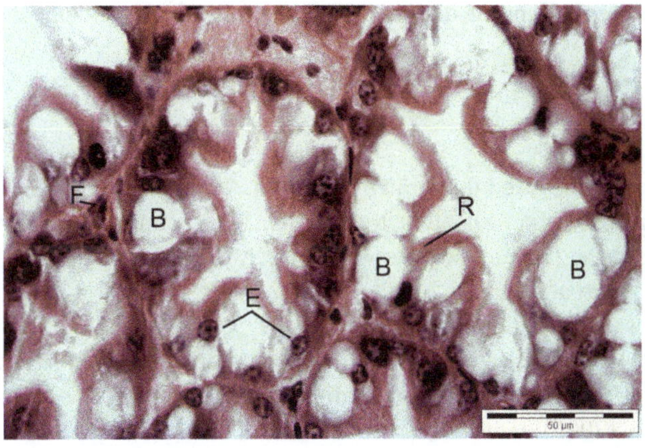

Fig. 1 Section of the hepatopancreas of giant freshwater prawn fed the control diet and not challenged with bacteria showing B cells (B), R cells (R), F cells and predominately E cells (E) within the tubules. H&E stain. Magnification ×40; bar 50 μm

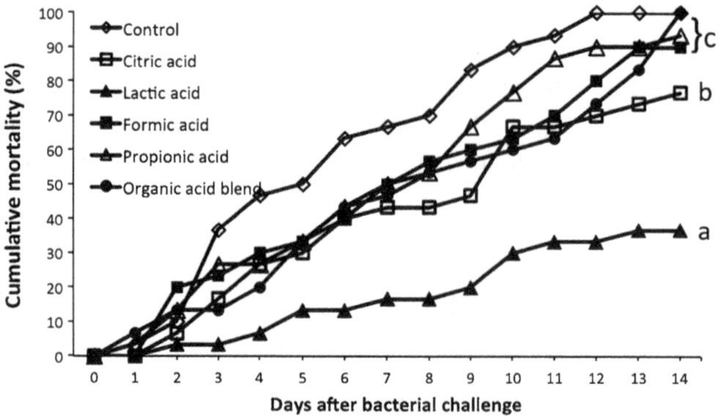

Fig. 2 The mean cumulative mortality (%) of the giant freshwater prawn fed diets with different added organic acids and subsequently challenged with *Vibrio harveyi* over a 14-day period. Values are mean of triplicate groups of ten prawns. Different letters indicate significant difference ($P < 0.05$)

mortality of prawns from the remaining dietary treatments was not significantly different from each other ($P > 0.05$).

Histology of the hepatopancreas after bacterial challenge

It was observed that substantial morphological changes occurred in all treatments; however, the amount of changes appeared to be dependent on the dietary treatment. It should be noted that since the control and OAB diets resulted in 100% mortality, near the end of the challenge period, freshly dead prawns were removed within about 1 h of death and only these prawns were used for histological analysis. Near the end of the disease challenge trial when there were only a few prawns remaining in the tanks, hourly observations were made to ensure that only freshly dead prawns were used. Results showed that prawns fed the control, PA, FA or OAB diets often had increased hemocytes within the interstitial spaces, separation of the myoepithelial layer and epithelium, a near collapse of the tubule lumen and, in some cases, tubule rupture. These effects appeared to be less pronounced for prawns fed the LA or CA diet since, aside from some cases of abnormally shaped lumen, the tubule structure appeared more intact and normal with less hemocyte infiltrations. The histopathology of the hepatopancreas of *M. rosenbergii* fed the control diet is shown in Fig. 3a, b and the LA diet in Fig. 3c, d.

After quantifying the various hepatopancreatic epithelial cells, it was found that prawns fed the LA or CA diets had significantly more R cells in their hepatopancreas than those fed the other diets while no significant changes were detected for the B cells or E cells among treatments (Table 2). Moreover, the R cells from prawns fed the LA or CA diet appeared to have a more rounded shape compared to the other treatments. The tubule diameter of the hepatopancreas of the prawns was not quantified since many individuals had tubules that were too severely compressed and/or distorted to accurately measure using a micrometer.

Discussion

In the present study, it was demonstrated that prawns fed the LA diet or the CA diet had significantly enhanced resistance to *V. harveyi* infection than all other treatments. Significantly lower prawn mortalities were also observed between prawns fed the LA diet compared to the CA diet. To the best of our knowledge, in terms of comparing different dietary organic acids on the disease resistance of a host, this is the first study to demonstrate this with a freshwater aquatic animal.

The use of different SCFA was shown to improve the resistance of *Artemia franciscana* to *V. campbelli* and was suggested that SCFA were ingested and thus inhibited the growth of *V. campbelli* (Defoirdt et al. 2006). This study by Defoirdt et al. (2006) demonstrated a significant step to the potential prophylactic protection of SCFA, but was noted that this would likely become cost-prohibitive in large-scale aquaculture operations if SCFA were to be added into the culture water. The use of organic acids, including SCFA, as prophylactics in

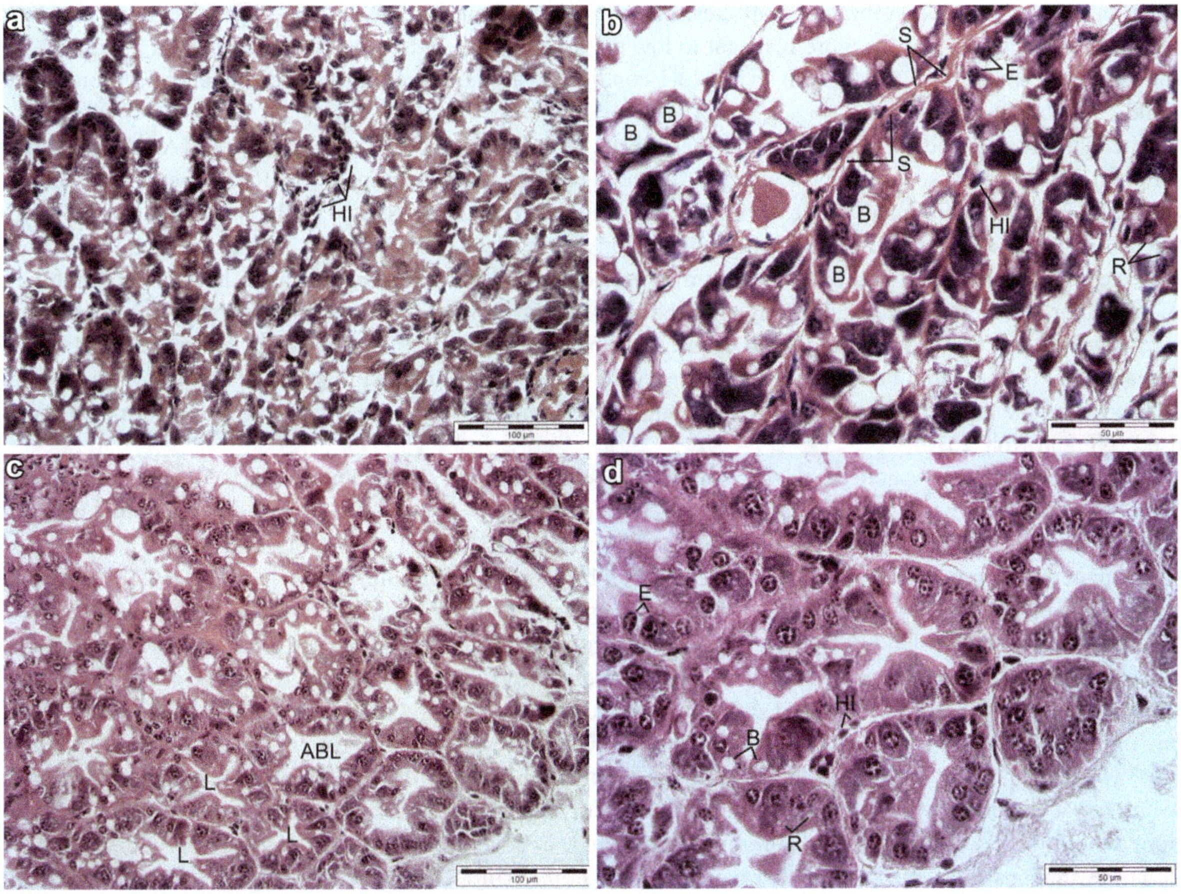

Fig. 3 Section of the hepatopancreas from the giant freshwater prawns fed the control diet (**a**, **b**) or LA diet (**c**, **d**) after being challenged with *Vibrio harveyi* for 2 weeks. The hepatopancreatic sections of the prawns fed the control diet show hemocyte infiltrations (HI) within the interstitial sinuses, a near collapse of the tubules (**a**) and separation of myoepithelial layer and epithelium (S) (**b**). The hepatopancreatic sections of the prawns fed the LA diet show a more normal tubule structure and lumen (L), but with some instances of abnormal lumen (ABL) (**c**) along with some hemocyte infiltrations (**d**). E cells (E), B cells (B), R cells (R). H&E stain. Magnification ×20; bar 100 μm (**a**, **c**). Magnification ×40; bar 50 μm (**b**, **d**)

Table 2 The prevalence of R cells, B cells and E cells (number/tubule) within the hepatopancreas of the giant freshwater prawn fed different organic acids added to the diets after 2 weeks of *Vibrio harveyi* challenge

Experimental diets[a]	Cell prevalence		
	R cell	B cell	E cell
Control	1.25 ± 1.42^b	7.41 ± 1.87	10.42 ± 2.30
CA	5.48 ± 2.45^a	6.54 ± 2.19	12.88 ± 3.11
LA	4.45 ± 1.85^a	6.42 ± 3.42	13.22 ± 4.08
FA	0.89 ± 0.28^b	7.22 ± 1.51	11.14 ± 2.77
PA	1.78 ± 0.34^b	8.14 ± 2.48	10.07 ± 1.18
OAB	1.29 ± 0.17^b	6.27 ± 1.85	13.85 ± 1.29

Cell prevalence is the mean number from 20 randomly selected tubules from three replicates and all values in the same column with different superscripts letters indicate significant differences ($P < 0.05$)

[a]Refer to Table 1 footnote for description of diets

the feeds of crustaceans has been met with good success recently and can be a cost-effective option (Ng et al. 2009; Park et al. 2011; Romano et al. 2015; Ng et al. 2015). It is believed that organic acids confer prophylactic protection from pathogenic bacteria by acting as antimicrobials, which includes their dissociation when entering the cell membranes of bacteria and lowering their cytoplasm pH (Ng and Koh 2016). Furthermore, the bacteria that are more resistant to low pH are more likely to survive. Such bacteria include lactic acid bacteria, since these are tolerant to low-pH conditions, and are known to act as beneficial probiotics to some aquatic animals.

Interestingly, Silva et al. (2013) found that the in vitro inhibitory abilities of LA or CA salts on three different *Vibrio* species, including *V. harveyi*, were significantly less compared to PA, butyrate and acetate acid salts. Dietary additions of butyrate and acetate acids were not investigated in the present study, although the fact that the LA and CA diets led to significantly higher resistance to *V. harveyi* challenge than the PA diet indicates that the antimicrobial effectiveness may rely on complex interactions with the host species and diet. In particular, different organic acids may alter the microbiota of the host animal in different ways. In addition, a review on the current knowledge and application of organic acids in the feeds of aquatic animals have revealed that the potential beneficial effects on growth performance and health status of fish and shrimp seem to depend on the aquatic animal species, type and concentrations of organic acids and the culture conditions used (Ng and Koh 2016).

The addition of organic acids to the diets of white and tiger shrimp significantly increased their resistance to *V. harveyi* challenge and was suggested that increased phenoloxidase activity as well as less hepatopancreatic damage were likely contributors to this finding (Ng et al. 2015; Romano et al. 2015). This is because the hepatopancreas of crustaceans is known to be targeted by *V. harveyi* (Lavilla Pitogo et al. 1998). Typically, *V. harveyi* is a brackish and marine bacterium, but this has been detected in diseased *M. rosenbergii* within some freshwater ponds in Thailand (Tonguthai 1992) and continues to be a major disease problem in *M. rosenbergii* hatcheries where the larval and post-larval stages are cultured in brackish water to complete their life cycle (Jayaprakash et al. 2006). Recently, it was demonstrated that *V. harveyi* is highly virulent to *M. rosenbergii*, but was dependent on the strain (Siripornadulsil et al. 2014). The present experiment also demonstrated that *V. harveyi* was pathogenic when present in sufficient numbers in *M. rosenbergii*, and caused substantial damage to the hepatopancreas which included a separation of myoepithelial layer and epithelium, hemocyte infiltrations within the interstitial sinuses, abnormal lumen and, in severe cases, tubule rupture and collapse.

These findings are consistent with reports of *V. vulnificus*-infected *M. rosenbergii* (Sharshar and Azab 2008) and *V. harveyi*-infected *P. monodon* (Soonthornchai et al. 2010). However, prawns fed the LA diet or the CA diet showed significantly higher survival to *V. harveyi* infection as well as substantially less hepatopancreatic damage/alterations and significantly more R cells compared to the other treatments. It is worthy to note that Siripornadulsil et al. (2014) showed that *V. harveyi*-infected *M. rosenbergii* had degraded lipid droplets in the hepatopancreas, which was suggested to be the result of phospholipase activity from the bacterium, compared to control prawns. Therefore, this greater protection to the hepatopancreas of prawns fed the CA or LA diet in the present study, along with indications of higher energy reserves, may have contributed to their enhanced bacterial resistance. Considering *Vibrio* spp. can lead to substantial mortalities, particularly in the hatchery and nursery stages (Tonguthai 1992; Jayaprakash et al. 2006), these findings may have important implications to controlling bacterial disease in this freshwater prawn aquaculture industry.

Finally, it has been reported that synergistic activities among organic acids may exist or provide a broader spectrum of antimicrobial protection than the sole use of one organic acid type (Thompson and Hinton 1997; Chaveerach et al. 2002). However, the inclusion of dietary OAB in the present study had little effect on the resistance of the prawns from *V. harveyi* infections despite our previous investigation demonstrating that a similar organic acid blend significantly improved the resistance of tilapia to *S. agalactiae* (Ng et al. 2009; Koh et al. 2016). This appears to indicate that the effectiveness of such blends are dependent on the host species and/or target pathogenic bacteria. Therefore, since CA and LA dietary additions to the diets of *M. rosenbergii* provided the best resistance to *V. harveyi* infections, the use of these organic acids at optimal levels should be further explored, especially for LA. The utilization of these two organic acids as a prophylactic against a potentially broader range of pathogenic bacteria should also be investigated. The results of the present study are certainly encouraging, and while more research directions should be made, dietary organic acids showed beneficial properties that may help reduce the reliance on antibiotics in prawn farming and thus produce a more environmentally friendly seafood product acceptable to a wider consumer market.

Acknowledgements This study was sponsored by a Research University Grant (1001/PBIOLOGI/811246) from Universiti Sains Malaysia. Dr. Romano was financially supported by a post-doctoral fellowship awarded by Universiti Sains Malaysia.

References

AOAC (Association of Official Analytical Chemists) (1997) In: Cunniff PA (ed) Official methods of analysis of AOAC International, 16th edn. AOAC International, Arlington

Briggs MRP, Jauncey K, Brown JH (1988) The cholesterol and lecithin requirements of juvenile prawn (*Macrobrachium rosenbergii*) fed semi-purified diets. Aquaculture 70:121–129

Chaveerach P, Keuzenkamp DA, Urlings HAP, Lipman LJA, van Knapen F (2002) In vitro study on the effect of organic acids on *Campylobacter jejuni/coli* populations in mixtures of water and feed. Poult Sci 81:621–628

D'Abramo LR, Sheen SS (1994) Nutritional requirements, feed formulation, and feeding practices for intensive culture of the freshwater prawn *Macrobrachium rosenbergii*. Rev Fish Sci 2:1–21

Defoirdt T, Halet D, Sorgeloos P, Bossier P, Verstraete W (2006) Short-chain fatty acids protect gnotobiotic *Artemia franciscana* from pathogenic *Vibrio campbelli*. Aquaculture 261:804–808

Defoirdt T, Boon N, Sorgeloos P, Verstraete W, Bossier P (2009) Short-chain fatty acids and poly-β-hydroxyalkanoates: (new) biocontrol agents for a sustainable animal production. Biotechnol Adv 27:680–685

FAO (Food and Agriculture Organization of the United Nations) (2016) Fisheries statistical collections. FIGIS data collection. Available via FIGIS from http://www.fao.org/fishery/statistics/globalproduction/en

Jayaprakash NS, Kumar VJR, Philip R, Singh ISB (2006) *Vibrios* associated with *Macrobrachium rosenbergii* (De Man, 1979) larvae from three hatcheries on the Indian southwest coast. Aquacult Res 37:351–358

Kim YC, Romano N, Lee KS, Teoh CY, Ng WK (2013) Effects of replacing dietary fish oil and squid liver oil with vegetable oils on the growth, tissue fatty acid profile and total carotenoids of the giant freshwater prawn, *Macrobrachium rosenbergii*. Aquacult Res 44:1731–1740

Kluge H, Broz J, Eder K (2006) Effect of benzoic acid on growth performance, nutrient digestibility, nitrogen balance, gastrointestinal microflora and parameters of microbial metabolism in piglets. J Anim Physiol Anim Nutr 90:316–324

Koh CB, Romano N, Siti-Zahrah A, Ng WK (2016) Effects of a dietary organic acids blend and oxytetracycline on the growth, nutrient utilization and total cultivable gut microbiota of the red hybrid tilapia, *Oreochromis* sp., and resistance to *Streptococcus agalactiae*. Aquacult Res 47:357–369

Lavilla Pitogo CR, Leaño EM, Paner MG (1998) Mortalities of pond-cultured juvenile shrimp, *Penaeus monodon*, associated with dominance of luminescent vibrios in the rearing environment. Aquaculture 164:337–349

Marshall BM, Levy SB (2011) Food animals and antimicrobials: impacts on human health. Clin Microbiol Rev 24:718–733

Ng WK, Koh CB (2011) Application of organic acids in aquafeeds: impacts on fish growth, nutrient utilisation and disease resistance. In: Luckstadt C (ed) Standards for acidifiers, principles for the use of organic acids in animal nutrition. Nottingham University Press, Nottingham, pp 49–58

Ng WK, Koh CB (2016) The utilization and mode of action of organic acids in the feeds of cultured aquatic animals. Rev Aquacult. doi:10.1111/raq.1241

Ng WK, Koh CB, Sudesh K, Siti-Zahrah A (2009) Effects of dietary organic acids on growth, nutrient digestibility and gut microflora of red hybrid tilapia, *Oreochromis* sp., and subsequent survival during a challenge test with *Streptococcus agalactiae*. Aquacult Res 40:1490–1500

Ng WK, Koh CB, Teoh CY, Romano N (2015) Farm-raised shrimp, *Penaeus monodon*, fed commercial feeds with added organic acids showed enhanced nutrient utilization, immune response and resistance to *Vibrio harveyi* challenge. Aquaculture 449:69–77

Park GH, Lee JH, Yun HH, Browdy CL, Bharadwaj AS, Bai SCC (2011) Effects of two different organic acid blends in olive flounder. Korean J Organ Agrocult 19:39–42

Paul M, Chanda M, Maity J, Sen SS, Patra BC, Dash G (2011) The use of antibiotics and disinfectants in the freshwater prawn (*Macrobrachium rosenbergii*) farms of north and south 24 Paraganas districts of West Bengal. Chron Young Sci 2:55–58

Romano N, Koh CB, Ng WK (2015) Dietary microencapsulated organic acids blend enhances growth, phosphorus utilization, immune response, hepatopancreatic integrity and resistance against *Vibrio harveyi* in white shrimp, *Litopenaeus vannamei*. Aquaculture 435:228–236

Sahoo PK, Pillai BR, Mohanty J, Kumari J, Mohanty S, Mishra BK (2006) In vivo humoral and cellular reactions, and fate of injected bacteria *Aeromonas hydrophila* in freshwater prawn *Macrobrachium rosenbergii*. Fish Shellfish Immunol 23:327–340

Sharshar KM, Azab EA (2008) Studies on diseased freshwater prawn *Macrobrachium rosenbergii* infected with *Vibrio vulnificus*. Pak J Biol Sci 11:2092–2100

Silva BC, Vieira FN, Mouriño JLP, Ferreira GS, Seiffert WQ (2013) Salts of organic acids selection by multiple characteristics for marine shrimp nutrition. Aquaculture 384–387:104–110

Siripornadulsil W, Thongserm M, Siripornadulsil S (2014) Pathogenicity of halophilic *Vibrio harveyi* in giant freshwater prawns (*Macrobrachium rosenbergii* De Man). Aquacult Res 45:1979–1988

Soonthornchai W, Rungrassamee W, Karoonuthaisiri N, Jarayabhand P, Klinbunga S, Söderhall K, Jiravanichpaisal P (2010) Expression of immune-related genes in the digestive organ of shrimp, *Penaeus monodon*, after oral infection by *Vibrio harveyi*. Dev Comp Immunol 34:19–28

Su X, Li X, Leng X, Tan C, Liu B, Chai X, Guo T (2014) The improvement of growth, digestive enzyme activity and disease resistance of white shrimp by the dietary citric acid. Aquacult Inter 22:1823–1835

Sukor SA, Taher S, Ehteshamei F, Arshad A, Ng WK, Romano N (2016) Effects of different dietary organic acids on the survival, growth and hepatopancreatic histopathology of the blue swimmer crab, *Portunus pelagicus*. J Shellfish Res 35:1–7

Thompson JL, Hinton M (1997) Antibacterial activity of formic and propionic acids in the diet of hens on salmonellas in the crop. Br Poult Sci 38:59–65

9

Production of fish protein hydrolysates step by step: technological aspects, equipment used, major energy costs and methods of their minimizing

Inna Petrova · Ignat Tolstorebrov · Trygve Magne Eikevik

Abstract Nowadays, there is a growing interest on how to utilize fish materials remaining from the main production and considered as unappropriated for a direct human consumption. There are numbers of possible solutions to recover valuable nutrients from that matter and one of the most efficient is the production of fish protein hydrolysate. This article is devoted to overview existing information about the production of dried fish protein hydrolysates with a focus on dehydration process during production and equipment used for moisture removal. Drying step of the production is considered as the most energy demanding and, therefore, described in detail. Questions considering energy demands of the drying are highlighted in the article together with the proposals for the improvement of energy efficiency. This work also describes source of the raw material, the main steps of the technological scheme with the equipment used and valuable information on the intermediate state of fish protein hydrolysate between the process operations.

Keywords Drying · Fish protein hydrolysate · Fish by-products · Industrial drying systems

Abbreviations

P Pressure (Pa)
pH Logarithmic measure of hydrogen ion concentration (dimensionless)
T Temperature (°C)

Introduction

Fish and marine products are one of the most used protein sources for human consumption. However, a big part of the catch is at its best utilized for production of cheap by-products such as fish flour, fish oil or animal feed or just wasted unused. At the same time, a growing amount of fishery and, consequently, of fishing wastes demands efficient solutions for their utilization since they are a good nutritional source, which is highly underestimated.

Fish protein hydrolysates (FPH) are a product made from fish or fish material by the method of protein hydrolysation (breakage of proteins from which fish tissues are constructed into smaller parts—peptides and

I. Petrova (✉) · I. Tolstorebrov · T. M. Eikevik
Department of Energy and Process Engineering, Norwegian University of Science and Technology, 7491 Trondheim, Norway
e-mail: inna.petrova@ntnu.no

finally into amino acids). Thus, FPH is a mixture of broken proteins (Damodaran et al. 2008). FPH are reported to have a number of improved properties compared to the protein of origin which are namely improved functional properties (He et al. 2013; Kristinsson and Rasco 2000) and bioactive properties such as anti-oxidative (He et al. 2013; Chalamaiah et al. 2012; Sarmadi and Ismail 2010) or anti-hypertensive activities (He et al. 2013). Recently, FPH is also started to be used as cryoprotectants for frozen fish products (Jenkelunas and Li-Chan 2018).

FPH is produced in two forms: liquid and dried. Liquid FPH is a watering mixture of hydrolyzed proteins, which contains up to 90% of moisture. FPH in a liquid form is highly unstable for a long-term storage and, moreover, it is difficult to be transported. Thus, dried FPH is preferable due to a longer shelf-life, easier storage and transportation. Nevertheless, at the same time, the removal of such a big amount of water from liquid FPH is a difficult and costly task, which is one of the challenges of dried FPH production.

Thus, FPH has a huge potential use as a protein source for human consumption, but the step of dehydration demands a big energy supply and is, consequently, very costly. This article is devoted to highlight the main steps of moisture removal from FPH during the production, equipment used for the dehydration processes and possibilities to save or recover the used energy to cut energy costs.

Technology and equipment of FPH production

Availability of raw material

Source of raw material

The very first task of the implementation of FPH technology is the analysis of main fishing sources, which are the most abundant and, possibly, underutilized.

According to FAO (2016), China following by Indonesia, USA and Russian Federation were the leaders of marine fishery production. The countries holding leadership in aquaculture were China with 45.5 million tons, or more than 60% of global farmed fish production, following by India, Viet Nam, Bangladesh and Egypt (FAO 2016).

The main marine species in worldwide capture over years were the families of Gardidae (Alaska Pollock, Atlantic Cod, Blue Whiting, Pacific Cod), Engraulidae (Anchoveta, Japaneese Anchovy), Scombridae (Skipjack Tuna, Chub Mackerel, Yellowfish Tuna, Atlantic Mackerel, Seerfishes Nei), Clupeidae (Sardinellas Nei, Atlantic Herring, European Pilchard, Araucanian Herring, European Sprat, Pacific Herring), Carangidae (Scads Nei), Trichiuridae (Largehead Hairtail), Ommastrephidae (Humboldt Squid, Argentine Shortfin Squid), Nemipteridae (Threadfin Breams Nei), Scomberesocidae (Pacific Saury), Portunidae (Gazami Crab), Akiami Paste Shrimp (FAO 2016). Worldwide, the most important fish families farmed in aquaculture are Cyprinidae (Carp), Salmonidae (Salmon, Trout), Serranidae (Seabass), Acipenseridae (Sturgeon), Scophthalmidae (Turbot), Sparidae (Sea Bream), Mytilidae (Mussels), Ostreidae (Oysters) and some families of clams (European Comission 2012). Thus, due to the abundance of aforeisaid species in worldwide fishing production, an attention must be paid on better utilization of a big amount of by-products rested after their processing.

Global fish capture and aquacultural production in 2014 reached the values of 93.4 and 73.8 million tons, respectively, and only 146.3 million tons were used for human consumption (FAO 2016). From the amount used for human consumption, a big part was used as by-products or simply wasted. For example, the amount of non-edible fishing part, remaining from the filleting of salmon, generally reached 45% according to FAO (2017). According to Seafoodsource (2016), if all fish globally are processed and all the by-products collected, the amount of 36 million tons left from the main fish processing would be available as raw material for further treatment. However, the authors of Seafoodsource (2016) claimed that only 5.7 million tons of by-products are currently processed and as much as 11.7 million tons are not collecting from processing plants for further utilization.

More precise distribution of fish by-products and unutilized wastes was published by Richardsen et al. (2015) for the Norwegian fishery sector. According to the authors, the main fishery groups produced in Norway in 2015 were aquaculture, pelagic and white fish production from which as much as 43%, 12% and 29% correspondingly were rested. The rests from pelagic sector were almost totally utilized for fish flour and other fish products. The amount of unutilized fishing rests from aquaculture facilities was moderate (9% of the total quantity) and the biggest amount was observed for the white fish industry (52% of the total quantity).

Thus, underutilized and underestimated fishing rests from the main production can be processed to get FPH, which can increase efficiency of fish processing by redirection of the by-products for human consumption. This is both more profitable and environmentally friendly than using fishing resting material for fish meal. In the case of FPH, the fishing industry avoids a two-stage circle of fishing rests consumption to feed animals, which will be consumed by humans and allows delivering a food resource straight to the final consumer.

In spite of economic benefits of FPH production, the main hindrance of complete by-product utilization is a logistic issue. There is a need for quick transportation of by-products in order to avoid spoilage. However, due to a general lack of well-organized model of collecting, storage and transportation, further utilization of fish by-products is complicated and available fish source may be even lost as wastes. An example of successful collaboration between fish processing plants and by-product utilization facilities is Iceland Ocean Cluster, which is made up of nearly 70 companies located in a common building of around 9000 square meters (SeaFoodSource 2016). Such models if realized globally may help better utilization of fish by-products due to easy transportation and high speed of fishing rest secondary processing.

Type of raw materials

Raw materials for FPH production can be both fish and fish left-overs. Since the usage of whole fish or only fish by-products may influence the processing, Table 1 is designated to roughly compare the chemical composition of whole fish and fish by-products exemplified by fish heads and backbones (salmon (FAO 2007; Ytrestøyl et al. 2014) and cod (Bikov et al. 1998) are taken for the comparison).

As it is seen from the table, some fluctuations and differences in the chemical content between different types of fish raw matter are possible due to the type of raw matter; however, similar fluctuations can be observed since seasoning, etc. In all cases, raw materials should be carefully evaluated at technological line in order to provide effective recovery of protein fraction. Further in the article, the terms of raw material will be corresponded to any type of raw materials, generally used for FPH production, including both whole fish and fish by-products.

Technological scheme

A principal technological scheme of FPH manufacturing is presented in Fig. 1. The process starts from a takeover of raw materials. Water, chemicals or enzymes are added to minced raw material to obtain chemical (acid or alkali) or enzymatic hydrolysis correspondingly. After a certain time, when a certain desired extent of hydrolysis is achieved, there is a need to stop the hydrolysation process by chemical or thermal treatments according to the hydrolysis method. After termination of hydrolysis, the protein mixture is delivered to solid

Table 1 Chemical composition of by-products and whole fish for salmon and cod

Nutrients	By-products			Whole body
	Heads	Backbones	Average	
Salmon				
Moisture	66	64.3	65.1	58.8
Lipids	16.7	15.2	16.0	21.3
Proteins	11.3	14.1	12.7	17.5
Ash	6.0	6.4	6.2	2.4
Cod				
Moisture	79.5	75.0	77.3	75.3
Lipids	0.3	0.4	0.4	5.7
Proteins	13.9	15.2	14.6	13.0
Ash	5.6	9.0	7.3	3.4

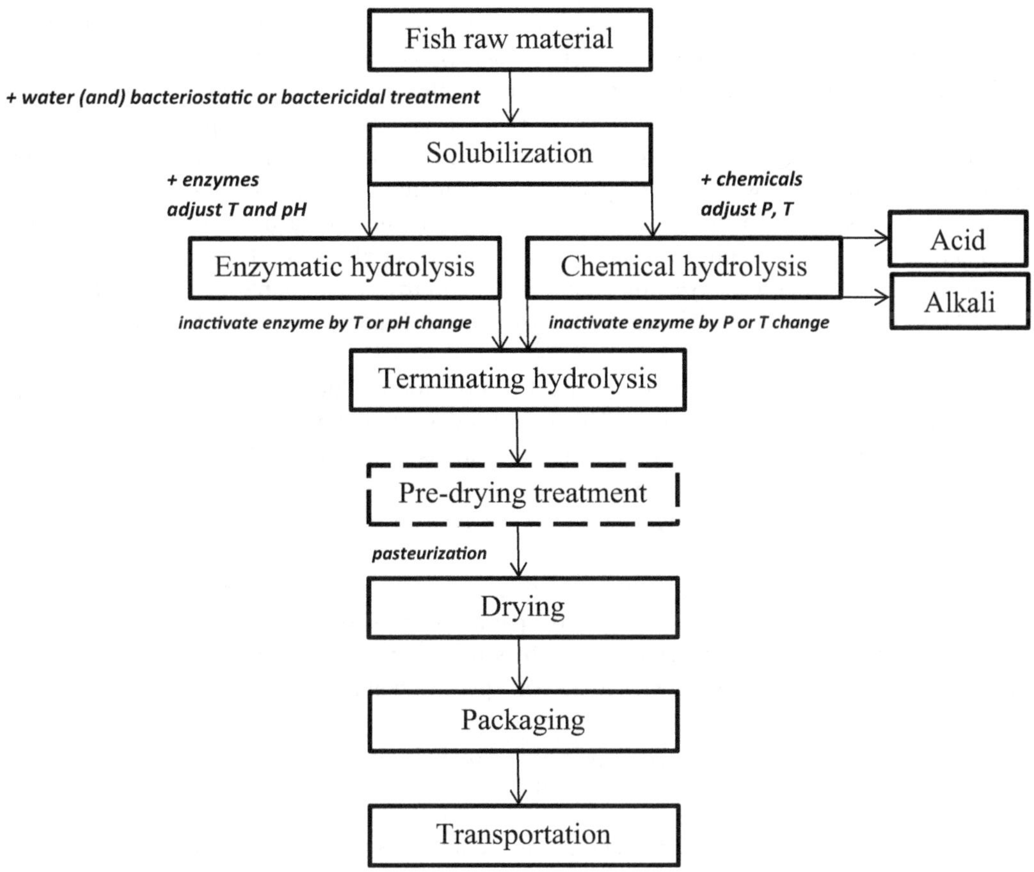

Fig. 1 Principal scheme of FPH production

separation where liquid part is separated from the solids. Then, a heating procedure can be applied if desired to treat the fish protein hydrolysate by temperature to reduce microbial activity (pasteurization). The term of pasteurization in FPH technology is different from that in dairy production due to higher temperatures used. In some cases when the final product has to have a certain quality, for example, to be used for biochemical purposes as a microbiological media, removal of salt can be provided. After the protein mixture can be concentrated to remove some water before the drying procedure. It makes the process of drying more efficient due to the decreased quantity of water to be removed. However, at the same time, preliminary water removal by concentration needs additional equipment and energy to provide the dehydration process, thus, the need of such an extra treatment is decided by a producer for each certain case. Liquid FPH slurry is then delivered for drying and dried FPH is sent for packaging and transportation. FPH are normally stored at 4 °C or lower, in some cases with vacuum packaging mainly to prevent lipid oxidation (He et al. 2013).

Hydrolysation

Hydrolysation of proteins can be generally provided by chemical (acid or alkali treatment) or by biochemical methods. Biochemical hydrolysis is provided by proteolytic enzymes, which are already present in fish tissues (autolysis) or by premixing of commercial enzymes in order to accelerate the process (enzymatic hydrolysis). Some literature on novel methods of fish protein hydrolysis is nowadays available in scientific publications, for example, the work of Hoeling and Volkov (2015) describing the method of hydrothermal extraction. However, such methods need to be well evaluated before they can be used at an industrial scale. Other extraction methods than those mentioned above do not hydrolyze proteins, but are used for protein concentration by removal of a part of water and/or fat (concentration or isolation of protein fraction) (Kristinsson and Rasco 2000).

All methods of hydrolysation are held with one final purpose—extraction of protein part by the deformation of fish proteins into smaller parts (peptides and amino acids) and their further separation. The peptides and amino acids, formed by protein degradation, have smaller molecular mass and can be absorbed by digesting system faster than proteins (Di Pasquale 2008). It is a particularly important property for those, whose organism is weakened by a disease or who need a quick nutritional supply (for examples, sportsmen).

Fish material is solubilized with water by the solid content of 8–20% (Pasupuleti and Braun 2010). Generally, water and fish material are mixed as 1:1. However, there is also a possibility of hydrolysation process without water addition (Himonides et al. 2011; Rebeca et al. 1991). Producers should evaluate the amount of added water for the hydrolysation step in order to provide a suitable extent of the protein breakdown and at the same time avoid extra costs for dehydration of unnecessary water fraction from FPH on the step of drying.

Hydrolysation is generally held in metallic tanks (hydrolysators), which are in most cases constructed and installed by producers of FPH by themselves. The hydrolysators should be constructed so, as heat is supplied to the tube shell of the tanks and, desirably, an agitation blade is installed into the tank to mix the hydrolysate mixture for uniform temperature and chemical distribution. The amount of heat provided as well as the temperatures used for the hydrolysation depends on the type of hydrolysis, nature of raw material, used chemicals and other factors.

During the treatment, one needs to know how the degradation of proteins should be extended. To evaluate the extent of hydrolysation and follow its kinetics, the parameter named "degree of hydrolysys" (DH) is used. DH can be measured by a range of methods, which is overviewed by Kristinsson and Rasco (2000) and expressed as the ratio of broken peptide bonds to the total peptide bonds in the mixture per unit weight (Eq. 1):

$$\mathrm{DH} = \frac{P_{\mathrm{br}}}{P_{\mathrm{tot}}} \cdot 100\% \tag{1}$$

where DH is the degree of hydrolysis in percent; P_{br} is the numbers of broken peptide bonds; P_{tot} is the total number of peptide bonds in the mixture.

DH is correlated with such functional properties of final FPH as solubility, emulsifying capability, foaming ability, fat absorption ability, and with such an important sensory property as bitterness (He et al. 2013; Kristinsson 1998; Kristensson and Rasco 2010; Kuehler and Stine 1974; Quaglia and Orban 1987, 1990). A number of researches showed that bitterness of protein fraction increases with the decrease in peptide size (Aliani and Eskin 2017; Aubes-Dufau et al. 1995). Slizyte et al. (2010) claimed that the presence of di- and tripeptides, which are the main protein fragments of FPH (Manninen 2004), will give far more bitter taste compared with FPH which mostly consists of larger protein fragments. However, Slizyte et al. (2010) also noticed that with further degradation of peptides into amino acids, the bitterness also decreases. In such a case, the process of hydrolysis shall be well adjusted to avoid undesirable sensory properties of final FPH and, therefore, to control not only DH, but also the obtained length of protein fragments. Chemical and enzymatic hydrolysation techniques allow a different amount of process control, which is extremely important for the choice of hydrolysation method.

Chemical hydrolysation

Chemical hydrolysis is provided by either acid or alkali. It is achieved by cleaving the bonds between different peptide groups in the protein sequence by chemical agents. Since this process is held at extreme working parameters (high concentration of acid or alkali, high temperature), the process of hydrolysis in this case is almost uncontrollable. As a result, the nutritional and functional properties of final FPH are reduced (Loffler 1986; Webster et al. 1982) and also varied due to the lack of control and traceability (Blenford 1994; Kristinsson and Rasco 2000; Skanderby 1994). Thus, FPH produced by chemical hydrolysis has a very limited spectrum of utilization.

Acid hydrolysation Acid hydrolysis is more widespread than that by alkali. Typically, fish is involved reacting with hydrochloric acid, but there are plenty of technologies involving sulfuric acid. High temperature and pressure of solution (121–138 °C and 220–310 mPa correspondingly) are held during several hours (typically 2–8) in order to achieve a certain DH (Pasupuleti and Braun 2010). The mixture is then neutralized to pH 6.0–7.0 and delivered for further dehydration (Kristinsson and Rasco 2000). The process of acid hydrolysation has several drawbacks such as high NaCl content which makes the final FPH inappropriate for some food and especially biochemical applications (Kristinsson and Rasco 2000; Pasupuleti and Braun 2010) or the destruction of tryptophan which is an essential amino acid (Jaswal 1990; Pasupuleti and Braun 2010).

Alkali hydrolysation Alkali hydrolysis is held at less elevated temperatures (typically 27–54 °C) in the presence of such alkaline agents as calcium, sodium or potassium hydroxide during several hours until the desired degree of hydrolysis is reached (Pasupuleti and Braun 2010). The use of alkali reactants, mainly sodium hydroxide, generally results in reduced functionality and nutritive value of final FPH. Moreover, a range of substances, which are not absorbed or even toxic for human organism, can be created during alkali hydrolysis (Kinsella and Melachouris 1976; Kristinsson and Rasco 2000; Lahl and Braun 1994; Linder et al. 1995). A number of amino acids as serine and threonine are destroyed during alkaline hydrolysis; however, tryptophan remains intact contrary to the acid hydrolysis (Pasupuleti and Braun 2010). Despite described drawbacks, limited alkali hydrolysis is used in the fish processing industry to recover a range of proteins from fish material and fish protein concentrates in order to hydrolyze a valuable protein fraction and, thus, improve the functionality (Kristinsson and Rasco 2000; Sikorski and Naczk 1981; Tannenbaum et al. 1970a, b).

Enzymatic hydrolysation

In this review, the attention is paid to the treatment when a specific enzyme or a mixture of enzymes is premixed to the hydrolysated slurry. This is more convenient and controllable method than that with autolytic enzymes due to a high variation in optimal conditions for each of enzyme in the autolytic enzymatic mixture and their variation influenced by species, season, gender, age, etc. (Kristinsson and Rasco 2000).

Enzymatic hydrolysis is carried out from one to several hours at mild conditions: slightly elevated temperatures (generally around 35–65 °C) and a certain pH according to the optimal requirements of the used enzymatic systems: alkali (for example, alcalase), neutral (papain, bromelain, alcalase, neutrase and flavourzyme) or acid (pepsins). Used for FPH production enzymes are of animal (pepsins), plant (papain, bromelain) or microbial origin (alcalase, neutrase, flavourzyme). Enzymes of microbial origin are considered as having greater pH and temperature stability (He et al. 2013). The application of enzymes allows obtaining a high-quality FPH avoiding such drawbacks as the developing of undesirable products of racemization as in a case of acid and alkali hydrolysis (Gonzalez-Tello et al. 1994). In addition, enzymatic hydrolysis allows a controllable distraction of a certain type of proteins by a certain chain cleavage due to enzymes' specificity with an easy process of enzyme inactivation after a certain DH is achieved (Pasupuleti and Braun 2010). The inactivation of enzymatic hydrolysis is typically accomplished by the elevated temperatures of around 75–100 °C applied for 5–30 min (Kristinsson and Rasco 2000).

Despite the aforesaid benefits, there are certain limitations when enzymatic hydrolysis is utilized. Among them are high costs of industrial enzymes, low yield, the need of a special treatment in order to deactivate the enzymatic hydrolysis, the difficulties of process control to obtain a certain molecular mass of the products of protein degradation and the bitterness of final FPH (He et al. 2013; Kristinsson and Rasco 2000). However, recent researches with different enzymes used for FPH production revealed the mixture of papain and bromelain as a very promising solution for avoiding bitterness at the production of FPH from herring (Slizyte et al. 2010). Table 2 shows a rough comparison of bitterness between different enzymatic systems based on the data published by Slizyte et al. (2010) for the production of FPH from herring. Exact values of relative bitterness as well as materials and methods of its evaluation can be found in the rapport of Slizyte et al. (2010). The values in the table are derived from the Fig. 19 (page 23) published by Slizyte et al. (2010).

Moreover, the resulting FPH has both an acceptable DH and yield (Slizyte et al. 2010.

Table 2 Comparison of bitterness for different enzymatic systems

Enzymatic system	Relative bitterness
Papain and bromelain	3.4
Autolytic enzymes	3.7
Flavourzyme	4.1
Bromelain	4.6
Protamex	4.7
Papain	4.8
Promod	5.1
Alcalase	5.5

Dehydration of FPH

Purification

To utilize hydrolyzed proteins as FPH, there is a need to separate protein fraction mainly from insoluble and fat fractions before the further dehydration. It is possible to be accomplished by a range of methods, mostly by centrifugation and plate and frame filtration.

In more advanced technologies, when a certain quality of final FPH is needed, the filtration of small chemical particles from the hydrolysation mixture can be provided (microfiltration, ultrafiltration and nanofiltration from bigger to smaller particle separation correspondingly) (Pasupuleti and Braun 2010). In some cases, electro-membrane filtration is also utilized for FPH fractionation, for example, in order to fractionate active peptides from complex hydrolysates (Suwal et al. 2018).

The temperature of purification mainly depends on the temperature of the delivered protein mixture and is not changed especially during the purification process by an additional heat supply.

After the separation of protein part, the mixture is generally pasteurized at high temperatures in order to eliminate possible microbial contamination. Pasteurization can be accomplished one or several times during the production line of FPH according to the technology used by a certain producer (Pasupuleti and Braun 2010).

Centrifugation

Generally, the process of centrifuging is used in order to separate the hydrolyzed mixture on several fractions for further recovery of the protein part.

Centrifuging is provided by different centrifuges, which are classified according to a certain property such as their intended use, continuous/batch operation, construction, speed, etc. A common future of all existed centrifuges is that the main element of the machine is a rotor which rotates at high velocities generally up to 20,000*g* (where *g* is the acceleration of gravity) if for industrial applications. The simplest principle scheme of the centrifuge is shown in Fig. 2, but can be modified according to the type of centrifugation.

Centrifugation in the processing of FPH can be done in one or several steps according to the desired quality of FPH. The centrifugation of mixture at 4000*g* for at least 20 min basically separates the mixture into three fractions: the sludge and solid layer at the bottom, the protein hydrolysate solution in the middle and the fish oil layer on the top (He et al. 2013). Cui (1996) reported more detailed fractionation of the mixture on sludge in the bottom, aqueous layer in the middle, lipid–protein fraction between aqueous layer and sludge, aqueous and oil layers and oil layer on the top; visualized picture of these layers can be found in the paper (Kristinsson and Rasco 2000).

The solids are separated from the liquid protein mixture mainly in order to purify the FPH from insolubles while the fish oil fraction is removed to avoid undesirable fat oxidation processes in the final FPH. The desirable fat content of final FPH is to be 0.5% (He et al. 2013; Spinelli et al. 1972) in order to avoid the influence of lipid changes on the quality of final FPH during shelf-life. However, Chalamaiah et al. (2012)

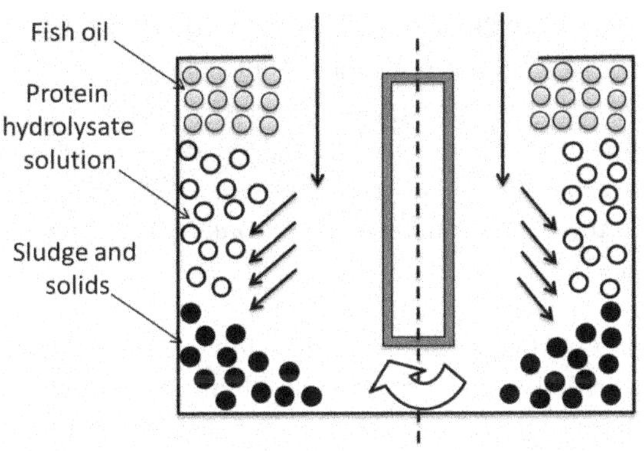

Fig. 2 Principal scheme of simplest centrifuge

overviewed a number of research works where the fat content reported to be at least less than 5% and even a few studies where the fat content was claimed to be less than 10%.

Plate and frame filtration

Plate and frame filtration presses working at the pressure of approximately 410–550 mPa are basic operation units, used by industrial FPH producers in order to purify protein fraction of hydrolysates from different impurities. As for investigations considering such a filtration for any studies with FPH, the information is very limited by a few publications (Chakraborty and Madhavan 1977; Loosen et al. 1994). It explains by the fact that plate and frame filtration is rather a technological step at a large production than a matter of investigation.

A simplest sketch of such a press is shown in Fig. 3. Plates equipped by filters consisted of special cloth material are constructed into a filter frame. Filter aid is applied on the surface of the filters in order to bind different impurities of the mixture and hold them into the filter cloth, or sometimes simply premixed to the mixture before filtration. The filtrated mixture enters the press and goes through the paths between the filter plates leaving impurities on the surface of the filters. Process recycles until a certain purity of the mixture is achieved.

The nature of impurities, which can be collected by the filters, depends on the nature of filter aid. Special attention must be paid on the nature of filter aid since it can also remove some desirable compounds from FPH slurry, such as some peptides and amino acids (Silvestre et al. 2009). Filter aids are generally made of porous

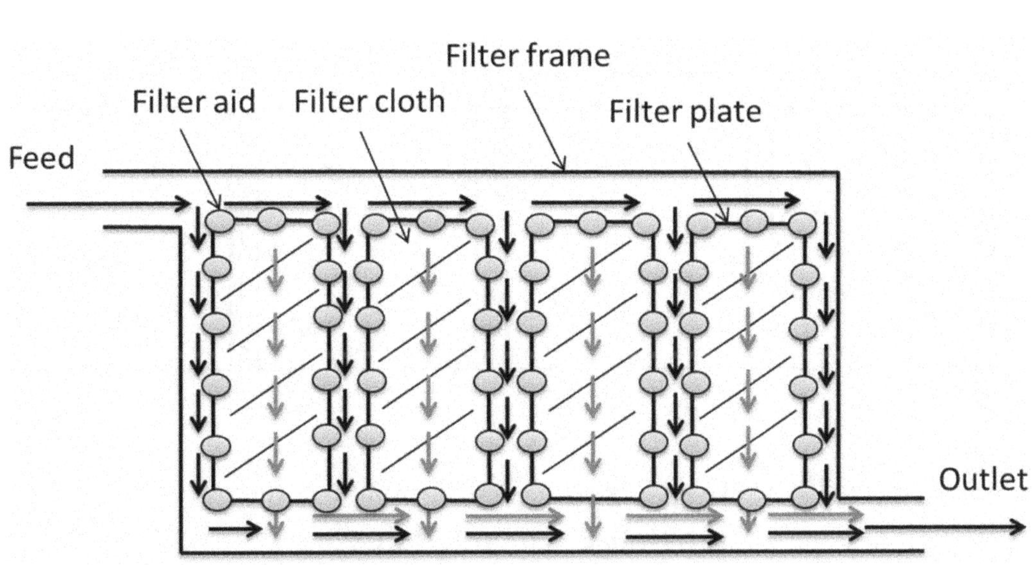

Fig. 3 Principal scheme of plate and frame filter press

inert materials such as silica-based rocks (diatomaceous earth), wood cellulose, perlite, etc. Filter aids for FPH usually consist of volcanic rocks or diatomaceous earth, but if the purpose is discoloration, or removing of hazard components, charcoal powder can be applied (Chae et al. 1998; John 1993).

The capabilities and operation time of the press are depended on the amount of solids in the filtrated mixture. Roughly, the capabilities are ranged between approximately 2–3 L per minute according to the initial solid content, which is estimated to be between 1 and 10% from the total amount (Kristinsson and Rasco 2000; Mine Engineer Product Manual 2009).

Micro-, ultra- and nano-filtration After plate and frame filtration, when bigger particles are caught by the filters, the protein mixture can be delivered to membrane filters for careful fractionation or concentration of FPH mixture. Membrane filtration is a useful tool for obtaining certain groups of peptides utilized for different purposes (for example, bioactive peptides) (Abejón et al. 2018).

Membrane filters are micro-, ultra- and nano-filters according to the aimed particle size: down to approximately 100, 10 and 1 nm correspondingly (Fig. 4). Peptides in FPH slurry can be also separated by membranes based on their charge and hydrophobicity characteristics (Abejón et al. 2018). Membranes are made of nylon, polyethersulfone, polyvinylidene difluoride, polytetrafluoroethylene, mixed cellulose ester, cellulose acetate, polypropylene and other material (Microlab Scientific 2018).

Microfiltration is not a popular method utilized by researchers for the filtration of FPH, probably, due to the aimed particles to be plugged (the most bioactive peptides) are usually with a smaller size ($\leq$ 4 kDa) (Abejón et al. 2018; Saidi et al. 2014; Vandanjon et al. 2009), than microfilters can retain. The filtration range of microfiltration is generally less than 100 kDa, which is far bigger than the aimed peptide size. By contrast, ultra- and nano-filtration is recently investigated in a large number of research articles on FPH (Abejón et al. 2016, 2018; Bourseau et al. 2009; Roslan et al. 2018; Saidi et al. 2014; Vandanjon et al. 2009) or was used as a filtration tool in other investigations of FPH (Abdelhedi et al. 2018; Halim et al. 2018).

Ultrafiltration is used in most studies for fractionation or separation, while nano-filtration is utilized for peptide or amino acid concentration (Saidi et al. 2014).

Coupling ultra- and nano-filters in various types of cascades is used by many researchers for better efficiency of the filtration process. (Abejón et al. 2016, 2018; Bourseau et al. 2009; Saidi et al. 2014; Vandanjon et al. 2009).

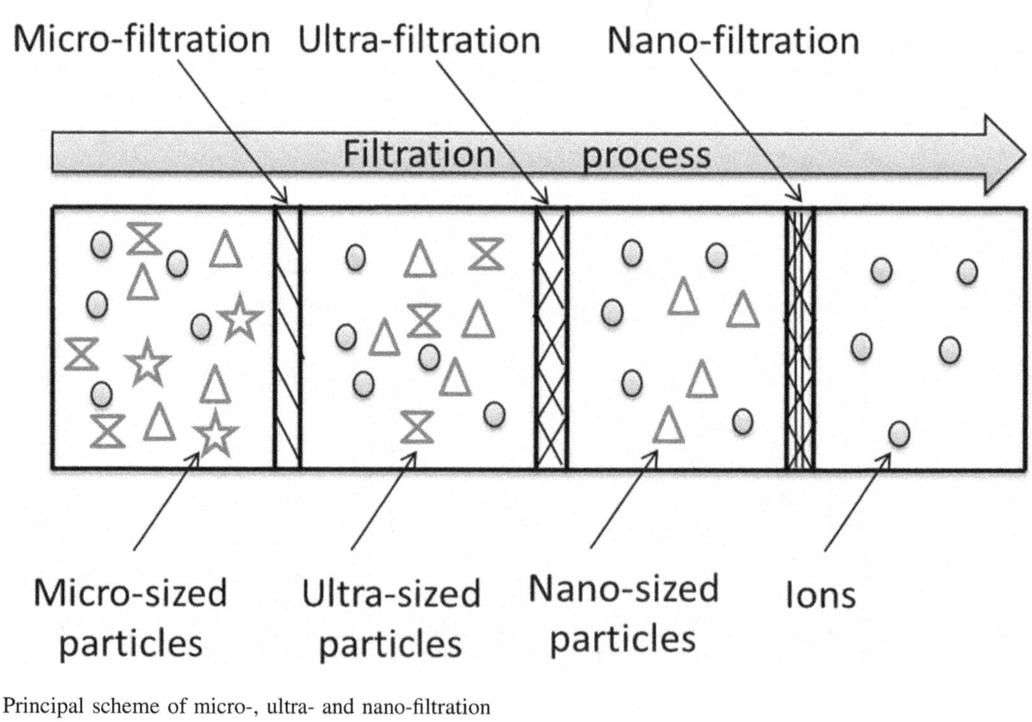

Fig. 4 Principal scheme of micro-, ultra- and nano-filtration

It is important to adjust the membrane to the desirable size of the particles to be separated in order to obtain an effective filtration process and avoid free flow of the aimed solids, which are too small through the membrane. At the same time, when deciding about the pore size, it is important to remember that all particles of the used size pore would be plugged. For example, if the aim is to remove endotoxins, the used membrane can cut off also a great amount of peptides with the same size (Pasupuleti and Braun 2010). Thus, the "golden mean" between the filtration and protein yield for FPH production should be found for a certain case.

When membrane separation is utilized, the process should be well adjusted to maximize the product purity and the process yield: pore size of membranes, pressure, flow rates, pH, salt concentration of the solution and other essential parameters should be considered. At the same time, total costs and environmental consideration must be also taken into account; in this case, water recovery may be a useful solution (Abejón et al. 2018). However, the simultaneous concurrence of high purity and yield values is not possible even considering all important factors, which influence the process of filtration (Abejón et al. 2016).

The effectiveness of peptide fractionation in FPH production can be troubled by disposition and, eventually, clogging of solid parts on membrane surface due to protein–protein or protein-membrane interaction. These phenomena are known as membrane fouling and are considered as a major problem leading to poor FPH separation and low peptide yield (Roslan et al. 2018). Membrane fouling can be controlled by various methods: applying external electric, magnetic or ultrasonic field, adjusting pH (Roslan et al. 2018; Zhu et al. 2017) or applying freshwater streams in order to control protein content in inlet and outlet streams of membrane modules (Abejón et al. 2018).

Concentration

Concentration is used to reduce water content in the protein mixtures before drying and, thus, reduce the energy costs of the drying step. Concentrators are apparatuses where the liquid solution to be concentrated goes through a large specific surface which is heated up. Thus, going through the heated surfaces of concentrators, moisture is evaporated and removed to the atmosphere by vapor.

The temperatures used for evaporation and evaporation rate vary for different FPH depending on the composition of the mixture and other properties influencing the boiling point such as DH, protein source or the choice of enzyme. At the same time, the productivity of the devices depends on their size and capability and can achieve 150 t/hour (GEA 2018).

Depending on the capabilities of the used concentrators and the allowable solid content in the mixture for further safe operation of dryers, the protein hydrolysate solution can be concentrated up to 50% solids (Pasupuleti and Braun 2010). Mostly used modifications of concentrators are falling and rising film evaporators which differ from each other by the way of product supply and, correspondingly, by the construction.

After evaporation, some producers add a final pasteurization before the concentrate is delivered to be dried.

Falling film evaporators A simple sketch of a falling film evaporator is shown in Fig. 5. The protein mixture starts to boil at the top of the apparatus. Then, it goes from the top of the apparatus through the heated tubes with help of gravitational force towards the bottom and the output valve. The surplus of moisture is removed through a special vapor valve.

Rising film evaporators In a rising film evaporator, the protein mixture, in contrast to a falling film evaporator, goes from the bottom to the top of the apparatus through heated tubes (Fig. 6). The mixture reaches the boiling temperature at the bottom and the liquid and vapor parts flow together upward towards the top against the gravitational force. The concurrent movement of liquid–vapor mixture upward creates a high degree of turbulence, which gives better efficiency of the evaporation process (NPTEL 2013). The surplus of moisture is then evaporated and removed through a special vapor valve.

Drying

Drying of FPH is a process, which allows ensuring a stable shelf-life of final FPH by water content decrease and making the transportation of FPH easier due to the reduced mass and volume of the final product.

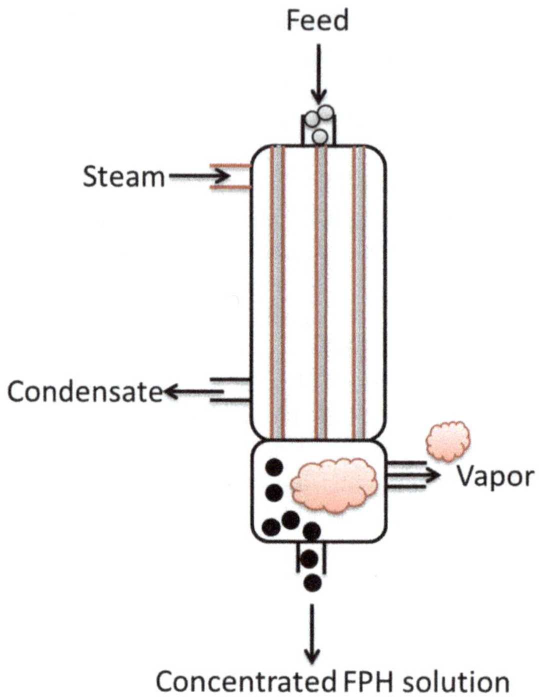

Fig. 5 Principal scheme of falling film evaporator

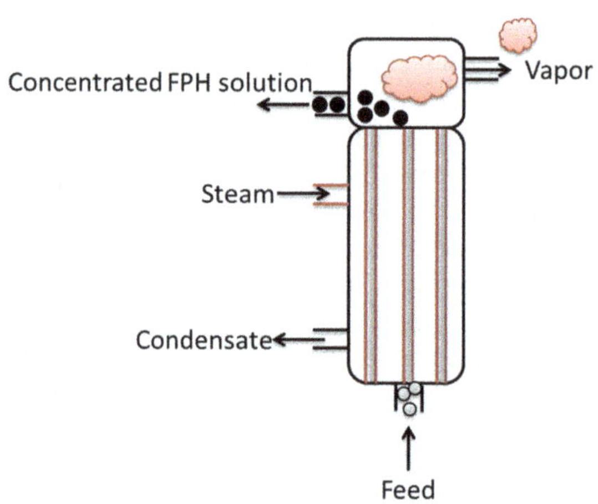

Fig. 6 Principal scheme of rising film evaporator

The drying process of FPH is usually accomplished by a limited number of drying apparatus: they are typically spray dryers which provide convectional drying, and vacuum freeze dryers and roller drum dryers which utilize contact heat supply. From the experience, the authors suggest convection spray dryers as the most used drying systems at large scale in FPH production due to a satisfactory productivity, reasonable quality of the final FPH and a relative easiness of operation. The term "spray dryer" is used in this article to describe convectional spray dryers; the authors do not devote the work to spray drying using radiation or mixed heat transfer.

The choice of drying method at laboratory scale generally depends on the availability of drying equipment at a certain laboratory and not depends on the purposes of the research or the properties of studied FPH. Thus, it is difficult to classify the used equipment according to the studied properties of FPH. Generally, the mostly used drying equipment for research purposes is spray and freeze dryers.

Spray drying A typical spray dryer is a drying apparatus, which consists of a cylindrical tube with a feeding system at the top and an outlet configuration normally equipped by a separator of dried FPH and a filter to clean the output air (Fig. 7). The design of the drying tube typically depends on rheological properties of droplets, an used droplet size, residence time of droplets in a drying chamber and desired moisture content of the final dried FPH. Rheological properties of dried droplets, which are important in a sense of spray drying, are solid content, density, surface tension and viscosity. The question of a spray dryer design, which is of interest, is uniform heat delivery to all droplets and efficient solid separation (Wisniewski 2015).

Liquid FPH is supplied to the feed system and sprayed into the tube chamber by a special device—atomizer. Atomizer can be made by a different configuration according to the dryer construction, the rheological properties of liquid and desired size of droplets.

Industrial spray dryers typically have rotary disk atomizers, or single fluid high pressure swirl nozzles. In rotary atomizers (Fig. 8a), liquid is centrifugally accelerated to high velocity before discharging into the hot gas in a co-current mode. Rotary atomizers are suitable for a large number of products. In a case with high pressure nozzles, a high pressure is applied to the liquid in the atomizer before forcing it out into the drying chamber in a co-current (Fig. 8b) or fountain mode (Fig. 8c). Pressure nozzles are typically applied for coarser particles than other atomizers (GEA Process Engineering A/S 2016).

Smaller spray drying apparatus may be equipped by swirl nozzles, typically two-fluid swirl nozzles, where the droplets are sprayed by a high velocity compressed gas jet. These atomizers can be constructed with co-current or fountain mode of droplet discharging. Small laboratory spray dryers generally have multi-fluid nozzles or ultrasonic spray nozzles (Wisniewski 2015).

Depending on the type of atomizer, the liquid is mixed with the hot drying gas and flows into the drying chamber towards the outlet. When liquid FPH is supplied into the drying chamber and sprayed by the atomizer, the fine particles of liquid start immediately to interact with hot inert drying gas, which is delivered from the heating system. Simultaneous heat and mass transfer occurs while heat is supplied to the droplets from the drying agent by convection. The temperature of the droplets is rapidly increased up to the wet bulb temperature and the moisture from the particles starts to evaporate with a certain almost constant rate. At the same time, the drying gas temperature decreases. This period is considered as a first stage of spray drying;

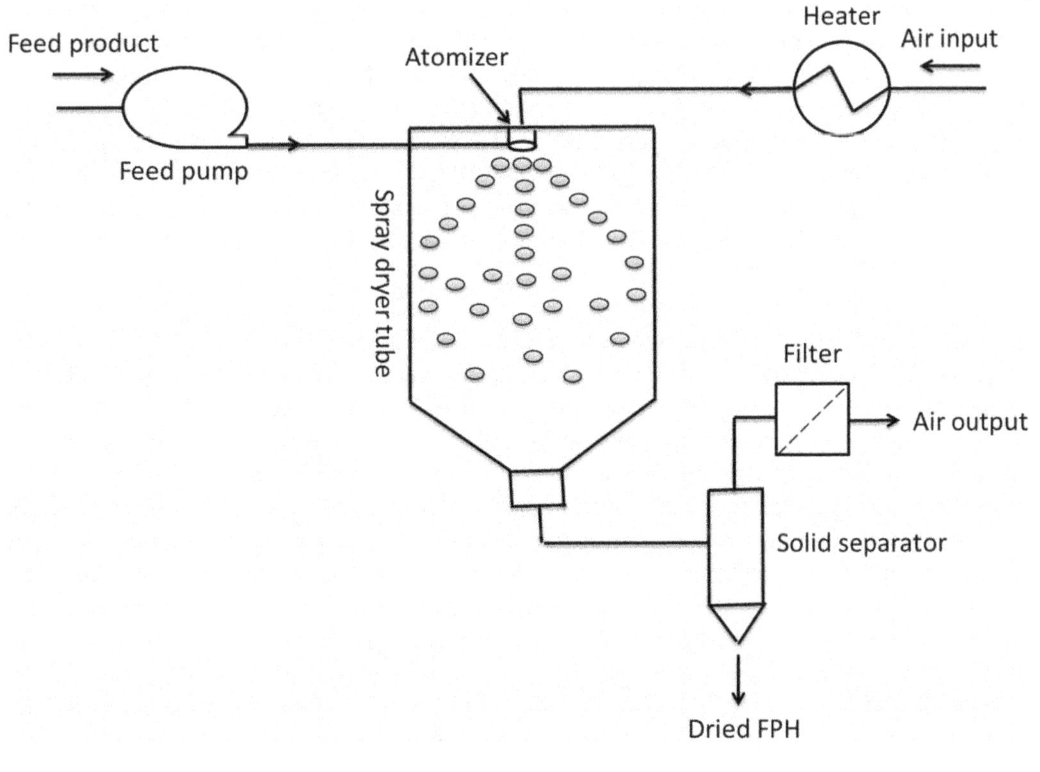

Fig. 7 Principal scheme of spray dryer

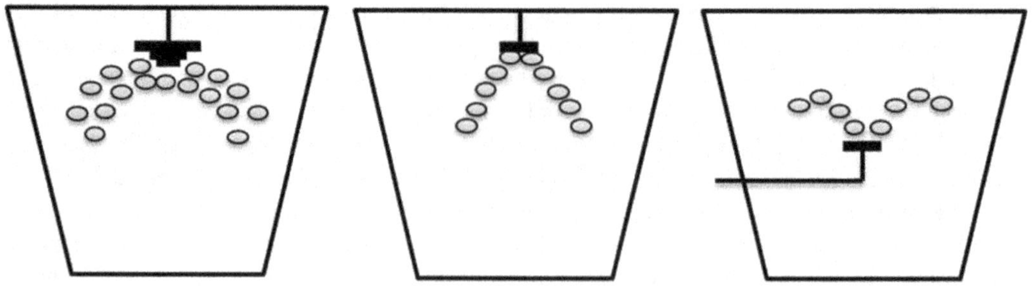

Fig. 8 Principal sketch of atomizer configuration: **a** Rotary atomizer. **b** Two-fluid nozzle atomizer with co-current mode. **c** Pressure nozzle atomizer with fountain mode

most water is removed during this stage. During the second stage, the temperature of the particles rises toward the air temperature at the outlet of the drying chamber (Huang et al. 2017, Wisniewski 2015). The final dried FPH is collected from the solid separator, which is mounted at the chamber's outlet and delivered to be packed and further transported.

Spray drying of FPH normally held at high temperatures in a diapason 150–220 °C. Exact temperature modes and residence time into the drying chamber are dependent on the rheological and sensitivity properties of the dried mixture and the desired moisture content in the final FPH. Higher inlet temperatures give higher evaporation rate, but can be damaging to the quality and bioactivities of final FPH. Thus, temperature mode is carefully adjusted in each case according to the purposes of production. Resistance time in spray dryer generally varies between 5 and 100 s (Masters 1991). Dried product rate of spray dryers is depended mainly on the working volume and capacity, rheological properties of solution, flow rate and a type of installed atomizer. Laboratory spray dryers such as model GA 32 (Yamato Scientific Co., Ltd.) have maximal water evaporation rate of 1.3 kg_{water}/h, while industrial spray dryers can evaporate much more than 1000 kg_{water}/h.

Spray drying is a popular method used at different productions over years due to its high productivity. However, for research purposes its use is limited by the availability at a certain research laboratory, where freeze dryers are rather available. Nevertheless, some research laboratories are equipped by spray driers and a number of research works on FPH have been published. Thus, Abdul-Hamid et al. (2002) studied nutritional quality of spray-dried FPH from Black Tilapia. Recently, Morales-Medina et al. (2016) used spray drying for the microencapsulation of fish oil by FPH made from sardine and horse mackerel. The new research of Silveira Alvares et al. (2018) used spray-dried FPH for the examination of its effect on vascular function of healthy individuals. Published articles on spray-dried FPH are generally aimed on studying of different properties of FPH using one-mode adjusted spray-drying just as an operation in the technological line. This gap should be filled by relative research devoted to the influence of spray-drying mode on the properties of FPH.

Freeze drying Freeze drying is a technique which utilizes the principle of moisture sublimation under lowered pressure. Because of that, the technique is often called vacuum freeze drying. Liquid FPH is frozen in a form of flat thin block at low temperatures (lower than − 35.0 °C). Then, the block of frozen FPH is transferred for drying into a vacuum chamber (Fig. 9) under low pressure (below 1.0 mbar). The energy for ice sublimation is provided through the shelves installed into the vacuum chamber. The pressure and temperature in the chamber are kept below the triple point of water; due to this, the sublimation of ice occurs in the frozen product, which forms fine porous structure of the dried layer. During the drying process, the water vapor crystallizes on the heat exchanger's surface. The temperature on the product's surface refers to the pressure in the drying chamber; it should be kept low enough to ensure only solid state of water at a given pressure (usually bellow − 56.0 C). The rate of moisture removal depends on thickness of frozen product and ice thickness on the heat exchanger's surface: increasing the product's thicknesses decreases the dehumidification rate significantly. Some part of the water is difficult to remove at low temperatures (primary drying). Thus, the drying temperature is increasing at the final stage of the drying process (secondary drying).

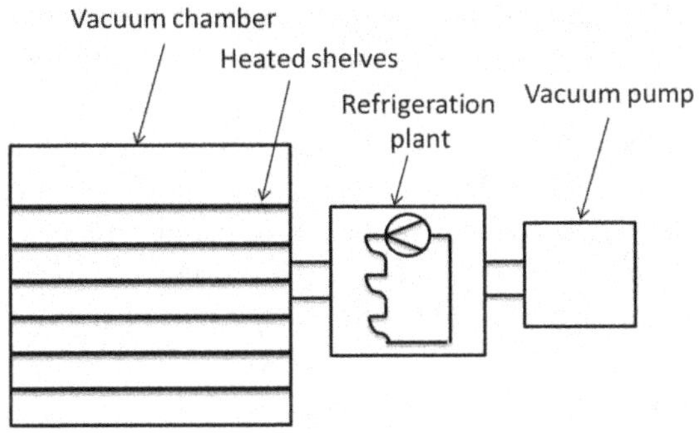

Fig. 9 Principal scheme of vacuum-freeze dryer

Freeze drying is a highly applicable method used for FPH drying at laboratory scale due to its simplicity and speed. A large number of authors have used freeze drying for the recovery of FPH in a number of researches, which studied different FPH purposes (Jenkelunas and Li-Chan, 2018; Noman et al. 2018; Halim et al. 2018) and many others. For example, the work of He et al. (2013) used freeze drying for the examination of functional properties of FPH. Elavarasan et al. (2016) used freeze drying in the evaluation of bioactive, clinical and structural properties of freeze-dried FPH from freshwater fish. Through a number of research work, freeze drying showed gentle treatment of the hydrolyzed mixture and a good protein recovery together with low moisture content in final dried FPH.

However, at large-scale production the cost and limited producing capacity of freeze-drying plays a negative role, where freeze drying is mostly used for the production of the highest quality FPH. When producing FPH for biochemical or medicine purposes, for example, as growing media for different microbiological cultures, freeze drying can be utilized at a certain extent.

Roller drum drying The principle of roller drum drying in the FPH production is to supply liquid FPH feed to the surface of horizontally mounted rotating hollow cylinder (single drum dryer) or two cylinders rotating towards each other (double drum dryer) or away from each other (twin drum dryer). The diameter of cylinders generally ranges from 0.5 to 6 m, while the length alters between 1 and 6 m (Juming and Shen 2003). Thus, thin coating of FPH (0.5–2 mm) on the heated surface is created. In Fig. 10, a simple sketch of a typical double drum dryer is shown. The heated surface at the temperature generally up to 200 °C provides necessary latent heat to the liquid product to be dried. The temperature of the product rises up to the temperature of free water boiling during a few moments after coating the drum. Most moisture evaporates at a water boiling

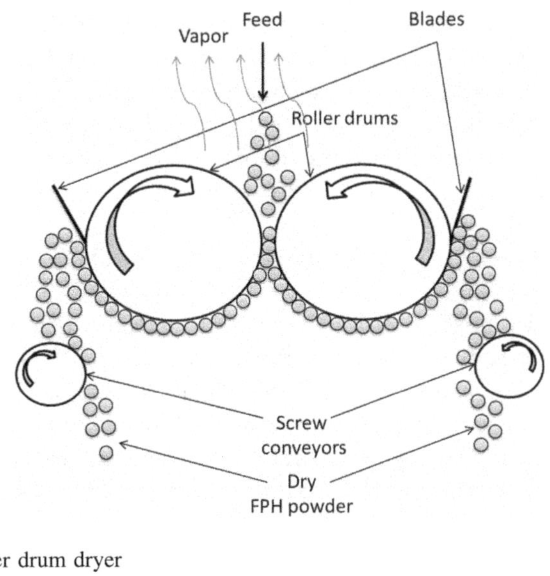

Fig. 10 Principal scheme of roller drum dryer

temperature during the following constant drying rate period; the rest is removed when the temperature of the dried mass goes up to the boiling temperature of bounded water just before the scrapping the dried mass by scrapping blades.

Such properties as feed method and rate, heat supply rate, roll speed, and thickness of the product film play a key role in the control and manage the roller drum drying (Juming and Shen 2003).

Drum drying is not a popular methods applied to the research on FPH, probably, due to the relative complexity of operation at small scale. The search of the works, which applied drum drying to FPH, gave the research of Zhou et al. (2016). The authors applied drum drying to FPH from squid and scallop.

Energy aspects of FPH production

Energy demands of main technological operations

There is a lack of information concerning drying of hydrolysates and energy efficiency analysis. However, other liquid food products have similar thermo-physical properties and moisture content; thus, the related energy efficiency values can be used as a reference for the evaluation.

Main energy demanding processes of FPH manufacture are hydrolysation, concentration and drying steps. Drying at industrial scale is mostly utilized by spray dryers and, thus, their performance must be evaluated for making a conclusion on energy status of the operation. Moisture evaporation rate for most spray dryers is found in the range between 600 and 41,000 kg/h (Baker and McKenzie 2005). Some pilot models of spray dryers have specific energy consumption in the range between 3 000 and 5500 kJ/kg of evaporated water (Al-Mansour et al. 2011), but the real heat consumption by industrial models of spray dryers is much higher. The average energy consumption for industrial spray dryers is in the range between 4 500 and 11 500 kJ/kg (Mujumdar 2007). The review, devoted to energy efficiency of spray dryers with respect to the industry, revealed the average energy consumption for food production as 4880 kJ/kg of evaporated water in the case of multi-stage spray dryers (Baker and McKenzie 2005).

Meanwhile, for the other heat demanding processes of FPH manufacture, heating loads are far lower. For example, for enzymatic hydrolysis the main energy consumption can be split into 3 stages: heating of the raw material and water to hydrolysation temperature, maintenance at hydrolysation temperature and final heating to terminate the hydrolysis. There is a necessity to reach working temperatures in a short period of time. Thus, the high heating loads are suspected from heating devices for the first and third stage due to the high specific heat capacity of the mixture. One can assume the heat capacity of the mixture in the range between 3.6 and 3.9 kJ/(kg*K) (ASHRAE 1999) considering the ratio of fish to water as 1:1. Thus, simple calculations give the approximate heat consumption for enzymatic hydrolysis in the range of 583–175, 0.032–3.8 and 136–252 kJ/kg for heating, hydrolysis procedure and termination of hydrolysis, respectively.

Concentration steps have higher heat consumption than for hydrolysis—around 5000 kJ/kg, but this step is an additional operation which may be added to prepare the mixture before the drying step and not an essential technological step. Moreover, it generally consumes much less energy than drying step of the FPH production.

Thus, the drying step of FPH production is the main energy consumer in the technological chain. Due to huge energy supply to the drying, energy saving in the drying process should become an important issue of FPH production. Energy effective solution provides not only huge economy savings, but also help to reduce the emission of the greenhouse effect gases, and, consequently, global warming (Mujumdar, 2007).

Possible solutions for the improvement in energy efficiency of dehydration procedure for FPH

Statistical data on convective dryers show that 20–60% of the supplied heat goes to moisture evaporation, 15–40% to heat losses with the exhaust air, 5–25% to product heating, 3–10% to heat losses through the walls and 5–20% to other losses (Danilov and Leontchik 1986; Mujumdar 2007). Thus, considering the main heat expenses, the efforts referring energy savings should be applied mainly to energy reduction during moisture evaporation and product heating; to recovery of the heat from exhaust air; and to decrease of the heat losses to the atmosphere. The following proposals referring the main heat expenses may help decreasing energy costs of FPH production and increasing its efficiency:

- Moisture evaporation.

Heat supply for the evaporation of moisture can be reduced by

- Product concentration before drying: in this case, solid content increases up to a certain value positively influence the drying efficiency (Masters 1991).
- Increase of the drying temperature or, consequently, the difference between the drying agent (surface) and the product: this would increase the drying rate, but also may influence badly on the quality of sensitive materials.
- Product distribution: this property contributes to uniform mass- and heat transfer when the product is distributed uniformly and in a properly adjusted amount into or over a certain volume of a spray drier or drum and vacuum-freeze driers, respectively.
- Other methods suitable for the operation of spray driers:

1. Multistage drying can be used as a solution for spray driers when the introduction of an additional unit into the system may give a positive result. In such a unit, the product is dried during the constant rate period, which is governed by internal moisture diffusion and, where moisture takes longer time to get out the product. An example of such a system is a fluidized bed (stationary or vibrated) installed at the lower part of a spray drier. This method with a proper operation can save up to around 35% of consumed energy (Mujumdar 2007).
2. Humidity control: for hygroscopic materials, air relative humidity of a drying agent can be a controlling factor. In this case, dehumidification of the drying agent can be considered. However, for the majority of application, air relative humidity at the entrance of spray drier is below the level that can inhibit completion of the drying process (Masters 1991).

- Recover of heat from exhaust air.

This part includes the recovery of the heat from drying air by the use of heat pumps or heat exchangers. The recovered heat can be used on different technological needs of the production.

- Product heating.

The preheating of the product can reduce energy consumption on product heating during drying. In this case, the viscosity of the product can be reduced which is desirable as for the better atomization of liquids in spray driers and for uniform distribution of the product on the surface of a drum drier and on the plates of a vacuum-freeze drier.

- Heat losses through the walls of equipment.

A proper insulation should be considered in order to decrease heat losses to the atmosphere through the walls of equipment. It can considerably reduce the energy consumption of the equipment with low investment costs and minimum labor efforts.

Conclusions

This paper gives a broad overview on the production of FPH from the step of raw material to the final drying of the hydrolysated product. The matters of type and availability of fish by-products were considered regarding the worldwide production. The stage of hydrolysis was evaluated according to the efficiency and safety. Enzymatic hydrolysis was chosen as the most promising solutions so far. Possible pre-treatment steps before the drying were considered in order to evaluate their efficiency in the decreasing of energy costs for the main drying step. Drying techniques used for FPH were highlighted together with their energy consumption and possible solutions for the effective energy management. Summarizing, the production of FPH ought to be adjusted so as to switch between available raw materials regarding its availability; the hydrolysation step can be carried out by enzymatic hydrolysis using the commercial mixture of Papain and Bromelain as a hydrolyzing agent; purification should be accomplished essentially by techniques which answer the producers' expectations to the final quality; the step of concentration may be added to the technological scheme, but not essentially; the drying step is still provided more efficiently by spray drying, but aforesaid improvements can be introduced to the existing installations.

Acknowledgements Thanks to the Department of Energy and Process Engineering and the Industrial Process Technology Group of NTNU for their valuable support and cooperation.

References

Abdelhedi O, Nasri R, Mora L, Jridi M, Toldrá F, Nasri M (2018) In silico analysis and molecular docking study of angiotensin I-converting enzyme inhibitory peptides from smooth-hound viscera protein hydrolysates fractionated by ultrafiltration. Food Chem 239:453–463. https://doi.org/10.1016/j.foodchem.2017.06.112

Abdul-Hamid A, Bakar J, Bee GH (2002) Nutritional quality of spray dried protein hydrolysate from Black Tilapia (Oreochromis mossambicus). Food Chem 78(1):69–74. https://doi.org/10.1016/S0308-8146(01)00380-6

Abejón R, Abejón A, Belleville MP, Sánchez-Marcano J, Garea A, Irabien Á (2016) Multiobjective Optimization of Membrane Networks for Fractionation of Protein Hydrolysate from Fish By-Products. In: Kravanja Z, Bogataj M (eds) Computer Aided Chemical Engineering. Elsevier, Amsterdam

Abejón R, Belleville MP, Sanchez-Marcano J, Garea A, Irabien A (2018) Optimal design of industrial scale continuous process for fractionation by membrane technologies of protein hydrolysate derived from fish wastes. Sep Purif Technol 197:137–146. https://doi.org/10.1016/j.seppur.2017.12.057

Aliani M, Eskin MNA (2017) Bitterness: perception, chemistry and food processing. Wiley, New Jersey

Al-Mansour HE, Al-Busairi BH, Baker CGJ (2011) Energy consumption of a pilot-scale spray dryer. Dry Technol 29(16):1901–1910. https://doi.org/10.1080/07373937.2011.595563

Alvares TS, Conte-Junior CA, Pierucci AP, de Oliveira GV, Cordeiro EM (2018) Acute effect of fish protein hydrolysate supplementation on vascular function in healthy individuals. Journal of Functional Foods 46:250–255. https://doi.org/10.1016/j.jff.2018.04.066

ASHRAE (1999) ASHRAE handbook. American Society of Heating, Refrigerating and Air Conditioning Engineers, Atlanta

Aubes-Dufau I, Capdevielle J, Seris JL, Combes D (1995) Bitter peptide from hemoglobin hydrolysate: isolation and characterization. FEBS Lett 364(2):115–119

Baker CGJ, McKenzie KA (2005) Energy consumption of industrial spray dryers. Dryi Technol 23(1–2):365–386. https://doi.org/10.1081/DRT-200047665

Bikov VP, Ionas GP, Golovkova GN, Didenko AP, Akulin VN, Perova LI, Odincov AB, Konstantinova LL, Dvinin UF, Christoferzen GS, Borisova LP (1998) Book of chemical content and technological properties of seawater and oceanic fishes. VNIRO Press, Moscow

Blenford DE (1994) Protein hyddrolysates: functionalities and uses in nutritional products. Int Food Ingr 3:45–49

Bourseau P, Vandanjon L, Jaouen P, Chaplain-Derouiniot M, Massé A, Guérard F, Chabeaud A, Fouchereau-Pérond M, Le Gal Y, Ravallec-Plé L, Bergé J-P, Picot L, Piot J-M, Batista I, Thorkelsson G, Delannoy C, Jakobsen G, Johansson I (2009) Fractionation of fish protein hydrolysates by ultrafiltration and nanofiltration: impact on peptidic populations. Desalination 244(1):303–320. https://doi.org/10.1016/j.desal.2008.05.026

Chae HJ, In MJ, Kim MH (1998) Process development for the enzymatic hydrolysis of food protein: effects of pre-treatment and post-treatments on degree of hydrolysis and other product characteristics. Biotechnol Bioproc E 3(1):35–39. https://doi.org/10.1007/BF02932481

Chakraborty PK, Madhavan P (1977) A pilot scale set up for the manufacture of fish hydrolysate. Fish Technol 14(2):159

Chalamaiah M, Dinesh Kumar B, Hemalatha R, Jyothirmayi T (2012) Fish protein hydrolysates: proximate composition, amino acid composition, antioxidant activities and applications: A review. Food Chem 135(4):3020–3038. https://doi.org/10.1016/j.foodchem.2012.06.100

Cui H (1996) The influence of degree of hydrolysis on the structural and functional properties of fish protein peptic hydrolysates. Masters thesis. University of Washington, Seattle

Damodaran S, Parkin KL, Fennema OR (2008) Fennema's Food Chemistry. CRC Press, Boca Raton

Danilov OI, Leontchik BI (1986) Energy Economic in Thermal Drying. Energoatomizdat, Moscow

Di Pasquale MG (2008) Amino acids and proteins for the athlete: the anabolic edge. CRC Press, Boca Raton

Elavarasan K, Shamasundar BA, Badii F, Howell N (2016) Angiotensin I-converting enzyme (ACE) inhibitory activity and structural properties of oven- and freeze-dried protein hydrolysate from fresh water fish (Cirrhinus mrigala). Food Chem 206:210–216. https://doi.org/10.1016/j.foodchem.2016.03.047

European Comission (2012) https://ec.europa.eu/fisheries/cfp/aquaculture/species_en. Accessed 26 Sept 2018

FAO (2007) http://www.fao.org/3/a-a1394e.pdf. Accessed 26 Sept 2018

FAO (2016) http://www.fao.org/3/a-i5555e.pdf. Accessed 26 Sept 2018

FAO (2017) http://www.fao.org/fi/static-media/MeetingDocuments/WorkshopAMR17/presentations/22.pdf. Accessed 26 Sept 2018

GEA (2018) http://www.gea.com/en/products/falling-film-evaporator.jsp Accessed 26 Sept 2018

GEA Process Engineering A/S (2016) GEA Niro spray drying. Small-scale solutions for R&D and production. GEA Process Engineering A/S, Danmark

Gonzalez-Tello P, Camacho F, Jurado E, Paez MP, Guadix EM (1994) Enzymatic hydrolysis of whey proteins: i. Kinetic models. Biotechnol Bioeng 44(4):523–528. https://doi.org/10.1002/bit.260440415

Halim NRA, Azlan A, Yusof HM, Sarbon NM (2018) Antioxidant and anticancer activities of enzymatic eel (monopterus sp) protein hydrolysate as influenced by different molecular weight. Biocatal Agric Biotechnol 16:10–16. https://doi.org/10.1016/j.bcab.2018.06.006

He S, Franco C, Zhang W (2013) Functions, applications and production of protein hydrolysates from fish processing co-products (FPCP). Food Res Int 50(1):289–297. https://doi.org/10.1016/j.foodres.2012.10.031

Himonides AT, Taylor AKD, Morris AJ (2011) Enzymatic Hydrolysis of Fish Frames Using Pilot Plant Scale Systems. Food Nutr Sci 2:586–593

Hoeling A, Volkov V (2015) Proteins from by-products - innovative components in sustainable industrial production. Proceedings of KGTU 38:83–92

Huang S, Vignolles ML, Chen XD, Le LY, Jan G, Schuck P, Jeantet R (2017) Spray drying of probiotics and other food-grade bacteria: a review. Trends Food Sci Tech 63:1–17. https://doi.org/10.1016/j.tifs.2017.02.007

Jaswal AS (1990) Amino Acid Hydrolysate from Crab Processing Waste. J Food Sci 55(2):379–380. https://doi.org/10.1111/j.1365-2621.1990.tb06768.x

Jenkelunas PJ, Li-Chan ECY (2018) Production and assessment of Pacific hake (Merluccius productus) hydrolysates as cryoprotectants for frozen fish mince. Food Chem 239:535–543. https://doi.org/10.1016/j.foodchem.2017.06.148

John TG (1993) Non-bitter protein hydrolyzates. US Patent 5266685, USA

Juming THF, Shen GQ (2003) Drum Drying. In: Encyclopedia of agricultural, food, and biological engineering. Marcel Dekker, Inc., New York, pp 211–214. https://doi.org/10.1081/e-eafe 120007091

Kinsella JE, Melachouris N (1976) Functional properties of proteins in foods: a survey. Crit Rev Food Sci 7(3):219–280. https://doi.org/10.1080/10408397609527208

Kristinsson HG (1998) Reaction kinetics, biochemical and functional properties of salmon muscle proteins hydrolyzed by different alkaline proteases. Masters thesis. University of Washington, Seattle

Kristinsson HG, Rasco BA (2000) Fish Protein Hydrolysates: production, Biochemical, and Functional Properties. Critical Rev Food Sci 40(1):43–81. https://doi.org/10.1080/10408690091189266

Kuehler CA, Stine CM (1974) Effect of enzymatic hydrolysis on some functional properties of whey protein. J Food Sci 39(2):379–382. https://doi.org/10.1111/j.1365-2621.1974.tb02899.x

Lahl WJ, Braun SD (1994) Enzymatic production of protein hydrolysates for food use. Food Sci 48:68–71

Linder M, Fanni J, Parmentier M, Sergent M, Phan-Tan-Luu R (1995) Protein recovery from veal bones by enzymatic hydrolysis. J Food Sci 60(5):949–952. https://doi.org/10.1111/j.1365-2621.1995.tb06268.x

Loffler A (1986) Proteolytic enzymes: sources and applications. Food Technol-Chicago 40(1):63

Loosen P, Bressollier P, Julien R, Pejoan CH, Verneuil BG (1994) Method for preparing an enzymatic hydrolysate. United States Patent 5356637

Manninen AH (2004) Protein hydrolysates in sports and exercise: a brief review. J Sport Sci Med 3(2):60–63

Masters K (1991) Spray drying handbook, 5th edn. Longman Scientific & Technical, New York

Microlab Scientific (2018) http://www.microlabscientific.com/Membrane-Filter.html. Accessed 26 Sept 2018

Mine Engineer Product Manual (2009) http://www.mine-engineer.com/mining/plate.htm Accessed 26 Sept 2018

Morales-Medina R, Tamm F, Guadix AM, Guadix EM, Drusch S (2016) Functional and antioxidant properties of hydrolysates of sardine (S. pilchardus) and horse mackerel (T. mediterraneus) for the microencapsulation of fish oil by spray-drying. Food Chem 194:1208–1216. https://doi.org/10.1016/j.foodchem.2015.08.122

Mujumdar AS (2007) Principles, classification and selection of dryers. In: Mujumdar AS (ed) Handbook of industrial drying. CRC Press, Boca Raton

Noman A, Xu Y, Al-Bukhaiti WQ, Abed SM, Ali AH, Ramadhan AH, Xia W (2018) Influence of enzymatic hydrolysis conditions on the degree of hydrolysis and functional properties of protein hydrolysate obtained from Chinese sturgeon (Acipenser sinensis) by using papain enzyme. Process Biochem 67:19–28. https://doi.org/10.1016/j.procbio.2018.01.009

NPTEL (2013) National programme on technology enhanced learning. Module 3: design of evaporator. Lecture 1: introduction and evaporator classifications. Government of India, Ministry of Human resource development, India

Pasupuleti VK, Braun S (2010) State of the art manufacturing of protein hydrolysates. In: Pasupuleti VK, Demain AL (eds) Protein hydrolysates in biotechnology. Springer, Dordrecht, pp 11–32. https://doi.org/10.1007/978-1-4020-6674-0_2

Quaglia GB, Orban E (1987) Enzymic solubilisation of proteins of sardine (sardina pilchardus) by commercial proteases. J Sci Food Agr 38(3):263–269. https://doi.org/10.1002/jsfa.2740380310

Quaglia GB, Orban E (1990) Influence of enzymatic hydrolysis on structure and emulsifying properties of sardine (sardina pilchardus) protein hydrolysates. J Food Sci 55(6):1571–1573. https://doi.org/10.1111/j.1365-2621.1990.tb03571.x

Rebeca BD, Peña-Vera MT, DiAz-Castañeda M (1991) Production of fish protein hydrolysates with bacterial proteases; yield and nutritional value. J Food Sci 56(2):309–314. https://doi.org/10.1111/j.1365-2621.1991.tb05268.x

Richardsen R, Nystøyl R, Marthinussen A (2015) Analyse marint resteråstoff, 2015. Rapport. Analyse av tilgang og anvendelse for marint restråstoff I Norge. SINTEF Fiskeri og Havbryk, Trondheim

Roslan J, Mustapa Kamal SM, Yunos KF, Abdullah N (2018) Evaluation on performance of dead-end ultrafiltration membrane in fractionating tilapia by-product protein hydrolysate. Sep Purif Technol 195:21–29. https://doi.org/10.1016/j.seppur.2017.11.020

Saidi S, Deratani A, Belleville M-P, Amar RB (2014) Production and fractionation of tuna by-product protein hydrolysate by ultrafiltration and nanofiltration: impact on interesting peptides fractions and nutritional properties. Food Res Int 65:453–461. https://doi.org/10.1016/j.foodres.2014.04.026

Sarmadi BH, Ismail A (2010) Antioxidative peptides from food proteins: a review. Peptides 31(10):1949–1956. https://doi.org/10.1016/j.peptides.2010.06.020

SeaFoodSource (2016) https://www.seafoodsource.com/news/supply-trade/entrepreneurs-getting-creative-with-seafood-byproducts. Accessed 26 Sept 2018

Sikorski ZE, Naczk M (1981) Modification of technological properties of fish protein concentrates. Crit Rev Food Sci 14(3):201–230. https://doi.org/10.1080/10408398109527305

Silvestre MPC, Vieira CR, Silva MR, Carreira RL, Silva VDM, Morais HA (2009) Protein extraction and preparation of protein hydrolysates from rice with low phenylalanine content. Asian J Sci Res 2:146–154

Skanderby M (1994) Protein hydrolysates: their functionality and applications. Food Technol Int 10:141–144

Slizyte R, Grimsmo L, Storrø I (2010) Raport. Processering av biråstoff fra sild til olje og proteinhydrolysat. Laboratorieforsøk med ulike proteaser og pilotforsøk med ultrafersk råstoff. Rapport. SINTEF Fiskeri og Havbryk, Trondheim

Spinelli J, Koury B, Miller R (1972) Approaches to the utilization of fish for the praparation of protein isolates. J Food Sci 37(4):599–603. https://doi.org/10.1111/j.1365-2621.1972.tb02702.x

Suwal S, Ketnawa S, Liceaga AM, Huang J-Y (2018) Electro-membrane fractionation of antioxidant peptides from protein hydrolysates of rainbow trout (Oncorhynchus mykiss) byproducts. Innov Food Sci Emerg 45:122–131. https://doi.org/10.1016/j.ifset.2017.08.016

Tannenbaum SR, Ahern M, Bates RP (1970a) Solubilization of fish protein concentrate. I. An alkaline process. Food Technol-Chicago 24:604–606

Tannenbaum SR, Ahern M, Bates RP (1970b) Solubilization of fish protein concentrate. II. Utilization of the alkaline-process product. Food Technol-Chicago 24:607

Vandanjon L, Grignon M, Courois E, Bourseau P, Jaouen P (2009) Fractionating white fish fillet hydrolysates by ultrafiltration and nanofiltration. J Food Eng 95(1):36–44. https://doi.org/10.1016/j.jfoodeng.2009.04.007

Webster JD, Ledward DA, Lawrie RA (1982) Protein hydrolysates from meat industry by-products. Meat Sci 7(2):147–157. https://doi.org/10.1016/0309-1740(82)90080-8

Wisniewski R (2015) Spray drying technology review. Paper presented at the 45th International Conference on Environmental Systems Bellevue, 12-16 July 2015. Washington

Ytrestøyl T, Aas TS, Åsgård T (2014). Recourse utilization of Norwegian salmon farming in 2012 and 2013. Report 36/2014. Nofima, Tromsø

Zhou Y, Thirumurugan R, Wang Q, Lee CM, Davis DA (2016) Use of dry hydrolysate from squid and scallop product supplement in plant based practical diets for Pacific white shrimp Litopenaeus vannamei. Aquaculture 465:53–59. https://doi.org/10.1016/j.aquaculture.2016.08.028

Zhu Z, Wu Q, Di X, Li S, Barba FJ, Koubaa M, Roohinejad S, Xiong X, He J (2017) Multistage recovery process of seaweed pigments: investigation of ultrasound assisted extraction and ultra-filtration performances. Food Bioprod Process 104:40–47. https://doi.org/10.1016/j.fbp.2017.04.008

10

Tetraselmis chuii biomass as a potential feed additive to improve survival and oxidative stress status of Pacific white-leg shrimp *Litopenaeus vannamei* postlarvae

Norazira Abdu Rahman · Helena Khatoon · Norhayati Yusuf · Sanjoy Banerjee · Noor Azilah Haris · Fathurrahman Lananan · Katayama Tomoyo

Abstract Shrimp is an important traded fishery commodity. When subjected to stress, shrimp usually suffers from oxidative stress, which leads to cell injury, senescence, and death. To maintain shrimp good health, performance and production, antioxidant and immune systems are important. Natural antioxidants found in microalgae may be used to increase the cell protection against oxidative damage, being a promising alternative to the carcinogenic synthetic antioxidants. In this study, *Tetraselmis chuii* was evaluated for its effect on survival, growth and oxidative stress biomarkers on *Litopenaeus vannamei* postlarvae (PL). The antioxidant properties of the formulated feed with *T. chuii* inclusion were determined using four antioxidant chemical assays. Meanwhile, the oxidative stress biomarkers on PL were analyzed by hydrogen peroxide, membrane stability and lipid peroxidation assays. Results showed that PL reared on diets supplemented with 50% *T. chuii* had a significantly higher ($P \leq 0.05$) survival (97.6 ± 1.4%) and lower oxidative stress in terms of hydrogen peroxide content (10.08 ± 0.4 mM g^{-1} FW) and electrolyte leakage (10.8 ± 0.3%). The result of this study also showed that shrimp PL reared on diets supplemented with microalgal, *T. chuii* have high resistance to reverse salinity stress test (76.7–100%). However, no significant differences ($P \geq 0.05$) were found in the growth and lipid peroxidation. Due to the positive effect on oxidative stress status, survival and resistance to salinity stress, the feeding of *L. vannamei* PL with diet containing at least 50% of *T. chuii* is recommended as a natural source of antioxidant for PL.

N. A. Rahman
School of Fisheries and Aquaculture Sciences, Universiti Malaysia Terengganu, 21030 Kuala Terengganu, Malaysia

H. Khatoon (✉) · N. A. Haris
Institute of Tropical Aquaculture, Universiti Malaysia Terengganu, 21030 Kuala Terengganu, Malaysia
e-mail: hlnkhatoon@gmail.com

N. Yusuf
School of Science and Technology, Universiti Malaysia Terengganu, 21030 Kuala Terengganu, Malaysia

S. Banerjee
Institute of Bioscience, Universiti Putra Malaysia, Serdang, Malaysia

F. Lananan
East Coast Environmental Research Institute, Sultan Zainal Abidin University, Gong Badak Campus, 21300 Kuala Terengganu, Terengganu, Malaysia

K. Tomoyo
Graduate School of Agricultural and Life Sciences, The University of Tokyo, 1-1-1, Yayoi, Bunkyo-ku, Tokyo 113-8657, Japan

Keywords Marine Microalgae · Antioxidant · Feed supplement · Oxidative stress biomarkers · Shrimp

Introduction

Litopenaeus vannamei (white-leg shrimp) is the most widely cultivated penaeid shrimp in many parts of the world, especially in the Southeast Asian region. However, shrimp aquaculture production is challenged by pathogenic infection, low survival, weak larvae, and environmental problems, which results in significant loss of production (Lightner 2011). In aquatic organisms, changes in the environmental condition, such as salinity, oxygen levels, and temperature, can enhance the generation of reactive oxygen species (ROS), causing oxidative damages in tissues (Lushchak 2011). Thus, physiological conditions such as antioxidant system are critical for shrimp health maintenance and healthy growth performance, especially under environmental stress (Castex et al. 2010). Failure in detoxifying the excess ROS production using their natural antioxidant defenses may lead to significant oxidative damages. Some of these include the inactivation of enzymes, degradation of the protein, DNA damages and lipid peroxidation (Di Giulio et al. 1995). These damages may lead to a disease outbreak and huge economic losses (Bachère 2000).

Commercial antioxidant supplements such as butylated hydroxytoluene (BHT) and butylated hydroxyanisole (BHA) have been incorporated into the feed to reduce the oxidative damage in aquatic animals. However, the utilization of both synthetic antioxidant has already been restricted due to their carcinogenic and toxic nature in animal models (Liu et al. 2007). This leads to the search for natural sources of antioxidant compounds instead of relying on the dangerous synthetic ones. In recent years, besides their important roles and application in aquaculture hatcheries, microalgae have also attracted increasing attention as animal feed supplement since they are a rich source of natural antioxidants, pigments, and other bioactive compounds. Microalgae of different genera and species have been found to exhibit significant antioxidant activity in vitro (Geetha et al. 2010; Rao et al. 2006) and, thus, they are considered as potential natural sources of antioxidants in vivo. In addition, most microalgal extracts showed a much higher inhibition of liver microsomal lipid peroxidation compared to BHT which is one of the most powerful synthetic antioxidant agents (AbouL-Enein et al. 2003).

Marine microalgal, *Tetraselmis* sp. are flagellated chlorophytes with rapid growth rate, and can stand a broad range of temperatures and pH values (Khatoon et al. 2014). *Tetraselmis* sp. is a widely used species in aquaculture as it contains adequate amounts of protein, lipid, carbohydrate and fatty acids which are essential for the cultured organisms. This species also serves as a rich source of a variety of bioactive compounds, such as vitamin E, carotenoids, phenolic compounds, and terpenes, which have antioxidant, anticancer and antimicrobial properties (Jahnke 1999; Carballo-Cárdenas et al. 2003; Ismaiel et al. 2016). Although there are already some species that has been produced commercially and studied for their potential as feed supplements, such as *Spirulina* and *Haematococcus pluvialis,* there is still need to search for the next best microalgae species. Many studies on microalgae as food additives in aquatic farmed animals focused on survival, growth, nutrition and pigmentation of the species (Ju et al., 2011, 2012; Macias-Sancho et al. 2014; Li et al. 2014). However, no studies have been reported on the efficacy of *Tetraselmis* sp. on the oxidative stress in *L. vannamei* postlarvae. The aim of the present study was to evaluate the effects of *T. chuii* supplementation on growth, survival and oxidative stress status in *L. vannamei* postlarvae. This study highlighted the potential of these microalgae when utilized as a natural source of exogenous antioxidant and their ability to enhance the shrimp defenses against oxidative stress and damage.

Materials and methods

Culture of microalgae

Tetraselmis chuii were obtained from the laboratory of the Institute of Tropical Aquaculture, Universiti Malaysia Terengganu. The pure *T. chuii* was cultured and maintained at 25 °C, 30 ppt salinity using Conway medium (Tompkins et al. 1995) and 24 h photoperiod. Conway medium was prepared by adding macronutrient (1 mL), trace metals (0.5 mL) and vitamins (0.1 mL) stock solution to 1 L of filtered and sterilized seawater (Tompkins et al. 1995). Subculturing of microalgal stock culture was done once in two weeks to maintain the health and purity of the stock. The stock culture was used for the culture experiments.

Tetraselmis chuii was cultured in 25 L closed cylinder photobioreactors under laboratory conditions using Conway medium according to Khatoon et al., (2014). The initial cell density for each culture was at 1×10^5 cells mL^{-1}. Cultures were maintained under a light intensity of 50 $\mu mol^{-2}\ s^{-1}$ at 25 °C and 24 h continuous light using white fluorescent light. Aeration was provided continuously throughout the culture. Once the culture reached their stationary phase (on day 8), the cells were harvested by centrifugation at 8000 rpm for 10 min using a tubular continuous centrifuge J-025 (Hanil Science Industrial Co.LTD). Samples were then freeze dried using Labconco Freezone 4.5 and kept at −20 °C until further use.

Feed formulation for *L. vannamei* postlarvae

The freeze-dried microalgal *T. chuii* was ground using a stand mixer (KitchenAid® Professional 600 Series, Michigan, USA). Then, formulated feed was prepared by incorporating commercial feed with microalgal *T. chuii* at 25% w/w, 50% w/w, 75% w/w and 100% w/w. A commercial feed commonly used for the production of *L. vannamei* at this rearing phase was used as the control diet. The formulated feed was then milled to 300 mm. The formulated feeds were then kept at −20 °C until further use.

Determination of antioxidant properties

Preparation of samples extracts

The formulated feed (with different percentages of dried microalgal *T. chuii*) was used in this experiment. Sample extracts were prepared by adding 12.5 mL distilled water to 0.1 g of dried formulated feed before it was incubated in 100 °C water bath for an hour (Gyamfi et al. 1999). Then, it was centrifuged for 10 min at 10,000 rpm using a high-speed refrigerated centrifuge (Hitachi CR22N, Japan). The supernatant was stored at 4 °C and used as the extract.

Antioxidant activity

Antioxidant activity was determined following the method of Yang et al. (2000). Linoleic acid emulsion (10 mM, pH 6.5) was prepared by dissolving 0.28 g linoleic acid (Sigma-Aldrich) and 0.28 g Tween 20 (Sigma-Aldrich) in 50 mL, sodium phosphate buffer (0.2 M, pH 6.5). Then, sample extract was added to 2 mL linoleic acid emulsion before it was incubated in the dark for 15 h at 37 °C to accelerate oxidation. Then, 6 mL of 60% methanol (HmbG Chemicals) was added before the absorbance was taken at 234 nm using a spectrophotometer (Shimadzu UV-1601, Japan) before and after oxidation.

Scavenging effect on diphenyl-1-picrylhydrazyl radical

Scavenging action of diphenyl-1-picrylhydrazyl (DPPH) radical was analyzed according to Shimada et al. (1992). One mL sample extract was added to 0.25 mL of DPPH radicals (0.2 mM) (Sigma-Aldrich) in methanol (HmbG Chemicals) before it was incubated for 30 min at room temperature. The absorbance was then taken at 517 nm using a spectrophotometer (Shimadzu UV-1601, Japan) to measure the reduction of the DPPH free radical. One mL Trolox (Sigma-Aldrich), Ascorbic acid (Sigma-Aldrich) and BHT (Sigma-Aldrich) were also used at 20 mM as positive controls. The scavenging effect on DPPH radical (%) was calculated with the following equation: $\text{Scavenging effect}(\%) = \frac{(A - A_1)}{A} \times 100$, where A was the absorbance of the negative control (DPPH solution only), and A_1 was the absorbance of the test sample.

Scavenging effect on superoxide radical

Superoxide anion was determined according to Robak and Gryglewski (1988) by adding 0.5 mL NADH (78 μM), 0.5 mL NBT (50 mM), 0.5 mL phenazine methosulfate (10 μM) and 1.0 mL of sample extract to

0.5 mL Tris–HCl buffer (16 mM, pH 8.0). The absorbance was then measured at 560 nm using a spectrophotometer (Shimadzu UV 1601). 20 mM Ascorbic acid, Trolox, and BHT were used as the positive controls. The inhibition ratio (%) was calculated with the following equation: Scavenging effect (%) $= \frac{(A - A_1)}{A} \times 100$, where A was the absorbance of the control, and A_1 was the absorbance of the test sample. All chemicals used in this analysis were obtained from Sigma-Aldrich.

Ferric reducing antioxidant potential assay

The ferric reducing antioxidant potential (FRAP) assay was carried out following the method of Hajimahmoodi et al. (2010). Firstly, FRAP reagent was prepared prior to use by mixing 5 mL of a TPTZ solution (2,4,6-tripyridyl-*S*-triazine) (Sigma-Aldrich) (10 mM L^{-1}) in HCl (40 mM L^{-1}) with 5 mL of $FeCl_3$ (20 mM L^{-1}) (Sigma-Aldrich) and 50 mL of acetate buffer (0.3 M, pH 3.6) before it was warmed to 37 °C. Then, a 100 μL sample extract was added to 3 mL of FRAP. The mixture was then incubated at 37 °C for 10 min, and the absorbance was taken at 593 nm. The results were expressed as micromole of Trolox equivalent.

Proximate analysis of the formulated diet

Protein analysis

Protein analysis was carried out according to Lowry et al. (1951). Each formulated feed (5–6 mg) was dissolved in 25 mL distilled water and 0.5 mL was taken from each diluted sample for protein analysis. Mixed reagent was prepared by adding 1 mL of Reactive 1 (1% Potassium sodium tartrate) to 50 mL of Reactive 2 (2 g of Sodium carbonate) in 100 mL of 0.1 M NaOH. Then, 0.5 mL of sample was added to 0.5 mL of 1 M Sodium hydroxide and it was incubated for 5 min in 100 °C water bath. After 10 min of cooling in water bath, 2.5 mL of the mixed reagent and 0.5 mL of Folin reagent were added to the mixed solution and were kept in the dark places for 30 min. The absorbance was measured using UV-1601 spectrophotometer (Shimadzu) at 750 nm.

Lipid analysis

Lipid analysis was conducted according to Marsh and Weinstein (1966). The carbonization was carried out using tripalmitin as the standard after lipid extraction was performed following Bligh and Dyer (1959). Sample extraction was carried out by adding 4.5 mL of chloroform: methanol (1:2) to the sample before it was centrifuged at 10,000 rpm for 10 min. The supernatant was separated and kept in a clean tube. The biomass was again extracted by centrifuging after adding 1.5 mL of chloroform and 1.5 mL of distilled water. Both supernatants were combined and evaporated with a water bath at 35 °C under vacuum after the polar phase was removed. Two mL of concentrated sulfuric acid was added after it completely dries and the mixture solution was then cooled down to 0 °C immediately. The absorbance was then taken at 375 nm after 3.0 mL of distilled water was added to the mixture.

Carbohydrate analysis

The sample solution was prepared by adding 5–6 mg of sample into 25 mL of distilled water (Dubois et al. 1956). Then, 1.0 mL of phenolic solution (5%) and 5.0 mL of sulfuric acid were added to the sample solution before the absorbance was taken at 488 nm, spectrophotometrically (Shimadzu UV-1601, Japan).

Experimental setup

The experiment was conducted at the Institute of Tropical Aquaculture, Universiti Malaysia Terengganu, Malaysia. Five treatments consist of 100% commercial feed (CF) as control; commercial feed with 25% *T. chuii* supplement (T25); commercial with 50% *T. chuii* supplement (T50); commercial feed with 75% *T. chuii* supplement (T75) and 100% *T. chuii* (T100). Fifteen 30 L rectangular aquaria filled with 20 L of filtered, dechlorinated seawater at 30 ppt salinity was prepared.

The *L. vannamei* PL stage five (PL5) with average initial weight of 0.45 mg was obtained from a commercial hatchery and stocked at a density of 50 PL L^{-1}. Constant aeration was provided to each aquarium and the hatchery tanks were maintained under a 12 h light: 12 h dark cycle. The shrimps were fed four times a day at 6 h intervals with the formulated diet and Artemia (Golden Dolphin, Malaysia). The shrimps were fed with artemia, ad libitum (Gamboa-Delgado and Le Vay 2009). All treatments were done in triplicates and were carried out for 12 days. Fecal matter and the remainder of the uneaten food were siphoned out daily at less than 10% of the culture volume to maintain water quality. On the final day of the culture, surviving PLs from every treatment tank were counted and weighed to estimate mean survival for each of the treatments and the control groups. The shrimp PL tissue samples were collected and kept at −80 °C for the oxidative stress biomarker analysis.

Physical and chemical analysis

Temperature, pH, salinity and dissolved oxygen in the culture tanks were measured daily using a portable multi-parameter probe (YSI 556 MPS, USA). Meanwhile, total ammonia nitrogen (TAN), nitrite nitrogen (NO_2-N) and phosphate phosphorous (PO_4-P) were determined every other day (Parsons et al. 1984).

Biological analyses

The specific growth rate (SGR) of PL was calculated from the body weight based on Ricker's model (1990): $SGR = \left(\frac{\ln_{wf} - \ln_{wi}}{\Delta t}\right) \times 100$, where wf is the weight at time (*t*) and wi is the initial weight of PL. At the end of the feeding experiment, the survival of PL for each treatment was calculated and reverse salinity stress test was conducted. The reverse salinity stress test is commonly practiced by commercial hatcheries to distinguish between healthy and weak PL (Khatoon et al. 2007). The stress test was carried out by transferring the shrimp PL from the culture tanks with 30 ppt salinity to freshwater (0 ppt) and the shrimp PL survival was recorded after 2 h.

Oxidative stress biomarker analysis

Hydrogen peroxide assay

Level of hydrogen peroxide (H_2O_2) was measured in the shrimp tissue collected on the final day of culture (Velikova et al. 2000). The fresh shrimp tissue sample (0.15 g) was homogenized at 0–4 °C in 1.0 mL of 0.1% trichloroacetic acid (TCA) (Sigma-Aldrich) before it was centrifuged for 10 min at 10,000 rpm. Then, 0.5 mL of supernatant was mixed with 0.5 mL of potassium phosphate buffer (10 mM, pH 7.0). One mL potassium iodide was added to the mixture before it was incubated under dim light for 10 min. The absorbance was measured at 390 nm, using UV1601 spectrophotometer (Shimadzu).

Lipid peroxidation assay

Lipid peroxidation level was determined following the method of Heath and Packer, (1968). Fresh shrimp sample (0.15 g) was homogenized at 0–4 °C in the presence of 1.0 mL of 0.1% TCA solution before it was centrifuged at 10,000 rpm for 5 min. Then, 2.25 mL TBA reagent was added to 0.75 mL of the supernatant. The TBA reagent was prepared by dissolving 0.5 g TBA (Sigma-Aldrich) in 100 mL of 20% TCA. The mixture was then placed in 95 °C water bath for 30 min and quickly cooled in ice bath for 15 min. Then, it was then centrifuged at 10,000 rpm for 10 min before the absorbance was measured at 532 nm and 600 nm. The level of lipid peroxidation [Malondialdehyde (MDA) equivalent] was then calculated as follows: $\frac{A_{532} - A_{600}}{15{,}5000} \times 10^6$, where A_{532} is absorbance at 532 nm and A_{600} is absorbance at 600 nm. The results were expressed as nanomole MDA/g fresh weight of tissue sample.

Membrane stability assay

The membrane stability index was measured as the relative percentage of total electrolyte leakage (Rady 2011). A total of 0.2 g shrimp tissue was washed with double distilled water and placed in a test tube containing 10 mL of double distilled water and incubated at 40 °C for 30 min, after which the first reading of conductivity was registered (C1). The sample was then boiled for 10 min at 100 °C and cooled at room temperature before conductivity readings were taken (C2) using EUTECH INSTRUMENT PC700 pH/mV/Conductivity/°C/°F meter. Percentage of leakage was calculated as follows: Relative leakage (%leakage) $= \frac{1-C_1}{C_2} \times 100$, where C_1 is conductivity before boiling and C_2 is conductivity after boiling.

Statistical analysis

The collected data from antioxidant chemical assays, proximate analysis, survival, growth and oxidative stress biomarkers assay were analyzed using one-way analysis of variance (ANOVA) and the significant differences among treatments were determined using Tukey test at 95% confidence interval level. All statistical analysis was carried out using the SPSS (Statistical Package for the Social Sciences) software.

Results and discussion

Antioxidant activity of the formulated diet

The antioxidant activity in all treatments is dependent on the concentration of *Tetraselmis chuii* in the diets (Table 1). T100 and T75 showed the highest antioxidant activity, scavenging effect on free radicals and FRAP compared to control (CF) and other treatments ($P \leq 0.05$). In addition, there were no significant differences ($P \leq 0.05$) between the antioxidant activity of T100 (81.0 ± 0.0%) with BHT (80.4 ± 1.2%), Trolox (91.1 ± 0.7%) and Ascorbic acid (85.2 ± 0.3%). The addition of 25% *T. chuii* (T25) (51.0 ± 0.0%) to commercial feed led to the increase of the antioxidant activity by 2.4 fold when compared to the feed without any *T. chuii* addition (21.4 ± 3.1%). A similar trend was also found regarding the scavenging effect on superoxide radical, DPPH radical and FRAP. The increase in the percentage of *T. chuii* into the feed led to the increase in the antioxidant properties of the feed. The addition of microalgae has been shown to increase the antioxidant compounds such as astaxanthin and lutein in the aquaculture feed (Li et al. 2015; Nonwachai et al. 2010). Due to the carcinogenic and tumorigenic nature of BHT, the potential of *T. chuii* as a natural source of antioxidants is very promising and can be considered as a substitute to replace the harmful synthetic substances.

Table 1 Antioxidant activity, free radical scavenging activities, and ferric reducing antioxidant potential (FRAP) of all formulated feed

Formulated feed	Antioxidant activity (%)	Scavenging effect on superoxide radical (%)	Scavenging effect on dpph radical (%)	FRAP (μm Trolox g^{-1})
CF	21.36 ± 3.11^{d}	35.12 ± 1.45^{e}	36.40 ± 3.7^{d}	33.40 ± 0.44^{e}
T25	50.94 ± 1.05^{c}	46.60 ± 3.74^{d}	55.91 ± 1.4^{c}	37.63 ± 0.34^{d}
T50	51.48 ± 6.92^{c}	60.22 ± 0.30^{c}	56.47 ± 0.2^{c}	49.19 ± 0.65^{c}
T75	67.94 ± 2.21^{b}	71.96 ± 0.30^{b}	71.25 ± 0.15^{b}	53.23 ± 0.73^{b}
T100	80.82 ± 1.86^{a,b}	77.85 ± 0.14^{b}	74.64 ± 0.35^{b}	60.28 ± 0.62^{a}
Trolox	91.11 ± 0.73^{a}	98.46 ± 0.06^{a}	99.01 ± 0.04^{a}	–
Ascorbic acid	85.18 ± 0.32^{a}	95.66 ± 0.26^{a}	97.85 ± 0.11^{a}	–
BHT	80.42 ± 1.24^{a}	97.84 ± 0.15^{a}	98.52 ± 0.32^{a}	–

Data shown are mean ± SE ($n = 3$). Means with the same letters are not significantly different

Proximate composition of the formulated diet

The dried *T. chuii* (T100) contained 36.86 ± 1.93% protein, 11.74 ± 0.49% lipid and 20.83 ± 0.65% carbohydrate (Table 2). For the formulated diet, the highest protein and lipid content was found in T50 with a total of 57.30 ± 1.30 and 13.98 ± 0.67%, respectively. Meanwhile, the highest carbohydrate content was shown by T100 and T75 which contained 20.83 ± 0.65 and 21.48 ± 0.73%, respectively.

The protein content (53–57%) of the treatment feed CF, T25 and T50 in this study is well within the range recommended for growth and survival of shrimp postlarvae which should be greater than 44% (Biedenbach et al. 1989). The total lipid content and carbohydrate content of the formulated diet with *T. chuii* were also within the ranges recommended for shrimp. According to Bautista (1986), a protein content between 40 and 50%, with 20% carbohydrate and 5–10% lipid, provided the best growth and survival to the shrimp, *Penaeus monodon*. Thus, the addition of *T. chuii* to the feed for growing shrimp PL could provide adequate amounts of nutrients needed, and be a good source of the antioxidant.

Effect of *T. chuii* as feed additives on physical and chemical parameter of water from culture tank

In shrimp hatchery, increased TAN and NO_2-N level with culture time are important factors that affect the survival, health and growth performance of shrimp postlarvae (Chin and Chen 1987). In this study, the inclusion of *T. chuii* into the feed leads to the improvement and maintenance of good water quality throughout the PL culture. Significantly lower TAN (Fig. 1a), NO_2-N (Fig. 1b) and PO_4-P (Fig. 1c) were found in water samples from the tank of PL supplemented with *T. chuii* compared to control during the final day of culture. This result was in accordance with Guedes and Malcata (2012), who showed that microalgae could stabilize and improve the water quality of culture. Although freeze-dried microalgae were used in this study, there are still some viable microalgal cells present even if it is at a very low level (Day 2007). Thus, there is a possibility of microalgal cells from leftover feed to start growing again in the tank. Throughout the shrimp culture period, a slow increase in green water concentration in the tank was observed. It was contributed by the increase in microalgae concentration which absorbs the nutrients available, especially in tank T50, T75 and T100. According to Chen et al. (2012), *T. chuii* showed the highest TAN uptake which led to lower nutrient toward the end of culture compared to the other microalgae species studied. Thus, minimal or no water exchange can be considered on PL fed with *T. chuii* inclusion which is beneficial due to the cost and labor usually needed for frequent water exchange in aquaculture to maintain a good water quality (Thompson et al. 2002). Meanwhile, there was no significant difference in the physical parameters (temperature, salinity, light and pH) of the water for all treatments during the whole experimental period (Table 3).

Growth, survival and oxidative stress biomarkers of shrimp PL supplemented with high antioxidant *T. chuii*

Antioxidant in microalgae is mainly contributed by its bioactive compounds such as carotenoid, amino acids, polyunsaturated fatty acids and others. Even though the nutritional and antioxidant composition of feed will be high with higher concentration of microalgae, there is only a certain level of microalgae inclusion that can be tolerated by shrimp. This is because, increasing microalgae level led to the increase in fiber content, as well as lowering the digestibility and palatability of the feed (Sudaryono et al. 1996). In the feeding trials with the *L.*

Table 2 Proximate composition (% dry weight) of formulated diets

Formulated feed	Total protein (% dw)	Total lipid (% dw)	Total carbohydrate (% dw)
CF	53.47 ± 0.75^{b}	11.95 ± 0.40^{b}	9.10 ± 0.60^{d}
T25	54.64 ± 1.09^{b}	13.48 ± 0.54^{a}	15.53 ± 0.72^{c}
T50	57.30 ± 1.30^{a}	13.98 ± 0.67^{a}	18.09 ± 0.72^{b}
T75	43.63 ± 2.32^{c}	11.89 ± 0.44^{b}	21.48 ± 0.73^{a}
T100	36.86 ± 1.93^{d}	11.74 ± 0.49^{b}	20.83 ± 0.65^{a}

Data shown are mean ± SE (n = 3). Means with the same letters are not significantly different ($P > 0.05$)

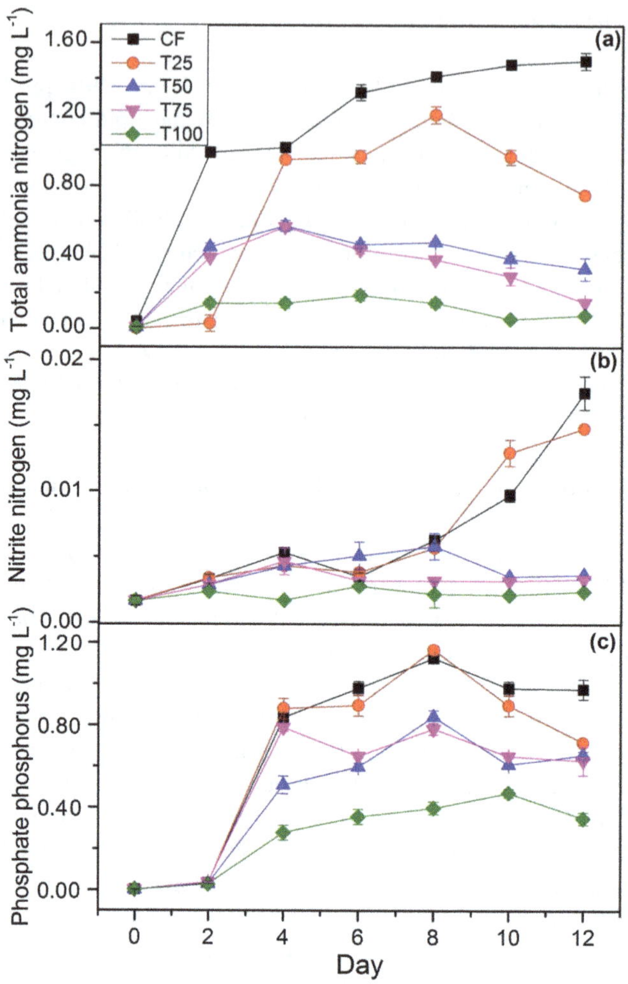

Fig. 1 **a** Total ammonia nitrogen, **b** nitrite nitrogen, **c** phosphate phosphorus concentrations of water in tanks with different formulated diet treatments compared to the control. Data shown are mean ± SE ($n = 3$)

Table 3 Temperature, dissolved oxygen, pH and salinity in control and treated tanks

Treatment	Parameter			
	Temperature (°C)	DO (mg L^{-1})	pH	Salinity (ppt)
CF	26.99 ± 0.16[a]	5.78 ± 0.10[a]	8.32 ± 0.03[a]	30.25 ± 0.07[a]
T25	27.10 ± 0.07[a]	5.73 ± 0.12[a]	8.28 ± 0.06[a]	30.31 ± 0.14[a]
T50	27.12 ± 0.17[a]	5.54 ± 0.14[a]	8.30 ± 0.05[a]	30.23 ± 0.08[a]
T75	27.05 ± 0.15[a]	5.83 ± 0.11[a]	8.31 ± 0.04[a]	30.21 ± 0.09[a]
T100	27.12 ± 0.26[a]	5.95 ± 0.22[a]	8.31 ± 0.07[a]	30.34 ± 0.10[a]

Data shown are mean ± SE ($n = 3$) for the whole 12 days of postlarvae culture. Means with the same letters are not significantly different ($P > 0.05$)

vannamei postlarvae, the inclusion of *T. chuii* into the diet has shown a significantly higher survival compared to PL fed with commercial feed (control) after 12 days of culture period (Fig. 2). In Malaysia, the survival rate of PL at commercial hatchery is usually between 40 and 70%, depending on the system used. In this study, the highest ($P \leq 0.05$) survival of shrimp was achieved in tanks of PL fed with T50 (97.6 ± 1.4%) and followed by T75 (85.6 ± 3.5%). Meanwhile, along with a lower survival rate, PL fed with CF and T25 also showed a lethargic movement and was less active than PL fed with T50, T75 and T100. In terms of SGR, no significant

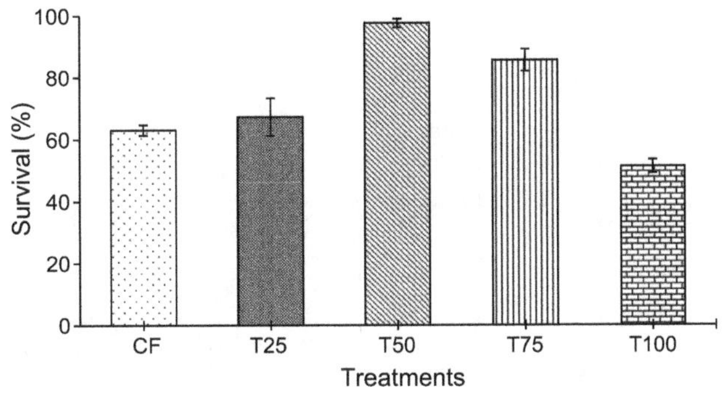

Fig. 2 Survival of postlarvae fed with formulated diets after a culture period of 12 days. Data shown are mean ± SE ($n = 3$)

Table 4 Specific growth rate (SGR) of shrimp postlarvae fed with formulated diets

Treatment	Specific growth rate, SGR
CF	27.92 ± 0.88^a
T25	26.76 ± 0.92^a
T50	26.60 ± 0.70^a
T75	22.16 ± 0.32^a
T100	23.20 ± 1.43^a

Data shown are mean ± SE ($n = 3$). Means with the same letters within column are not significantly different at $P < 0.05$

difference ($P > 0.05$) was observed among all treatments (Table 4). The performance and health of the aquatic organism are directly related to the excessive production of ROS and protection of the cell membrane by antioxidant against the production of free radical (Mourente et al. 2002). In this study, the higher survival of PL supplemented with *T. chuii* can be linked to the hydrogen peroxide content (H_2O_2), lipid peroxidation and electrolyte leakage which act as the biomarkers of the oxidative stress in the shrimp tissue.

Postlarvae fed with the addition of *T. chuii* in their diet showed lower H_2O_2 compared to CF (Fig. 3a). The H_2O_2 content in PL fed with T75 (10.22 ± 0.3 mM g^{-1} FW) and T50 (10.08 ± 0.4 mM g^{-1} FW) significantly decreased compared to CF, T25, and T100. This suggests that the inclusion of *T. chuii* into the PL diet up to 75% *T. chuii* presented a protective effect against oxidative stress in the shrimp under both stressful situations and also during development to protect the cell. This result is in accordance with Amar et al. (2004), who reported that the antioxidant defense in an aquatic organism depends on the nutritional factors. High supplementation of dietary antioxidant caused a reduced need for the endogenous antioxidant enzyme (such as catalase and superoxide dismutase) in scavenging hydrogen peroxide (H_2O_2) and superoxide radical (Lygren et al. 1999). In accordance with the study by Sheikhzadeh et al. (2012) and Da Silva et al. (2015), supplementation of the high astaxanthin *H. pluvialis* into feed also leads to better performance with increased antioxidant activity in juvenile *L. vannamei*.

The degree of cell membrane injury induced by oxidative stress was measured as the relative percentage of total electrolyte leakage from the cells (Fig. 3b). In the present study, the lowest electrolyte leakage was found in PL fed with T50 ($10.8 \pm 0.3\%$) which was significantly lower ($P \leq 0.05$) than CF, T25, and T100. There were no significant differences ($P > 0.05$) in the electrolyte leakage between PL fed with T50 ($10.8 \pm 0.3\%$) and T75 ($12.7 \pm 1.0\%$). It was found that with the inclusion of *T. chuii* to the PL diet, the electrolyte leakage decreased compared to control. However, the addition of *T. chuii* of more than 75% led to the increase in the electrolyte leakage of the PL. In cells, stress-induced electrolyte leakage is caused by the accumulation of ROS such as H_2O_2 and often results in decreased membrane permeability and programmed cell death (Neill et al. 2002; Demidchik et al. 2014). Both animal and plant showed similar mechanism with the ROS-activated K^+ permeable cation channel in the cell membrane, which is a major parameter in regulating the programmed cell death (Demidchik et al. 2014). The addition of *T. chuii* is capable of protecting the cells from cell damage and death by neutralizing the ROS into safer intermediates before it can cause further damages to the cell.

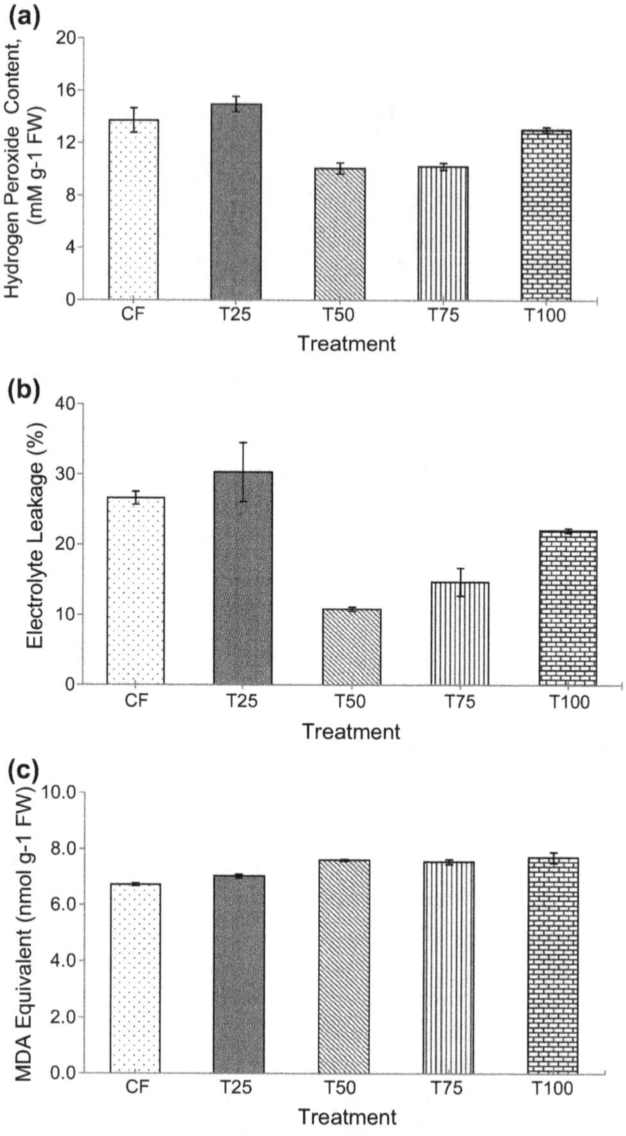

Fig. 3 **a** Hydrogen peroxide content, **b** electrolyte leakage, **c** lipid peroxidation of postlarvae fed with formulated diets after a culture period of 12 days. Data shown are mean ± SE ($n = 3$)

Lipid peroxidation was measured by determination of malondialdehyde (MDA) as an end product of lipid peroxidation. In this study, the lipid peroxidation in the PL did not differ significantly ($P > 0.05$) between all treatments (Fig. 3c). According to Tirmenstein and Reed (1988), during exposure to stress, lipid peroxidation can cause subcellular organelles and biomembrane damage. Although no significant difference was observed in the lipid peroxidation among all treatment, based on the H_2O_2, electrolyte leakage and the survival of the PL, the results suggest that without the enrichment of the diet with the high antioxidant *T. chuii*, the PL seem to be more highly susceptible to oxidative stress, which led to cell damage and death. The lower survival of CF and T25 treatment (Fig. 2) is probably due to the high rate of cell damage and death that occurred throughout the culture period. Survived PL in the CF tank may have adapted or had more intact cell membranes by the end of culture period, leading to lower lipid peroxidation in the PL. Ruff et al. (2001) also reported no significant difference in the lipid peroxidation of PL supplemented with dietary tocopherol and ascorbic acid to the lipid peroxidation. Thus, the present study showed that the inclusion of *T. chuii* led to better protection against oxidative stress and oxidative damages by improving the cells scavenging effect on ROS, which resulted in higher survival with low H_2O_2 content and electrolyte leakage compared to CF.

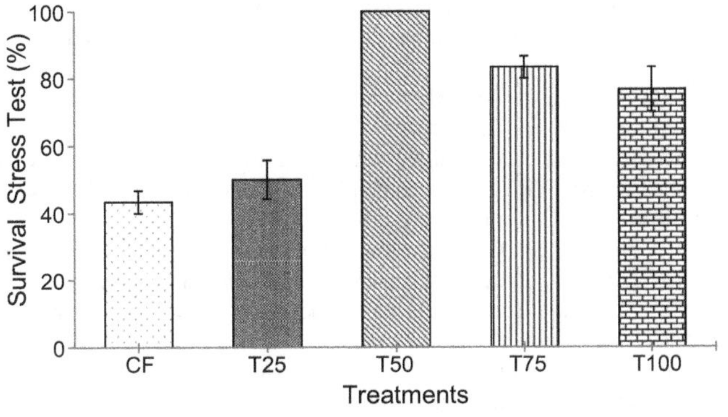

Fig. 4 Survival after salinity stress test in larval rearing tanks after a culture period of 12 days. Data shown are mean ± SE ($n = 3$)

Postlarvae fed with *T. chuii* also showed a higher resistance against exposure to salinity stress (Fig. 4). In aquatic organisms, changes in salinity may cause a variety of physiological stress, energy metabolism, electrolyte equilibrium and enhanced ROS generation which leads to oxidative damage (Choi et al. 2008). After 2 h of exposure to freshwater, PL fed a diet enriched with *T. chuii* showed a significantly higher survival against salinity stress compared to the control. Interestingly, 100% survival was exhibited by PL fed with 50% *T. chuii* (T50) and followed by T75. In fact, all PL fed with T50 were still alive and active even after 4 h of exposure to freshwater. No survival was observed after 4 h of exposure in PL fed with CF and T25.

The present study showed that the addition of *T. chuii* with high antioxidant properties into the PL diet at 50–75% is very beneficial and it enhances the PL defense against excessive oxygen radicals during normal physiological conditions and upon exposure to stress. This is because, under reduced salinity, there is a potential for excessive production of oxygen radicals in shrimp to maintain osmotic stability. The present study showed that the supplementation of *T. chuii* with high antioxidant properties could increase the defense of the PL against oxidative stress and damage, thereby leading to a higher survival rate and a higher tolerance against stress. In addition, supplementation with *T. chuii* also provides adequate nutrient to the PL and improvement in water quality throughout the culture period.

Conclusions

The present study demonstrated that supplementation with 50% *T. chuii* which have high antioxidant properties could improve the survival of the shrimp PL as well as maintain the water quality. In addition, it also enhances the shrimp defense against oxidative stress and damages by improving the cells scavenging effect on ROS, which led to a higher PL survival, with a lower H_2O_2 content, electrolyte leakage and higher tolerance against stress. Therefore, the present findings indicate the potential application of this microalgal species as a valuable natural antioxidant source that can be utilized for *L. vannamei* postlarvae.

Acknowledgements This study was supported by the Ministry of Higher Education, Malaysia, through the Fundamental Research Grant Scheme (FRGS) Project no. FRGS/1/2015/WAB01/UMT/03/6.

References

AbouL-Enein AM, El-Baz FK, El-Baroty GS, Youssef AM, Abd El-Baky HH (2003) Antioxidant activity of algal extracts on lipid peroxidation. J Med Sci 3:87–98

Amar EC, Kiron V, Satoh S, Watanabe T (2004) Enhancement of innate immunity in rainbow trout (*Oncorhynchus mykiss Walbaum*) associated with dietary intake of carotenoids from natural products. Fish Shellfish Immun 16:527–537

Bachère E (2000) Shrimp immunity and disease control. Aquaculture 191:3–11

Bautista MN (1986) The response of *Penaeus monodon* juveniles to varying protein/energy ratios in test diets. Aquaculture 53:229–242

Biedenbach JM, Smith LL, Thomsen TK, Lawrence AL (1989) Use of the Nematode *Panagrellus redivivus* as an artemia replacement in a larval penaeid diet. J World Aquacult Soc 20:61–71

Bligh EG, Dyer WJ (1959) A rapid method of total lipid extraction and purification. Can J Biochem Physiol 37:911–917

Carballo-Cárdenas EC, Tuan PM, Janssen M, Wijffels RH (2003) Vitamin E (α-tocopherol) production by the marine microalgae *Dunaliella tertiolecta* and *Tetraselmis suecica* in batch cultivation. Biomol Eng 20:139–147

Castex M, Lemaire P, Wabete N, Chim L (2010) Effect of probiotic *Pediococcus acidilactici* on antioxidant defences and oxidative stress of *Litopenaeus stylirostris* under *Vibrio nigripulchritudo* challenge. Fish Shellfish Immun 28:622–631

Chen SY, Pan LY, Hong MJ, Lee AC (2012) The effects of temperature on the growth of and ammonia uptake by marine microalgae. Bot Stud 53:125–133

Chin TS, Chen JC (1987) Acute toxicity of ammonia to larvae of the tiger prawn, *Penaeus monodon*. Aquaculture 66:247–253

Choi CY, An KW, An MI (2008) Molecular characterization and mRNA expression of glutathione peroxidase and glutathione S-transferase during osmotic stress in olive flounder (*Paralichthys olivaceus*). Comp Biochem Phys A 149:330–337

Da Silva FO, Tramonte VLCG, Parisenti J, Lima-Garcia JF, Maraschin M, da Silva EL (2015) *Litopenaeus vannamei* muscle carotenoids versus astaxanthin: a comparison of antioxidant activity and in vitro protective effects against lipid peroxidation. Food Biosci 9:12–19

Day JG (2007) Cryopreservation of microalgae and cyanobacteria. In: Day JG, Stacey GN (eds) Cryopreservation and freeze-drying protocols, 2nd edn. Humana, Totowa, pp 39–58

Demidchik V, Straltsova D, Medvedev SS, Pozhvanov GA, Sokolik A, Yurin V (2014) Stress-induced electrolyte leakage: the role of K+-permeable channels and involvement in programmed cell death and metabolic adjustment. J Exp Bot 65:1259–1270

Di Giulio RT, Benson WH, Sanders BM, Van Veld PA (1995) Biochemical mechanisms: Metabolism, adaptation and toxicity. Fundamentals of aquatic toxicology, 2nd edn. Taylor and Francis, Washington, pp 523–561

Dubois M, Gilles KA, Hamilton JK, Rebers PA, Smith F (1956) Colorimetric method for determination of sugars and related substances. Anal Chem 28:350–356

Gamboa-Delgado J, Le Vay L (2009) Artemia replacement in co-feeding regimes for mysis and postlarval stages of *Litopenaeus vannamei*: nutritional contribution of inert diets to tissue growth as indicated by natural carbon stable isotopes. Aquaculture 297:128–135

Geetha BV, Navasakthi R, Padmini E (2010) Investigation of antioxidant capacity and phytochemical composition of sun *Chlorella*—an in vitro study. J Aquac Res Development 1:1–7

Guedes AC, Malcata FX (2012) Nutritional value and uses of microalgae in aquaculture. In: Muchlisin Z (ed) Aquaculture. INTECH Open Acess Publisher, pp 59–78

Gyamfi MA, Yonamine M, Aniya Y (1999) Free-radical scavenging action of medicinal herbs from Ghana. Gen Pharmacol-Vasc S 32:661–667

Hajimahmoodi M, Faramarzi M, Mohammadi N, Soltani N, Oveisi M, Nafissi-Varcheh N (2010) Evaluation of antioxidant properties and total phenolic contents of some strains of microalgae. J Appl Phycol 22:43–50

Heath RL, Packer L (1968) Photoperoxidation in isolated chloroplasts. I. Kinetics and stoichiometry of fatty acid peroxidation. Arch Biochem Biophys 125:189–198

Ismaiel MM, El-Ayouty YM, Piercey-Normore M (2016) Role of pH on antioxidants production by *Spirulina* (*Arthrospira*) *platensis*. Braz J Microbiol 47(2):298–304

Jahnke LS (1999) Massive carotenoid accumulation in *Dunaliella bardawil* induced by ultraviolet-A radiation. J Photochem Photobiol B 48:68–74

Ju ZY, Deng DF, Dominy WG, Forster IP (2011) Pigmentation of Pacific white shrimp, *Litopenaeus vannamei*, by dietary Astaxanthin extracted from *Haematococcus pluvialis*. J World Aquac Soc 42:633–644

Ju ZY, Deng DF, Dominy WG (2012) A defatted microalgae (*Haematococcus pluvialis*) meal as a protein ingredient to partially replace fishmeal in diets of Pacific white shrimp (*Litopenaeus vannamei*, Boone, 1931). Aquaculture 354–355:50–55

Khatoon H, Yusoff FM, Banerjee S, Shariff M, Mohamed S (2007) Use of periphytic cyanobacterium and mixed diatoms coated substrate for improving water quality, survival and growth of *Penaeus monodon Fabricius* postlarvae. Aquaculture 271:196–205

Khatoon H, Abdu Rahman N, Banerjee S, Harun N, Suleiman SS, Zakaria NH, Lananan F, Abdul Hamid SH, Endut A (2014) Effects of different salinities and pH on the growth and proximate composition of *Nannochloropsis* sp. and *Tetraselmis* sp. isolated from South China Sea cultured under control and natural condition. Int Biodeterior Biodegradation 95:11–18

Li M, Wu W, Zhou P, Xie F, Zhou Q, Mai K (2014) Comparison effect of dietary astaxanthin and *Haematococcus pluvialis* on growth performance, antioxidant status and immune response of large yellow croaker *Pseudosciaena crocea*. Aquaculture 434:227–232

Li Y, Xiao G, Mangott A, Kent M, Pirozzi I (2015) Nutrient efficacy of microalgae as aquafeed additives for the adult black tiger prawn, *Penaeus monodon*. Aquac Res 47(11):1–11

Lightner DV (2011) Virus diseases of farmed shrimp in the Western Hemisphere (the Americas): a review. J Invertebr Pathol 106:110–130

Liu Y, Wang WN, Wang AL, Wang JM, Sun RY (2007) Effects of dietary vitamin E supplementation on antioxidant enzyme activities in *Litopenaeus vannamei* (Boone, 1931) exposed to acute salinity changes. Aquaculture 265:351–358

Lowry OH, Rosebrough NJ, Farr AL, Randall RJ (1951) Protein measurement with the Folin phenol reagent. J Biol Chem 193:265–275

Lushchak VI (2011) Environmentally induced oxidative stress in aquatic animals. Aquat Toxicol 101:13–30

Lygren B, Hamre K, Waagbo R (1999) Effects of dietary pro- and antioxidants on some protective mechanisms and health parameters in Atlantic salmon. J Aquat Anim Health 11:211–221

Macias-Sancho J, Poersch LH, Bauer W, Romano LA, Wasielesky W, Tesser MB (2014) Fishmeal substitution with Arthrospira (*Spirulina platensis*) in a practical diet for *Litopenaeus vannamei*: effects on growth and immunological parameters. Aquaculture 426–427:120–125

Marsh JB, Weinstein DB (1966) Simple charring method for determination of lipids. J Lipid Res 7(4):574–576

Mourente G, Díaz-Salvago E, Bell J, Tocher D (2002) Increased activities of hepatic antioxidant defence enzymes in juvenile gilthead sea bream (*Sparus aurata* L.) fed dietary oxidised oil: attenuation by dietary vitamin E. Aquaculture 214:343–361

Neill S, Desikan R, Hancock J (2002) Hydrogen peroxide signalling. Curr Opin Plant Biol 5:388–395

Nonwachai T, Purivirojkul W, Limsuwan C, Chuchird N, Velasco M, Dhar AK (2010) Growth, nonspecific immune characteristics, and survival upon challenge with *Vibrio harveyi* in Pacific white shrimp (*Litopenaeus vannamei*) raised on diets containing algal meal. Fish Shellfish Immunol 29:298–304

Parsons TR, Maita Y, Lalli CM (1984) A manual of chemical and biological methods for seawater analysis. Pergamon Press, New York, p 173

Rady MM (2011) Effect of 24-epibrassinolide on growth, yield, antioxidant system and cadmium content of bean (*Phaseolus vulgaris* L.) plants under salinity and cadmium stress. Sci Hortic-Amsterdam 129:232–237

Rao AR, Sarada R, Baskaran V, Ravishankar GA (2006) Antioxidant activity of *Botryococcus braunii* extract elucidated in vitro models. J Agr Food Chem 54:4593–4599

Ricker WE (1990) Growth rates and models. In: Hoar WS, Randal DJ, Brett JR (eds) Fish Physiology, vol VIII. Academic Press, USA, Bio- energetic and Growth, p 230

Robak J, Gryglewski RJ (1988) Flavonoids are scavengers of superoxide anions. Biochem Pharmacol 37:837–841

Ruff N, Lavens P, Huo JZ, Sorgeloos P, Nelis HJ, De Leenheer A (2001) Antioxidant effect of dietary tocopherol and ascorbic acid on growth and survival of *Litopenaeus vannamei* postlarvae. Aquac Int 9:115–126

Sheikhzadeh N, Tayefi-Nasrabadi H, Khani Oushani A, Najafi Enferadi M (2012) Effects of *Haematococcus pluvialis* supplementation on antioxidant system and metabolism in rainbow trout (*Oncorhynchus mykiss*). Fish Physiol Biochem 38:413–419

Shimada K, Fujikawa K, Yahara K, Nakamura T (1992) Antioxidative properties of xanthan on the autoxidation of soybean oil in cyclodextrin emulsion. J Agr Food Chem 40:945–948

Sudaryono A, Tsvetnenko E, Evans LH (1996) Digestibility studies on fisheries by product based diets for *Penaeus monodon*. Aquaculture 143:331–340

Thompson FL, Abreu PC, Wasielesky W (2002) Importance of biofilm for water quality and nourishment in intensive shrimp culture. Aquaculture 203:263–278

Tirmenstein MA, Reed DJ (1988) Characterization of glutathione-dependent inhibition of lipid peroxidation of isolated rat liver nuclei. Arch Biochem Biophys 261:1–11

Tompkins J, Deville MM, Day JG, Turner MF (1995) Catalogue of Strains. Titus Wilson and Son Ltd. Kendall. UK, p 166

Velikova V, Yordanov I, Edreva A (2000) Oxidative stress and some antioxidant systems in acid rain-treated bean plants protective role of exogenous polyamines. Plant Sci 151:59–66

Yang JH, Mau JL, Ko PT, Huang LC (2000) Antioxidant properties of fermented soybean broth. Food Chem 71:249–254

11

Peroxiredoxin 6 from cuttlefish (*Sepiella maindroni*): molecular characterisation during development and the response to thermal stress

Weiwei Song · Chunlin Wang · Ronghua Li · Changkao Mu

Abstract The aim of this work was to investigate the function of the peroxiredoxin family member Peroxiredoxin 6 in *Sepiella maindroni* (*Smprx6*) during embryonic development and the response to thermal stress. The 1064 bp full-length coding sequence of *Smprx6* includes an open reading frame of 660 bp encoding a polypeptide of 219 amino acids with a predicted molecular mass of 24.4 kDa. The deduced protein sequence shares significant homology with other reported Prx6 enzymes. Phylogenetic analysis classified SmPrx6 into the 1-Cys-Prx cluster, and quantitative real-time PCR (qPCR) demonstrated constitutive mRNA expression in all tested tissues and during five different embryonic developmental stages, as previously documented in other cephalopods. Thermal stress induced *Smprx6* expression, which peaked at 6 h then returned to control levels. Recombinant SmPrx6 was active at different concentrations and was thiol-dependant. The results suggest that SmPrx6 plays a major role in the development of *Sepiella maindroni* and its response to oxidative and thermal stress.

Keywords *Sepiella maindroni* · Peroxiredoxin 6 (Prx6) · Embryonic development · Thermal stress · Phylogenetic analysis · qPCR

Introduction

Cephalopods have high rates of growth and food conversion, which for aquaculture results in a short culture cycle, a high ratio of production to biomass, and high cost-effectiveness. However, at present, only small-scale culturing of a limited number of species is possible (Castellanos-Martínez and Gestal 2013). Furthermore, most reported culturing procedures rely on captured wild individuals (Vidal et al. 2014). Our research group previously developed a system for breeding *Sepiella maindroni*, a commercially important fishery species accounting for a large proportion of cuttlefish aquaculture in China, by creating artificial culture cycles capable of rearing at least six generations in captivity (Wei-Wei and Chun-Lin 2009). However, despite the

W. Song · C. Wang (✉) · R. Li · C. Mu
Key Laboratory of the Ministry of Education for Applied Marine Biotechnology, School of Marine, Ningbo University, Fenghua Rd. 818, Ningbo 315211, Zhejiang, People's Republic of China
e-mail: wangchunlinnbdx@163.com

W. Song · C. Wang · R. Li · C. Mu
Collaborative Innovation Center for Zhejiang Marine High-Efficiency and Healthy Aquaculture, Ningbo University, Ningbo, People's Republic of China

development of an artificial rearing method, *S. maindroni* suffers from the predicament of a high summer mortality. Cephalopods are susceptible to environmental changes, especially water temperature fluctuations, but if they survive such changes, they readily adapt to new environmental conditions (Gestal and Castellanos-Martinez 2015). Thermal fluctuations in a marine environment can induce physiological changes through oxidative stress-related mechanisms that generate reactive oxygen species (ROS) to modulate gene expression (Bogdan et al. 2000).

Although low concentrations of ROS are essential for normal physiological signalling (Poynton and Hampton 2014), excess ROS and oxidative stress can lead to cell damage, mutations, and even death (Zheng et al. 2010). Antioxidants are known to play a critical role in regulating ROS levels by detoxifying deleterious oxygen intermediates without disrupting normal cellular functions (Wood et al. 2003b). Peroxiredoxins (Prx) are a ubiquitous family of antioxidant enzymes that prevent oxidative damage (Robinson et al. 2010) and serve as effective medicinal agents for treating ROS-related diseases (Poynton and Hampton 2014). The Prx family has six isoforms classified into two sub-groups, including five 2-Cys-Prxs (Prx1 to Prx5) and one 1-Cys Prx (Prx6). The 2-Cys Prxs are divided into 'typical' and 'atypical' classes, depending on whether the conserved cysteine residues form intermolecular or intramolecular disulfide bridges, respectively (Wood et al. 2003b).

Prx6 is a bifunctional enzyme possessing both peroxidase and phospholipase A2 (PLA2) activities (Lee et al. 2015). Prx6 has been studied in various aquatic organisms including rock cod (*Trematomus bernacchii*) (Tolomeo et al. 2016), Japanese eel (*Anguilla japonica*) (Priyathilaka et al. 2016), turbot (*Scophthalmus maximus*) (Zheng et al. 2010), Chinese mitten crab (*Eriocheir sinensis*) (Mu et al. 2009), bivalve (*Laternula elliptica*) (David et al. 2007), disk abalone (*Haliotis discus discus*) (Nikapitiya et al. 2009), oyster (David et al. 2007), black tiger shrimp (*Penaeus monodon*) (Qiu et al. 2010) and Chinese shrimp (*Fenneropenaeus chinensis*) (Zhang et al. 2007). Prx6 has a single conserved catalytic cysteine residue in the N-terminal catalytic motif (PVCTTE) that is responsible for the peroxidase activity (Manevich and Fisher 2005). Evidence suggests Prx6 is distributed in all major tissues that are sensitive to oxidative stress (Manevich and Fisher 2005), and its expression is induced by stimuli such as bacteria (Zheng et al. 2010), viruses (Nikapitiya et al. 2009), pathogen-associated molecular patterns (PAMPs) (De Zoysa et al. 2012), chemicals (Wang et al. 2008), thermal stress (Park et al. 2008; Tolomeo et al. 2016) and environmental pollutants (David et al. 2007).

Antioxidant enzymes operating within the oxidant system of *S. maindroni* are significantly affected by thermal stimulation, particularly those related to scavenging ROS, as demonstrated by cDNA microarray studies based on an EST library (in review). A better understanding of the mechanisms underpinning the responses to oxidative stress could help to improve squid rearing and cephalopod aquaculture. In this study, we identified a peroxiredoxin 6 homolog in *S. maindroni* (*Smprx6*), measured *Smprx6* mRNA levels in the liver during oxidative and thermal stress, and produced and characterised the recombinant protein.

Materials and methods

Animals, embryos, acute thermal temperature challenge and sample collection

S. maindroni weighing 158.2 ± 2.8 g were collected in Ningbo, China, and acclimated in a blue plastic cylinder tank (diameter 1 m, depth 0.6 m) with a salinity of 28–30% at 22 ± 2 °C for 1 week before processing. Cuttlefish (120; length = 3.1 ± 0.5 cm) were randomly divided into six groups containing 20 individuals for the thermal stress experiment. After acclimation for 2 days, three groups were exposed to thermal stress conditions by gradually increasing the temperature by 1 °C per h to 32 °C, which is the thermal limit for survival based on breeding experiments (Jiang et al. 2010). Additionally, three control groups were maintained at 22 °C, which is the normal temperature for rearing in seawater. After treatment, five individuals were randomly sampled at 0, 0.5, 2, 6, 12, 24, 48 and 72 h, and the liver was collected and preserved in RNAlater (TaKaRa, Tokyo, Japan) for RNA extraction. Eight tissues (muscle, ink sac, liver, pancreas, ovary, testis, stomach, and gill) were collected from *S. maindroni* adults, and total RNA was extracted using RNAiso Plus (TaKaRa, Tokyo, Japan) according to the manufacturer's protocol.

Several clutches of fertilised eggs were collected and acclimated at 20 ± 1 °C. Embryonic development of *S. maindroni* can be divided into 12 stages (Jiang et al. 2010), with the early stages combined into five major developmental phases: (I) the zygote stage; (II) the pre-embryonic stage, which occurs over an extremely short period and includes cleavage, blastula, and gastrula stages; (III) the organogenesis stage, when the main body parts and organs such as the wrist, eye, heart and ink sac are formed; (IV) the morphological integrity stage, which corresponds to further ongoing development of bodily structures, with the yolk sac still present; (V) the pre-hatching stage, during which the yolk sac disappears in preparation for hatching. Embryos were collected at these five stages and used for RNA extraction.

All aspects of the experiment were conducted according to the guidelines provided by the Ethical Committee of Experimental Animal Care at Ningbo University (Ningbo, Zhejiang, China).

Identification and cloning of the full-length cDNA of SmPrx6

A squid cDNA library was constructed from the ink sac using the Creator Smart cDNA Library Construction Kit (TaKaRa, Tokyo, Japan). cDNA clones were selected randomly from the cDNA library, and single-pass sequenced using a random T3 primer on an ABI 3730 Genetic Analyzer (Applied Biosystems, Foster City, USA). A large-scale expressed sequence tag (EST) sequencing project for *S. maindroni* was initiated to identify and functionally annotate as many unique transcripts as possible. BLAST searches were performed to compare processed cDNA sequences against ESTs and genes available in the GenBank database (http://www.ncbi.nlm.nih.gov/blast) (Wei-Wei et al. 2012). Intriguingly, EST GT618012 (Contig 485; length = 982 bp) was homologous to Prx6 in *Rattus norvegicus* (RNO17295).

Amplification of flanking sequences was carried out using a 5′- and 3′-Full RACE kit (TaKaRa, Tokyo, Japan) to obtain the complete sequence of EST GT618028 following the manufacturer's instructions. Purified genes were cloned into the pMD-18T vector (Takara, Tokyo, Japan) and transformed into *Escherichia coli* Top10F-competent cells. Finally, M13-47 and RV-M universal primers (Table 1) were used to screen for three positive recombinants that were subsequently sequenced.

Sequence analysis, multiple sequence alignment and phylogenetic analysis

The sequence similarity of nucleotide and protein sequences was compared using BLAST (http://www.ncbi.nlm.gov/blast). Amino acid sequences were deduced by the Expert Protein Analysis System (http://www.expasy.org). Protein motif features were predicted by the Simple Modular Architecture Research Tool (http://smart.emblheidelberg.de/) (Schultz et al. 1998). Signal peptide prediction was performed by SignalP 4.1 (http://www.cbs.dtu.dk/services/SignalP). Multiple sequence alignment of *Smprx6* was performed using ClustalW (http://www.ebi.ac.uk/clustalw/). The neighbour-joining method was used to perform multiple sequence alignment of *Smprx6* and its homologs in other species and construct a phylogenetic network with

Table 1 Oligonucleotide primers used in this experiment

Primer	Sequence(5′-3′)	Sequence information
P1 (reverse)	AACTGCACGCGCGGTTTTCA	*Sm*Prx6-specific primer
P2 (forward)	GTAAGAATGCCAGTCGTGTT	*Sm*Prx6-specific primer
P3 (reverse)	CATGGCTACATGCTGACAGCCTA	Universal primer
P4 (forward)	TACCGTCGTTCCACTAGTGATTT	Universal primer
P5 (reverse)	ACTGAGCATCCTCTATCCAG	Real-time *Sm*Prx6 primer
P6 (forward)	TATCACCAACCTTCCAGTCG	Real-time *Sm*Prx6 primer
P7 (reverse)	GACTCCTACGTAGGAGACGA	Real-time β-actin primer
P8 (forward)	CGTTGAAGGTCTCGAACATGA	Real-time β-actin primer
P7 (reverse)	CATATGGCTAACCTTGGAGATATTTTCCCA	Recombinant primer (for expression)
P8 (forward)	CTCGAGTTAGTGGTGGTGGTGGTGGTGTTCTGGA TGGGGTGTCATACGA	Recombinant primer (for expression)
T3	AATTAACCCTCACTAAAGGG	Vector primer
T7	GTAATACGACTCACTATAGGGC	Vector primer
M13-47	CGCCAGGGTT TTCCCAGTCACGAC	Vector primer
RV-M	GAGCGGATAACAATTTCACACAGG	Vector primer

MEGA 6.0. Tertiary structures of *Smprx6* were predicted using the SWISSMODEL prediction algorithm (http://swissmodel.expasy.org).

Transcriptional analysis of SmPrx6 by quantitative real-time RT-PCR

Quantitative real-time RT-PCR was performed using an ABI 7300 (Applied Biosystems, Foster City, USA). Reactions were carried out in a final volume of 25 μL containing 12.5 μL of SYBR Green Real-time PCR Master Mix (TaKaRa, Tokyo, Japan), 9.5 μL of water, 0.5 μL of each primer (P5 and P6 for SmPrx6, and P7 and P8 for β-actin; see Table 1), and 2 μL of cDNA. Reactions were incubated at 95 °C for 3 s, followed by 40 cycles at 95 °C for 5 s, 56 °C for 5 s, and 72 °C for 20 s, and a final extension was performed at 72 °C for 1 min. Reactions were performed in a 96-well plate, each sample was repeated in triplicate, and an internal control gene (β-actin) was included. Dissociation curve analysis of amplification products was performed at the end of each reaction to confirm that only one PCR product was amplified and detected. PCR data were analysed using the ABI 7300 SDS software (Applied Biosystems, Foster City, USA). To maintain consistency, the baseline was set automatically by the software. The comparative C_t method was used to analyse the relative expression levels of *Smprx6*. C_t values for the amplified target products and the internal control (β-actin) were determined for each sample, and the difference in C_t for the target and internal control was calculated to normalise the differences in template and the efficiency of RT-PCR. Liver from squid in the control group was used as the reference sample for calibration. ΔC_t values for each sample were subtracted from the ΔC_t of the calibrator to generate $\Delta\Delta C_t$ values. The expression level of *Smprx6* was calculated using the $2^{-\Delta\Delta Ct}$ method, and values represent the n-fold difference compared with the calibrator (Livak and Schmittgen 2001).

Expression of recombinant SmPrx6

A PCR fragment encoding the mature SmPrx6 polypeptide was amplified with gene-specific primers P7 and P8 (Table 1) and cloned into the pMD18-T simple vector (TaKaRa, Tokyo, Japan), digested with restriction enzymes *Nde*I and *Xho*I (New England Biolabs, MA, USA), then subcloned into the *Nde*I and *Xho*I sites of the expression vector pET-21a(+) (Novagen, Shanghai, China). The recombinant plasmid was transformed into *E. coli* Origami (DE3) (Novagen, Shanghai, China) and subjected to DNA sequencing. The parent vector without the inserted fragment was used as a negative control. After sequencing to confirm in-frame insertion, positive transformants and the negative control were incubated in LB medium containing 50 μg/mL ampicillin at 37 °C with shaking at 220 rpm. When the culture reached an OD_{600} of 0.5–0.7, IPTG was added to a final concentration of 1 mM and culturing continued for an additional 4 h under the same conditions.

Cells were harvested by centrifugation at 8000×*g* for 2 min and suspended in 50 mM Tris containing 5 mM EDTA, 50 mM NaCl, and 5% glycerol (pH 7.9). After sonication for 60 s on ice, the rSm*Prx6* and negative control samples were purified using a HisTrap Chelating Column (Amersham Biosciences, Stockholm, Sweden) according to the manufacturer's instructions. Purified recombinant proteins were analysed by 15% SDS-PAGE to check the molecular mass and purity. Assaying of the antioxidant activity was performed using the Total Antioxidant Capacity Assay Kit (Jiancheng Bioengineering Institute, Nanjing, China) using the 2,2′-azino-bis(3-ethyl-benzothiazoline-6-sulphonic acid) ABTS method (Miller et al. 1993). Peroxidase activity was measured based on the change in absorbance at 420 nm that accompanies the breakdown of H_2O_2. One unit of activity was defined as the amount of enzyme catalysing the breakdown of 1 mg of substrate by 1 g of fresh tissue in the reaction system at 37 °C. Peroxidase activity was calculated using the formula supplied with the Peroxidase assay kit (Jiancheng Bioengineering Institute, Nanjing, China).

Statistical analysis

All data were expressed as the mean ± standard deviation ($n = 5$), and mean values were subjected to one-way analysis of variance (ANOVA). SPSS20 software was used to determine statistical significance ($p < 0.05$), and Origin 8.0 was used to prepare figures.

Results

Cloning and analysis of *SmPrx6*

The complete cDNA sequence of *Smprx6* was obtained by overlapping EST GT618012 with amplified fragments. The sequence has been deposited in Genbank under accession number HQ662844. The full-length mRNA of *Smprx6* is 1064 bp and includes an open reading frame of 660 bp encoding a polypeptide of 219 amino acids with a predicted molecular mass of 24.4 kDa and a theoretical isoelectric point of 6.17 (Fig. 1). The deduced amino acid sequence of *Smprx6* is shown in Fig. 1. The protein includes a typical Prx domain (from Leu4 to Ser160) and an alkyl hydroperoxide reductase (AhpC) domain (from Leu4 to Ser140) as identified by the SMART program. A signal peptide was not predicted by SignalP.

```
   1 GGGGGGTTCACATGTCAAAACTTTAAAAGGCGGACATGCCGATTGTCACTTCATTCTGTT
   1                                             M  A  N  L  G  D
  61 TCTTTTTCCCCAGTCGTCAGATTGTAAATTCGATTGCAAGGATGGCTAACCTTGGAGAT
   7 I  F  P  N  F  D  A  D  T  T  A  G  H  F  N  F  H  E  F  I
 121 ATTTTCCCAAACTTTGATGCTGACACCACAGCGGGCCATTTCAACTTTCATGAATTCATC
  27 N  N  S  W  A  I  L  F  S  H  P  A  D  Y  T  P  V  C  T  T
 181 AATAATTCTTGGGCCATCCTCTTCTCTCATCCTGCTGATTACACACCTGTCTGCACAACG
  47 E  L  T  R  V  V  Q  L  I  P  E  F  K  K  R  G  V  K  L  C
 241 GAGCTGACGAGAGTTGTCCAGTTGATTCCAGAATTCAAGAAAAGGGGAGTGAAACTGTGT
  67 A  L  S  C  D  D  V  E  S  H  H  G  W  I  K  D  I  I  A  Y
 301 GCCTTGTCATGTGATGATGTTGAAAGTCATCATGGTTGGATAAAAGATATTATTGCTTAC
  87 G  K  L  A  E  F  P  Y  P  I  I  S  D  K  T  R  K  L  A  V
 361 GGCAAACTTGCTGAGTTTCCATATCCAATCATCTCTGACAAAACCCGAAAGCTCGCAGTC
 107 S  L  G  M  V  D  P  D  E  K  D  L  A  G  M  P  L  T  A  R
 421 AGTCTGGGTATGGTGGATCCTGATGAAAAGGACCTGGCTGGAATGCCCCTAACTGCACGC
 127 A  V  F  I  I  G  P  D  K  K  L  K  L  S  I  L  Y  P  A  T
 481 GCGGTTTTCATCATTGGCCCTGACAAGAAATTAAAACTGAGCATCCTCTATCCAGCAACC
 147 T  G  R  N  F  D  E  I  L  R  V  I  D  S  L  Q  L  T  A  K
 541 ACTGGGAGAAATTTTGATGAAATTTTGAGAGTGATTGATTCCCTCCAATTAACGGCTAAG
 167 H  K  V  A  T  P  V  D  W  K  V  G  D  K  C  M  V  V  P  S
 601 CACAAGGTGGCCACTCCTGTCGACTGGAAGGTTGGTGATAAATGCATGGTTGTCCCCTCC
 187 V  K  V  E  D  E  K  E  I  F  P  K  G  V  E  T  I  P  V  P
 661 GTCAAAGTTGAAGACGAAAAAGAAATCTTCCCCAAGGGGGTTGAAACAATCCCAGTGCCT
 207 S  G  K  S  Y  L  R  M  T  P  H  P  E  *
 721 TCAGGAAAATCCTATCTTCGTATGACACCCCATCCAGAATAAAGAAATGCAGAAAGGGCA
 781 GGACTGCTGACCTTTTTTGCTTTTGACCTCTTTGAGATAAATTCAAAGGTGGGGGGGCAG
 841 GGGGCTACTCTAAGAACAAAATACTGCTGTAATTAATTGTAAGAATGCCAGTCGTGTTTG
 901 AAATCTATTTTTTTTTTCTACTCATAACGAATTGTTTAAAATCTGATTGTATAAATCATT
 961 TTGAAAGTAAAAACAGTTTTGTATTTGAGTCGTTCTACCGACTGTAATTTGGATGAAGGG
1021 ATGAATTTGATTTCAATAAAGATTTCTAAAGGAAAAAAAAAAAA
```

Fig. 1 Nucleotide and deduced amino acid sequences of SmPrx6. The *asterisk* (*) indicates the stop codon. The polyadenylation signal is shown in ***bold italics***. The active cysteine residue is in **bold** and is underlined. The catalytic motif is *shaded*. The conserved positively charged His 36 and Arg 126 are enclosed by *solid lines*

Multiple sequence alignment and phylogenetic analysis

The deduced amino acid sequence of *Smprx6* shares significant homology with other reported Prx6 proteins, with 71% identity with *Saccostrea glomerata* (ACX50969), and 70.9% identify with *Haliotis discus discus* (ABO26614). Multiple alignments revealed that Cys44 is conserved in all analysed Prx VI sequences. Moreover, the predicted SmPrdx6 protein has a unique cysteine residue (C44) in the peroxidase motif (42PVCTTE47) and a catalytic serine residue (S29) in the lipase motif (27NNSWA31) at the N-terminus (Fig. 2). Both of them are essential amino acid residues in the active sites that mediate peroxidase and PLA2 activity, respectively.

A phylogenetic tree was constructed based on Prx amino acid sequences (Fig. 3). Three distinct groups (1-Cys, 2-Cys, and an atypical 2-Cys subgroup) formed distinct clusters, with all Prx5 proteins part of the atypical 2-Cys subgroup, and Prx1, Prx2, Prx3, and Prx4 clustered in the 2-Cys subgroup. The *Smprx6* sequence and other Prx6 sequences from invertebrates were placed in the 1-Cys subgroup.

Expression of *Smprx6* in different tissues

To determine tissue-specific mRNA expression profiles of *Smprx6*, quantitative RT-PCR was carried out using various tissues from unchallenged (control) cuttlefish. Expression levels were normalised against β-actin mRNA levels which were tested in different tissues and following treatments to provide an internal control. The results revealed no significant differences in cycle threshold (C_t) values for β-actin in different experiments (Fig. 4; $p > 0.05$). The expression level of *Smprx6* in the ink sac was defined as 1, and expression was 460.71-fold higher in the pancreas, followed by liver (148.76-fold), gill (90.15-fold), ovary (26.6-fold), testis (11.53-fold), muscle (7.10-fold), and stomach (1.01-fold).

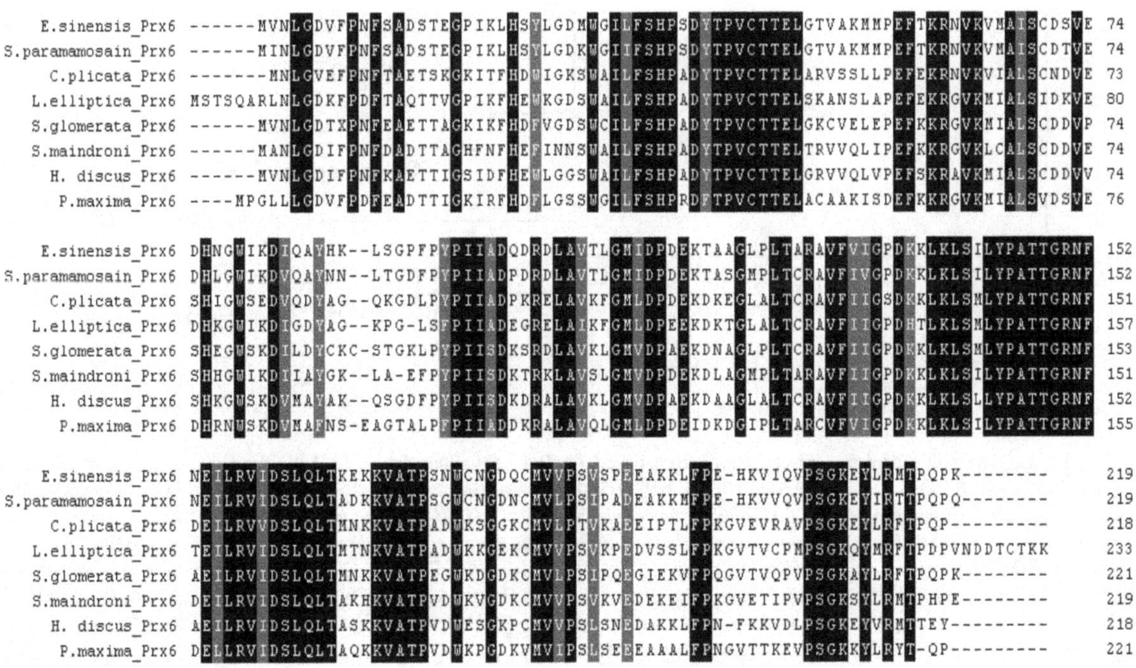

Fig. 2 Multiple alignments of the deduced amino acid sequence of SmPrx6 with other Prx6 proteins deposited in GenBank. *Black shading* indicates conserved amino acid residues. *Gaps* are indicated by *dashes* to improve the alignment. Abbreviations are as follows ACX50969, *Saccostrea glomerata*_Prx6; ABO26614, *Haliotis discus discus*_Prx6; ACF35639, *Eriocheir sinensis*_Prx6; ADN06076, *Cristaria plicata*_Prx6; ACJ53746, *Scylla paramamosain*_Prx6; ADJ57694, *Psetta maxima*_Prx6; ACE76885, *Laternula elliptica*_Prx6. Prx6 from *S. maindroni* is designated as *Smprx6*

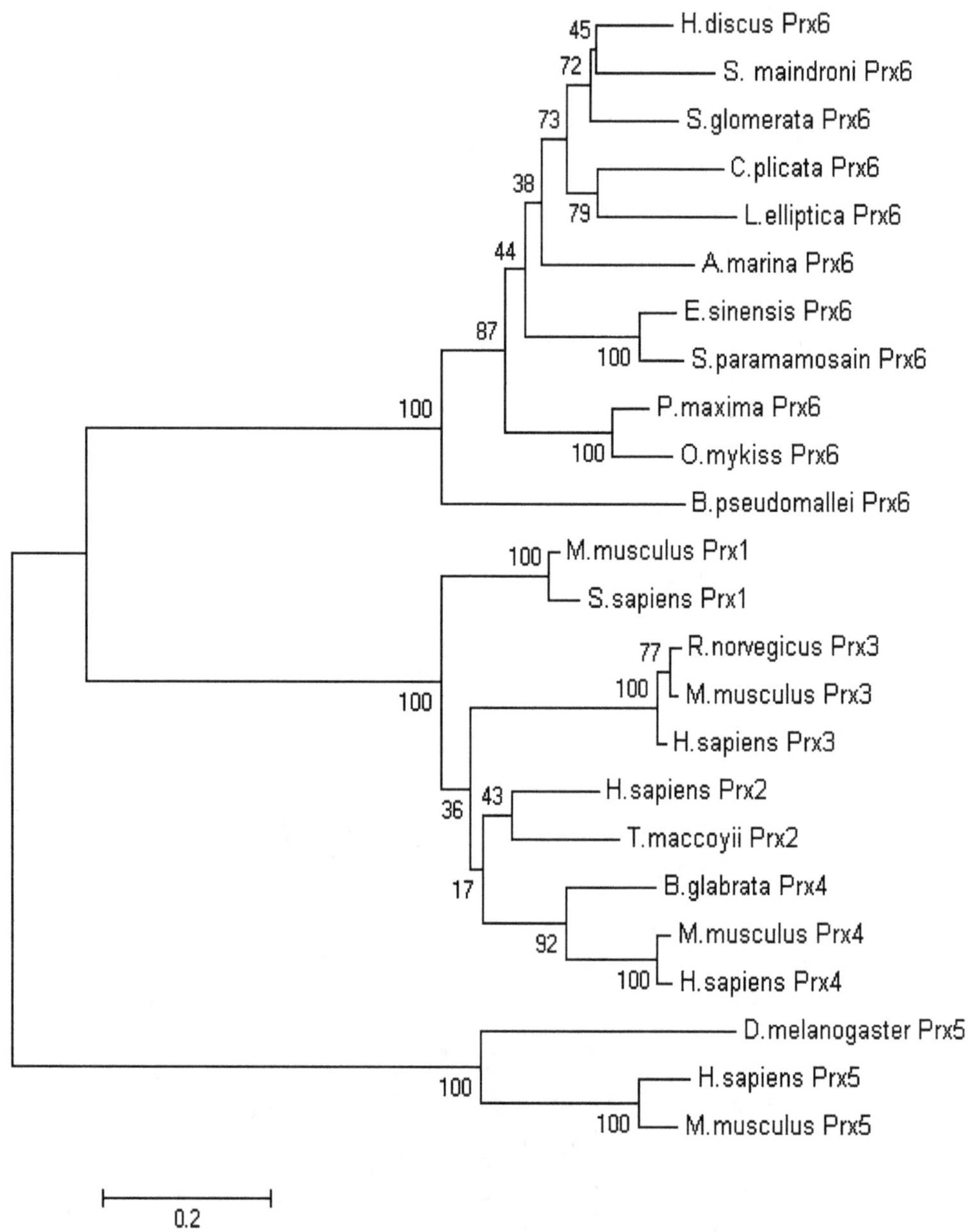

Fig. 3 Consensus neighbour-joining phylogenetic tree based on Prx sequences from the following species: CAM16508, *Mus musculus*_Prx1; NP_859048, *Homo sapiens*_Prx1; NP_005800, *Homo sapiens*_Prx2; ABW88997, *Thunnus maccoyii*_Prx2; AAH60567, *Rattus norvegicus*_Prx3; AAH05626, *Mus musculus*_Prx3; CAG29340, *Homo sapiens*_Prx3; NP_058044, *Mus musculus*_Prx4; ACI42880, *Biomphalaria glabrata*_Prx4; CAG46506, *Homo sapiens*_Prx4; AAF55497, *Drosophila melanogaster*_Prx5; CAG33484, *Homo sapiens*_Prx5; AAG13450, *Mus musculus*_Prx6; ACX50969, *Saccostrea glomerata*_Prx6; ABO26614, *Haliotis discus*_Prx6; ACF35639, *Eriocheir sinensis*_Prx6; ADN06076, *Cristaria plicata*_Prx6; AAY96294, *Arenicola marina*_Prx6; ACJ53746, *Scylla paramamosain*_Prx6; ADJ57694, *Psetta maxima*_Prx6; NP_001158604, *Oncorhynchus mykiss*_Prx6; ACE76885, *Laternula elliptica*_Prx6

Expression of *Smprx6* during embryonic development

Smprx6 mRNA levels in embryos were measured by quantitative RT-PCR and transcripts were detected in all stages of development, with maximum expression during the pre-embryonic stage (1369.11-fold), but expression then decreased markedly during the organogenesis stage (15.2-fold), and only trace amounts were detected in the pre-hatching stage ($p < 0.05$; Fig. 5).

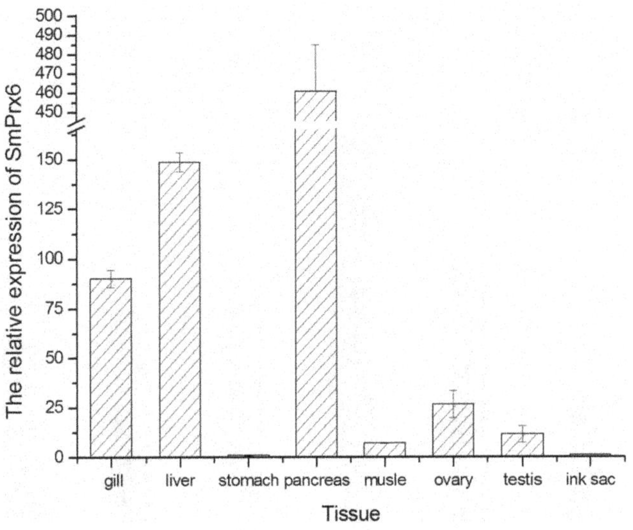

Fig. 4 Tissue distribution of *Smprx6* transcripts measured by SYBR Green RT-PCR. Muscle, ink sac, liver, pancreas, ovary, testis, gastric and gill were collected from three individual cuttlefish. *Vertical bars* represent the mean ± SD ($p < 0.05$)

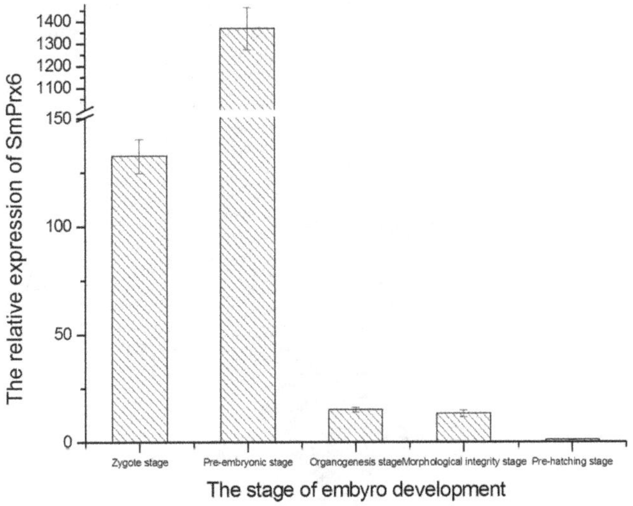

Fig. 5 *Smprx6* mRNA expression in embryos measured by SYBR Green RT-PCR. The β-actin gene was used as an internal control for all samples. *Vertical bars* represent the mean ± SD, and *bars* with different *letters* are significantly different ($p < 0.05$)

Expression of SmPrx6 in liver after thermal treatment

To investigate *Smprx6* expression following an external temperature stimulus, *S. maindroni* were exposed to 32 °C for 0.5, 2, 6, 12, 24, 48 and 72 h, while controls were maintained at 22 °C. Expression of *Smprx6* was assayed by quantitative real-time RT-PCR with β-actin as an internal control. At 1 h after thermal stimulation, *Smprx6* mRNA levels were slightly up-regulated, but the difference from the control group was not significant ($p > 0.05$). *Smprx6* mRNA levels then increased rapidly and peaked at 6 h, representing an increase of 3.72-fold ($p < 0.05$), before declining and eventually returning to control levels (Fig. 6).

Expression, purification and characterisation of recombinant SmPrx6

The recombinant plasmid pET-21a–SmPrx6 was transformed into *E. coli* Origami (DE3) as described above, and rSm*Prx6* was expressed and purified under native conditions. SDS-PAGE confirmed the presence of an overexpressed protein in the whole cell lysate following IPTG induction, and the purified protein ran as a

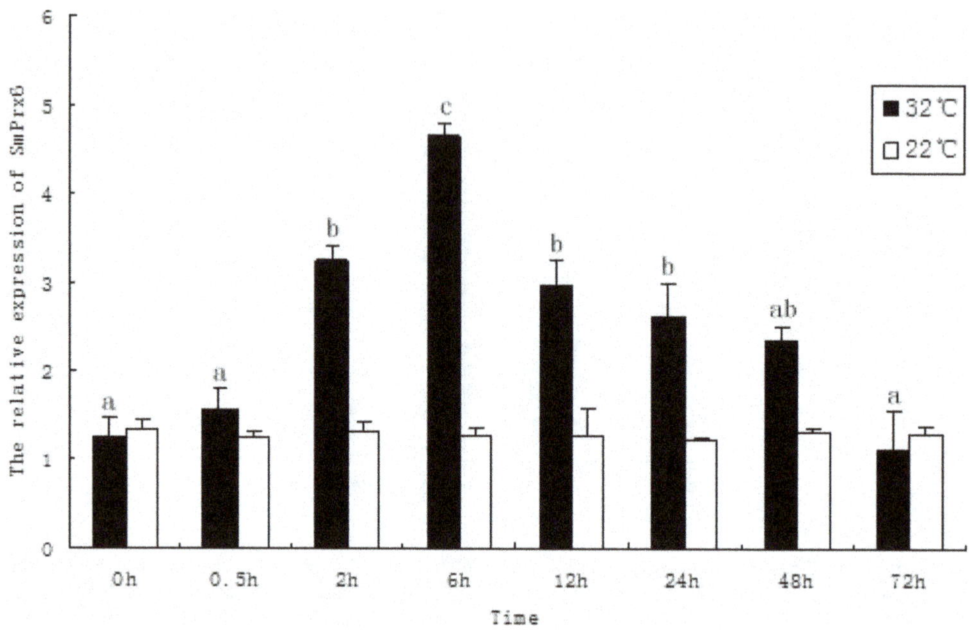

Fig. 6 Temporal expression profile of *Smprx6* transcripts in liver following thermal stress measured by SYBR Green RT-PCR. *S. maindroni* adults were incubated at 32 °C for 0, 0.5, 2, 6, 12, 24, 48 and 72 h. The control group was maintained at 22 °C. *Vertical bars* represent the mean ± SD, and *bars* with different *letters* are significantly different ($p < 0.05$)

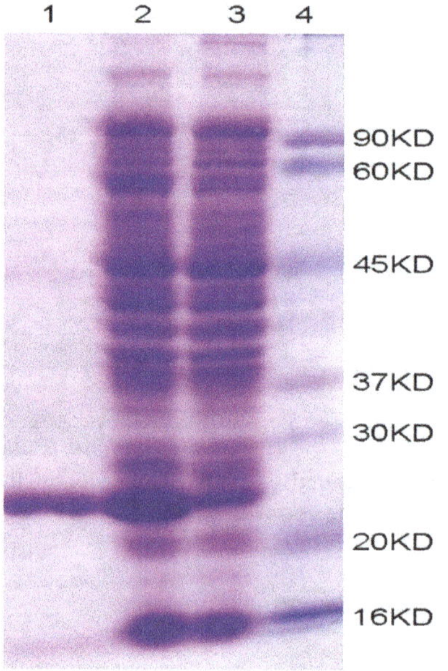

Fig. 7 SDS-PAGE of SmPrx6. *Lane 1* purified rSmPrx6; *lane 2* expression of rSmPrx6 after IPTG induction; *lane 3* uninduced control without IPTG induction; *lane 4* protein molecular standards

signal band with a molecular mass (~25 kDa) comparable to that predicted for rSm*Prx6* (Fig. 7). The purified rSm*Prx6* enzyme displayed antioxidant activity of 6.17 U/mg and POD activity of 10.17 U/mg.

Discussion

SmPrx6 shares common structural features with other invertebrate and vertebrate Prx6 proteins. The phylogenetic tree showed that Smprx6 falls into the 1-cys-Prx cluster, members of which contain a consensus sequence with a conserved cysteine residue (Cys 44) in the N terminus, and a AhpC domain with a peroxidase catalytic motif (PVCTT). In mammals, Cys47, His39 and Arg132 form the catalytic triad responsible for peroxidase activity, with Cys47 bound to His39 via hydrogen bonding, and electrostatically activated by Arg132 (Manevich and Fisher 2005). Interestingly, a catalytic triad (Cys44, His36 and Arg 126) is also present in squid and other mollusks (Nikapitiya et al. 2009). The catalytic motif responsible for PLA2 activity (27NNSWA31) was observed in the deduced amino acid sequence, with Ser29 the catalytic nucleophile forming part of the NNSWG consensus sequence responsible for the PLA2 hydrolase activity (Lee et al. 2015). Ser is replaced with a Met, Arg or Gly in *E. sinensis* (Mu et al. 2009), *Xenopus laevis* (Genbank accession no. NP_001084316) and *Suberites domuncula* (Genbank accession no. CAC38779), respectively. The effect this replacement has on the catalytic activity warrants further study.

Smprx6 was constitutively transcribed in all examined tissues, suggesting it is ubiquitously expressed in squid, and levels were highest in pancreas and liver, which is different from other aquatic animals. Ploch et al. (1999) reported that liver cells from a variety of fish species are relatively resistant to DNA modification by ROS generated by various compounds. Based on the observed high Prx6 expression levels in pancreas and liver, we propose that Prx6 in squid possesses other activities such as the PLA2 activity described above. Indeed, ubiquitous expression of EsPrx6 mRNA in different tissues has been linked to an involvement in numerous physiological functions including basal metabolism in *E. sinensis* (Mu et al. 2009). In channel catfish, Prx6 is expressed in liver, head, kidney, intestine, and skin, but not spleen (Yeh and Klesius 2007). However, Zheng and colleagues reported that turbot Prx6 is more highly expressed in blood and heart than liver in unchallenged fish, in contrast with rock bream and channel catfish that display tissue-specific expression profiles (Zheng et al. 2010). This apparent variation in the transcription of Prx6 in different organisms may be due to the expression of multiple isoforms of Prx family members in different organs (Huang et al. 2011).

The consumption of yolk stores during a specific period of development involves governing regulatory signals and controlling ROS by-products (Shafer et al. 2011). Most antioxidant enzymes such as catalase and glutathione peroxidase are expressed at high levels during yolk consumption. Understanding the sub-cellular management of metabolism before, during, and after yolk consumption is, therefore, an intriguing area of investigation. As observed in many other organisms, SmPrx6 mRNA was detected throughout embryonic development. Expression was highest during the pre-embryonic stage (II), which was 1358.26-fold higher than the pre-hatching stage ($p < 0.05$), and expression then declined gradually from stage III to stage V. The early embryonic period can be considered a time of substantial proliferative stress because early-stage embryogenesis involves massive cellular proliferation over a very short time. Correspondingly, SmPrx6 has been implicated in complex cellular processes including cell proliferation (Wood et al. 2003a), differentiation (Rabilloud et al. 1995) and apoptosis (Zhou et al. 2000; Bernier-Villamor et al. 2004), with the highest expression in early-stage embryos. Our present results indicate that SmPrx6 plays a major role in embryonic development in *S. maindroni*, especially during the early stages of tissue differentiation.

Oxidative stress often accompanies thermal stress. Although low levels of ROS modulate a variety of cellular functions such as cell proliferation, high concentrations of ROS cause cell damage or apoptosis (Park et al. 2008). Among the antioxidant systems that protect organisms from excessive ROS, the Prx family of antioxidant enzymes has received much attention during the past decade. The significant upregulation of Prx transcript levels after a high-temperature stimulus indicates the induction of gene expression (Park et al. 2005). Expression of SmPrx6 increased rapidly in *S. maindroni* and peaked at 6 h post-stimulus (3.72-fold; $p < 0.05$) before declining and eventually returning to control levels. This indicates that SmPrx6 may play a major role as an antioxidant protein to protect organisms from excessive ROS following thermal stress (Tolomeo et al. 2016).

Cephalopods are highly adaptable marine organisms, and can adjust their biology and life cycles in response to the prevailing environmental and ecosystem conditions. These organisms can presumably avoid increased levels of oxidative damage normally during short-term thermal stress, possibly by altering metabolic processes.

In conclusion, we identified and characterised Prx6 from *S. maindroni*, and detected *Smprx6* mRNA transcripts in all tested tissues and developmental stages. The observed upregulation of Prx6 expression following a high-temperature stimulus indicates that its peroxidase activity may be important for the response to oxidative stress. Therefore, Prx6 may play a major role in maintaining proteins in a reduced state by scavenging toxic ROS.

Acknowledgements The authors are grateful to all the laboratory members for their continuous technical advice and helpful discussion. This research was supported by the National Natural Science Foundation of China [Grant numbers 41176124, 41206114], the Public Welfare Project of Zhejiang Province [Grant number ZX2013000403], the Ningbo Agricultural Major Project [Grant number 201401C1111001], the Open Fund of Zhejiang Provincial Top Key Discipline of Aquaculture at Ningbo University, and the K. C. Wong Magana Fund of Ningbo University.

References

Bernier-Villamor L, Navarro E, Sevilla F, Lazaro JJ (2004) Cloning and characterization of a 2-Cys peroxiredoxin from *Pisum sativum.* J Exp Bot 55:2191–2199

Bogdan C, Rollinghoff M, Diefenbach A (2000) Reactive oxygen and reactive nitrogen intermediates in innate and specific immunity. Curr Opin Immunol 12:64–76

Castellanos-Martínez S, Gestal C (2013) Pathogens and immune response of cephalopods. J Exp Mar Biol Ecol 447:14–22

David E, Tanguy A, Moraga D (2007) Peroxiredoxin 6 gene: a new physiological and genetic indicator of multiple environmental stress response in Pacific oyster *Crassostrea gigas.* Aquat Toxicol 84:389–398

De Zoysa M, Ryu JH, Chung HC, Kim CH, Nikapitiya C, Oh C, Kim H, Saranya Revathy K, Whang I, Lee J (2012) Molecular characterization, immune responses and DNA protection activity of rock bream (*Oplegnathus fasciatus*), peroxiredoxin 6 (Prx6). Fish Shellfish Immunol 33:28–35

Gestal C, Castellanos-Martinez S (2015) Understanding the cephalopod immune system based on functional and molecular evidence. Fish Shellfish Immunol 46:120–130

Huang CF, Sun ZJ, Zhao YF, Chen XM, Jia J, Zhang WF (2011) Increased expression of peroxiredoxin 6 and cyclophilin A in squamous cell carcinoma of the tongue. Oral Dis 17:328–334

Jiang XM, Lu ZR, He HJ, Ye PL, Ying Z, Wang CL (2010) Effects of several ecological factors on the hatching of *Sepiella maindroni* wild and cultured eggs. Ying Yong Sheng Tai Xue Bao 21:1321–1326

Lee S, Jia B, Liu J, Pham BP, Kwak JM, Xuan YH, Cheong GW (2015) A 1-Cys peroxiredoxin from a thermophilic archaeon moonlights as a molecular chaperone to protect protein and DNA against stress-induced damage. PLoS One 10:e0125325

Livak KJ, Schmittgen TD (2001) Analysis of relative gene expression data using real-time quantitative PCR and the 2(−Delta Delta C(T)) Method. Methods 25:402–408

Manevich Y, Fisher AB (2005) Peroxiredoxin 6, a 1-Cys peroxiredoxin, functions in antioxidant defense and lung phospholipid metabolism. Free Radic Biol Med 38:1422–1432

Miller NJ, Rice-Evans C, Davies MJ, Gopinathan V, Milner A (1993) A novel method for measuring antioxidant capacity and its application to monitoring the antioxidant status in premature neonates. Clin Sci (Lond) 84:407–412

Mu C, Zhao J, Wang L, Song L, Zhang H, Li C, Qiu L, Gai Y (2009) Molecular cloning and characterization of peroxiredoxin 6 from Chinese mitten crab *Eriocheir sinensis.* Fish Shellfish Immunol 26:821–827

Nikapitiya C, De Zoysa M, Whang I, Kim CG, Lee YH, Kim SJ, Lee J (2009) Molecular cloning, characterization and expression analysis of peroxiredoxin 6 from disk abalone *Haliotis discus discus* and the antioxidant activity of its recombinant protein. Fish Shellfish Immunol 27:239–249

Park M, Shin HJ, Lee SY, Ahn TI (2005) Characterization of a cDNA of peroxiredoxin II responding to hydrogen peroxide and phagocytosis in *Amoeba proteus.* J Eukaryot Microbiol 52:223–230

Park H, Ahn IY, Kim H, Cheon J, Kim M (2008) Analysis of ESTs and expression of two peroxiredoxins in the thermally stressed Antarctic bivalve *Laternula elliptica.* Fish Shellfish Immunol 25:550–559

Poynton RA, Hampton MB (2014) Peroxiredoxins as biomarkers of oxidative stress. Biochim Biophys Acta 1840:906–912

Priyathilaka TT, Kim Y, Udayantha HM, Lee S, Herath HM, Lakmal HH, Elvitigala DA, Umasuthan N, Godahewa GI, Kang SI et al (2016) Identification and molecular characterization of peroxiredoxin 6 from Japanese eel (*Anguilla japonica*) revealing its potent antioxidant properties and putative immune relevancy. Fish Shellfish Immunol 51:291–302

Qiu L, Ma Z, Jiang S, Wang W, Zhou F, Huang J, Li J, Yang Q (2010) Molecular cloning and mRNA expression of peroxiredoxin gene in black tiger shrimp (*Penaeus monodon*). Mol Biol Rep 37:2821–2827

Rabilloud T, Berthier R, Vincon M, Ferbus D, Goubin G, Lawrence JJ (1995) Early events in erythroid differentiation: accumulation of the acidic peroxidoxin (PRP/TSA/NKEF-B). Biochem J 312(Pt 3):699–705

Robinson MW, Hutchinson AT, Dalton JP, Donnelly S (2010) Peroxiredoxin: a central player in immune modulation. Parasite Immunol 32:305–313

Schultz J, Milpetz F, Bork P, Ponting CP (1998) SMART, a simple modular architecture research tool: identification of signaling domains. Proc Natl Acad Sci USA 95:5857–5864

Shafer ME, Willson JA, Damjanovski S (2011) Expression analysis of the peroxiredoxin gene family during early development in *Xenopus laevis.* Gene Expr Patterns 11:511–516

Tolomeo AM, Carraro A, Bakiu R, Toppo S, Place SP, Ferro D, Santovito G (2016) Peroxiredoxin 6 from the Antarctic emerald rockcod: molecular characterization of its response to warming. J Comp Physiol B 186:59–71

Vidal EA, Villanueva R, Andrade JP, Gleadall IG, Iglesias J, Koueta N, Rosas C, Segawa S, Grasse B, Franco-Santos RM et al (2014) Cephalopod culture: current status of main biological models and research priorities. Adv Mar Biol 67:1–98

Wang Q, Chen K, Yao Q, Zhao Y, Li Y, Shen H, Mu R (2008) Identification and characterization of a novel 1-Cys peroxiredoxin from silkworm, *Bombyx mori*. Comp Biochem Physiol B Biochem Mol Biol 149:176–182

Wei-Wei S, Chun-Lin W (2009) Genetic diversity of *Sepiella maindroni* in cultured and natural populations. Oceanol et Limnol Sin 40:590–595

Wei-Wei S, Chang-Kao M, Chu S, Yu-Yu Z, Ping-Ping Z, Chun-Lin W (2012) Expression screening and annotation of *Sepiella maindroni* ink sac cDNA library. J Genet 91:e103–e108

Wood ZA, Poole LB, Karplus PA (2003a) Peroxiredoxin evolution and the regulation of hydrogen peroxide signaling. Science 300:650–653

Wood ZA, Schroder E, Robin Harris J, Poole LB (2003b) Structure, mechanism and regulation of peroxiredoxins. Trends Biochem Sci 28:32–40

Yeh H-Y, Klesius P (2007) cDNA cloning, characterization, and expression analysis of channel catfish (*Ictalurus punctatus* Rafinesque, 1818) Peroxiredoxin 6 gene. Fish Physiol Biochem 33:233–239

Zhang Q, Li F, Zhang J, Wang B, Gao H, Huang B, Jiang H, Xiang J (2007) Molecular cloning, expression of a peroxiredoxin gene in Chinese shrimp *Fenneropenaeus chinensis* and the antioxidant activity of its recombinant protein. Mol Immunol 44:3501–3509

Zheng WJ, Hu YH, Zhang M, Sun L (2010) Analysis of the expression and antioxidative property of a peroxiredoxin 6 from *Scophthalmus maximus*. Fish Shellfish Immunol 29:305–311

Zhou Y, Kok KH, Chun AC, Wong CM, Wu HW, Lin MC, Fung PC, Kung H, Jin DY (2000) Mouse peroxiredoxin V is a thioredoxin peroxidase that inhibits p53-induced apoptosis. Biochem Biophys Res Commun 268:921–927

12

Development of a prototype 16S rRNA gene-based microarray for monitoring planktonic actinobacteria in shrimp ponds

Jingjing Wang · Wei Zhao · Yajie Liang · Yifan Han · Xingbiao Wang · Xiaoxia Zhang · Xiaomei Ma · Demin Zhang · Zhiyong Huang

Abstract Many reports have shown that the composition of the bacterioplankton community can serve as a biological indicator to evaluate the occurrence of shrimp diseases. However, the distribution, diversity, and function of planktonic actinobacteria in shrimp ponds are still poorly understood. In this study, a prototype of a 16S rRNA gene-based taxonomic microarray was developed and evaluated for monitoring of planktonic actinobacteria in shrimp ponds. The prototype microarray is composed of 30 probes that target ten dominant families of planktonic actinobacteria found in shrimp ponds. The specificity of the actinobacterial microarray was validated by a set of control hybridizations with 16S rRNA genes clones. The prototype microarray was subsequently tested with two seawater samples from ponds with diseased shrimp populations (PDS) and ponds with healthy shrimp populations (PHS). The actinobacteria hybridization profiles revealed a lower abundance of Microbacteriaceae and a higher abundance of Mycobacteriaceae in PDS than in PHS. The changes in planktonic actinobacterial communities were validated by pyrosequencing data. These results support the utility of the microarray for monitoring planktonic actinobacteria in shrimp ponds and aquaculture environments.

Keywords Microarray · Planktonic actinobacteria · Shrimp ponds · 16S rRNA gene · Aquaculture environment

Introduction

Predicting the occurrence of shrimp disease outbreaks in aquaculture farms can be of considerable value to the long-term sustainable development of the industry (Leung and Tran 2000). Monitoring multiple geochemical factors has been widely employed to generate water quality indices for disease forecasting (Ferreira et al. 2011; Ma et al. 2012). However, in practice, it is generally difficult to set a threshold value for a specific abiotic parameter to evaluate the disease risk precisely (Xiong et al. 2014b). Thus, the application of only traditional physicochemical variables to assess shrimp health status may be inadequate (Rao et al. 2000). Recently, a number of studies have revealed that the bacterioplankton community can serve as a biological

J. Wang · W. Zhao · Y. Liang · Y. Han · X. Wang · X. Zhang · X. Ma · Z. Huang (✉)
Tianjin Key Laboratory of Industrial Biological Systems and Process Engineering, Tianjin Institute of Industrial Biotechnology, Chinese Academy of Sciences, No. 32, West 7th Road, Tianjin Airport Economic Area, Tianjin 300308, China
e-mail: huang_zy@tib.cas.cn

D. Zhang
School of Marine Sciences, Ningbo University, Ningbo 315211, China

indicator to evaluate the occurrence of shrimp diseases (Xiong et al. 2014b; Zhang et al. 2014). Actinobacteria, which are a prolific source of secondary metabolites with antibacterial, immunosuppressive, and antitumor activity (Becerril-Espinosa et al. 2013), are widely distributed in aquatic environments, where they play important roles (Subramani and Aalbersberg 2012; Ward and Bora 2006). However, the distribution, diversity, and function of planktonic actinobacteria in shrimp ponds are still poorly understood.

Bacterioplankton communities can be estimated easily and rapidly using techniques, such as high-throughput sequencing and phylogenetic microarray (phylochip; Ma et al. 2016; Zhang et al. 2013). However, data analysis of high-throughput sequencing techniques requires highly trained specialists, which has restricted its application in large-scale biomonitoring efforts (Seong Woon et al. 2010). Less expensive and simple analyses using phylochips make this a powerful method for routine monitoring of selected target sequences using hundreds of samples on a large scale (Seong Woon et al. 2010). The most common target of the phylochip is the 16S rRNA gene because of its ubiquity in the prokaryotic kingdom and its large presence in databases, which enables access to almost all bacteria by PCR (Bruce et al. 1992; Woese 1987). Although a few oligonucleotide microarrays based on the 16S rRNA gene have been developed for monitoring bacteria in aquatic environments (Bianca et al. 2004; Desantis et al. 2007; Jörg et al. 2004), the full-scale application of microarray technology for aquaculture environmental monitoring is scarce. This is in part because bacterial communities in the aquatic environments are distinct from those in aquaculture environments (Xiong et al. 2015).

This study was conducted to develop a prototype 16S rRNA gene-based microarray for monitoring planktonic actinobacteria in shrimp ponds. To accomplish this, 30 probes (30 nt) targeting ten dominant families of planktonic actinobacteria in shrimp ponds were designed. This prototype microarray was validated using 16S rRNA gene clones and tested with seawater samples from shrimp ponds. The results were compared to those obtained from the pyrosequencing of the same samples.

Materials and methods

DNA samples

Six 16S rRNA gene clones belonging to different families of actinobacteria were collected from marine environments (Table 1). All sequences were then sequenced and submitted to the NCBI database. Plasmid DNA was extracted from the clones using the Axygen Plasmid Miniprep Kit (Axygen, Hangzhou, China).

On the sampling day, approximately 1 L of seawater samples was collected from ponds with diseased shrimp populations (PDS) and ponds with healthy shrimp populations (PHS) located in Zhanqi, Ningbo, eastern China (29°32′N, 121°31′E; Zhang et al. 2014). After pre-filtering through nylon mesh (100 μm pore size) and 0.2 μm polycarbonate membrane (Millipore), water genomic DNA was extracted using a Power Soil® DNA Isolation Kit (MO BIO Laboratories, Carlsbad, CA, USA) according to the manufacturer's protocols. DNA was quantified with a NanoDrop ND-2000 spectrophotometer (NanoDrop Technologies, Wilmington, DE, USA) and stored at −80 °C until amplification.

Table 1 Clones of 16S rRNA gene used to validate the microarray

Clones	Similar organisms	Identity (%)	Group	Accession number	Environmental source
C1	*Ilumatobacter sp.*	98	Acidimicrobiaceae	KP688285	East China Sea
C2	Uncultured Iamiaceae	99	Iamiaceae	KP688286	East China Sea
C3	*Mycobacterium smegmatis*	99	Mycobacteriaceae	KP688287	East China Sea
C4	Uncultured Cellulomonadaceae	99	Cellulomonadaceae	KP688288	South China Sea
C5	*Microbacterium sp.*	96	Microbacteriaceae	KP688290	East China Sea
C6	*Micrococcus* sp.	99	Micrococcaceae	KP688292	South China Sea

PCR amplification of 16S rRNA gene and labeling

The pMD®19-T vector specific primers M13-47/RV-M (Zhang et al. 2008) and the universal bacterial primers 27F/1492R (Tanner et al. 1999) were used to amplify 16S rRNA genes from clones and seawater samples, respectively. The PCR reaction mixture (total volume 25 μL) contained 12.5 μL of Green PCR Master Mix (2×) (Thermo Fisher Scientific, Shanghai, China), 1 μL of each primer (2.5 μM), 1 μL of genomic DNA or 16S rRNA gene clone (~20 ng), and 9.5 μl of ddH_2O. PCR consisted of denaturation at 94 °C for 3 min, followed by 30 cycles of 1 min denaturation at 94 °C, 1 min annealing at 55 °C, 1 min elongation at 72 °C, and a final elongation step of 10 min at 72 °C. PCR products (5 μL) were denatured with 3 μL of random primers Cy3-NNN NNN NNN (Sangon BioTech, Shanghai, China) and 9 μL of ddH_2O at 95 °C for 3 min, then treated with an ice bath for 3 min. Next, 2.5 μL 5 × Klenow Buffer, 2.5 μL NTP (2.5 mM), and 1 μL Klenow enzyme (NEB, Beijing, China) were added to the mixture, followed by incubation at 37 °C for 90 min and 70 °C for 10 min (Schenk et al. 2000).

Oligonucleotide probe design and microarray manufacturing

The phylogenetic software package ARB (http://www.arb-home.de) under Ubuntu with the ARB 16S rRNA database (ssu_jan04_corr_opt.arb) was used to design probes (Ludwig et al. 2004). The parameters of the probe design function were set as follows: 30 nt, 40 < GC% < 60, position 1–10 000 in *E. coli*, max target out of group of 0, min group = 10–50%. All probes were tested with the probe match function of ARB against the total ARB 16S rRNA database. The best probes, which had a weighted mismatch (WMM) value of <2 with the targeted taxon and >2 with non-targets, were selected from the probe results window. The predicted melting temperature Tm (according to the nearest neighbor method) was calculated using Oligo7 (Molecular Biology Insights, West Cascade, CO, USA) with the default settings. The probes were chosen preferably when they had the same melting temperature (Tm), no hairpin, and a stable homoduplex. The specificity of the probes was then checked using BLAST searches of the NCBI database (http://blast.ncbi.nlm.nih.gov/Blast.cgi?PROGRAM=blastn&PAGE_TYPE=BlastSearch&LINK_LOC=blasthome). Probes having higher similarity (>90%) with the non-target group were removed from the probe set.

Probes were designed to target planktonic actinobacteria at the family level based on their predictive accuracy (Xiong et al. 2014b). Of the 42 families recognized in actinobacteria, the probes targeted ten families (Acidimicrobiaceae, Iamiaceae, Mycobacteriaceae, Cellulomonadaceae, Intrasporangiaceae, Microbacteriaceae, Micrococcaceae, Nocardioidaceae, Propionibacteriaceae, and Nitriliruptoraceae) shown to be dominant in shrimp ponds in the previous studies (Xiong et al. 2014a). Three probes targeted the same family, complementing several unique regions of the 16S rRNA gene (Loy et al. 2002).

Probes were synthesized (Sangon BioTech, Shanghai, China) with a 5′C6-NH_2 group for covalent attachment onto aldehyde slides AL (CapitalBio, Beijing, China), then spotted onto slides using a SmartArrayer 48 (CapitalBio, Beijing, China).

Hybridization protocols

The labeled DNA (15 μL) was mixed with 5 μL of hybridization buffer (5× SSC, 0.02% SDS, 5% formamide), denatured for 5 min at 95 °C, and immediately placed on ice. The slides were then placed in a hybridization chamber (CapitalBio, Beijing, China) and covered with a Hybri-slip (CapitalBio, Beijing, China). The labeled DNA was transferred onto the slide through holes in the Hybri-slip. Hybridization was conducted at 50 °C for 2 h, after which the slides were immediately washed by shaking in 2× SSC containing 0.02% SDS at 50 °C for 4 min, fresh 0.2× SSC at 50 °C for 4 min and 95% ethanol at room temperature for 4 min, successively. Finally, the slides were quickly dried by centrifugation at 1500 rpm for 4 min (Sanguin et al. 2006).

Scanning and data analysis

The slides were scanned at 532 nm at a resolution of 10 μm using a NimbleGen MS200 scanner (Roche NimbleGen). Images were analyzed with the GenePix 4.01 software (Axon, Union City, CA, USA). The spot quality was visually checked, and spots of poor quality were excluded from further analyses. A given spot was considered to provide strong positive hybridization if the spot pixels had an intensity two times higher than the median pixel intensity of the local background plus twice the standard deviation of the local background. A given probe was only considered to be truly hybridized when at least three of four replicate spots provided a strong positive hybridization (Sanguin et al. 2006). The pixel intensity of positive hybridization was then normalized relative to a mean hybridization signal observed from the set of positive control (PC) probes to provide the best array-to-array consistency, because PC probes were universal for every hybridized PCR amplicon (Liles et al. 2010).

Pyrosequencing analysis

An aliquot (50 ng) of DNA from seawater samples (PDS and PHS) was used as the template for amplification. The V1–V3 hyper variable region of bacterial 16S rRNA genes was amplified using the primer set 27F and 519R (Xiong et al. 2014a). The bacterial 16S rRNA gene products were sequencing using a Roche FLX 454 pyrosequencing machine (Roche Diagnostics Corporation, Branford, CT, USA). The sequencing reads were analyzed by the Quantitative Insights Into Microbial Ecology (QIIMEv1.5.0) workflow (Caporaso et al. 2010; Zhang et al. 2014).

Results and discussion

Probe design and microarray format

Probes can be designed at various taxonomic levels quickly and easily with the ARB software using 16S rRNA gene databases (Ludwig et al. 2004). In this study, probes were designed to target planktonic actinobacteria at the family level, as previous work showed that the bacterioplankton communities had a higher predictive accuracy of shrimp health and disease at family level than other taxa (Xiong et al. 2014b).

The specificity and sensitivity of probes varying from length among 9 and 70 nucleotides (nt) were evaluated. Short probes (25 nt) were significantly more specific than longer ones (35 nt), but the latter were more sensitive. Probes with a length of 30 nt are often selected as the best compromise between required specificity and sensitivity (Relogio et al. 2002). To maintain homogenous hybridization conditions, probes with identical melting temperatures and GC content should be selected (Sanguin et al. 2006). When these requirements could not be met, probes with suboptimal conditions were accepted to provide adequate phylogenetic coverage. Finally, 30 probes targeting ten families of actinobacteria were designed. The probes were 29–31 nucleotides long (80% of the probes were 30 nucleotides long), with a G + C content between 48.3 and 65.5% (average 57.2%). The melting temperature was between 72.8 and 83.3 °C (90% of probes fell within the 80 ± 5 °C temperature ranges; Table 2). As hybridizations can occur even in case of non-perfect matches (Kyselková et al. 2008; Loy et al. 2002), the specificity of probes was assessed using both Probe Match and BLAST.

A schematic diagram of the probe positions on the microarray is presented in Fig. 1. One probe has spotted as a positive hybridization control (PC), one as a negative hybridization control (NC), and one as a printing control (CK). PC targeted the V3 region of the 16S rRNA gene. The 30 polyT corresponding to the sequences not amplified during PCR was spotted as NC, and the 30 polyT labeled by HEX was spotted as CK. Each probe was printed in quadruplicate. The basic probe pattern on the microarrays consisted of two PC, six NC, and four CK, which could be used as the position reference and landmarks for image analysis. The 5′ end of each oligonucleotide probe was tailed with ten dTTP molecules (T-spacer) to increase the on-chip accessibility of spotted probes to target DNA.

Validation of probe set with pure cultures

The specificity of the individual probe was evaluated with six clones belonging to six families, respectively. Highly specific signals were observed, and all targets could be clearly differentiated by the signal patterns

Table 2 Oligonucleotide probes used in this study

Probe	Specificity	Sequence (5′–3′)	*L* (bp)	Tm (°C)	GC%
ID 1	Acidimicrobiaceae	CTTGGCAAGTCGGATGTGAAATCTCCAGGC	30	77.2	53.3
ID 2	Acidimicrobiaceae	GAGTCCGGTAGAGGATCGTGGAATTCCTGG	30	77.1	56.7
ID 3	Acidimicrobiaceae	GAGTGCGGTAGGGGAGCGGGGAATTCCTAG	30	80.6	63.3
ID 4	Iamiaceae	CGAAGACCGGGATAACCCTCCGAAAGGAGA	30	78.4	56.7
ID 5	Iamiaceae	TGACTTGAGTCCGGTAGGGGAGCACGGAA	29	80.1	58.6
ID 6	Iamiaceae	GACACCGCGAGGTTGAGCGAATCCCACAAA	30	80.1	58.6
ID 7	Mycobacteriaceae	ACAGCTTAACTGTGGGCGTGCGGGCGATAC	30	81.4	58.6
ID 8	Mycobacteriaceae	TACTAGGTGTGGGTTTCCTTCCTTGGGATC	30	75.2	48.3
ID 9	Mycobacteriaceae	GTTCCCTTGTGGCCTGTGTGCAGGTGGTG	29	81.0	62.1
ID 10	Cellulomonadaceae	CCGGAAAAGTGCAGAGATGTGCTCCCCGTAA	31	80.0	58.6
ID 11	Cellulomonadaceae	ACCTCAGGCTCAACCTGGGGCTTGCAGTGGG	31	81.8	62.1
ID 12	Cellulomonadaceae	GTACACCGACCTTGCGGGGCAACCATCTCT	30	80.3	62.1
ID 13	Intrasporangiaceae	ATCCGGGGCTCAACCCCGGACTTGCAGTGG	30	83.3	65.5
ID 14	Intrasporangiaceae	AATCCGGGGCTCAACCCCGGACTTGCAGTG	30	83.2	63.3
ID 15	Intrasporangiaceae	GACATACACCGGACACATCCAGAGATGGGTG	31	77.2	53.3
ID 16	Microbacteriaceae	CAGTATCCCATGAGTTCCCACCATTACGTG	30	74.7	50.0
ID 17	Microbacteriaceae	TAGATACCCTGGTAGTCCACCCCGTAAACG	30	76.4	50.0
ID 18	Microbacteriaceae	CAGAGTCAAGGGCAGATTGCTCACGTGTTA	30	76.3	50.0
ID 19	Micrococcaceae	CTGTCGTGAAAGTCCGGGGCTTAACCCCG	29	80.1	62.1
ID 20	Micrococcaceae	CACACAAGGTGGTTAGGCCATCGGCTTCGG	30	79.0	58.6
ID 21	Micrococcaceae	GTCCGAGGCTCAACCTCGGATCTGCGGTGG	30	81.3	65.5
ID 22	Nocardioidaceae	CACCACCTGTACACCAGTATCAAAGAGACC	30	72.8	48.3
ID 23	Nocardioidaceae	CGAAGGCGGTTCTCTGGGCATTACCTGACG	30	78.4	58.6
ID 24	Nocardioidaceae	AATCCGTGGAAAGGACCCCACACCTAGCGC	30	79.6	58.6
ID 25	Propionibacteriaceae	TGCTTTCGATACGGGTTGACTTGAGGAAGG	30	74.7	48.3
ID 26	Propionibacteriaceae	AGCTCGTAGGTGGTTGATTGCGTCGGAAGT	30	78.0	55.2
ID 27	Propionibacteriaceae	TACAAAGAGTTGCGAGCCTGTGAGGGTGAG	30	77.4	51.7
ID 28	Nitriliruptoraceae	CCGTAGAGATGCGGTGGGTTCGTCCGTGCT	30	81.5	65.5
ID 29	Nitriliruptoraceae	CATAGGGTCCGAGCGTTGTCCGGAATCATT	30	77.8	55.2
ID 30	Nitriliruptoraceae	ACCTCCAGAAGAAGGACCGGCCAACTATGT	30	77.5	55.2
PC	Positive control	ACTCCTACGGGAGGCAGCAGTGGGGAATAT	30	80.0	58.6
NC	Negative control	TTTTTTTTTTTTTTTTTTTTTTTTTTTTTT	30	58.6	0
CK	Printing control	TTTTTTTTTTTTTTTTTTTTTTTTTTTTTT-HEX	30	58.6	0

obtained (Fig. 2). Each of the expected-positive groups produced a hybridization signal with each of the positive control rRNAs. None of these probes showed false-negative or false-positive signals.

The full-scale application of 16S rRNA-based taxonomic microarray relies on a highly specific probe set (He et al. 2004). Identifications can be made with a high degree of confidence using an intelligent combination of several probes (Loy et al. 2005). Three probes that targeted separate signature sites of one family were constructed in our study. If they bound to the same colonies, or to the same fragment or fraction of DNA, the possibility of false positives is virtually eliminated. For example, in the hybridization results, two of three probes of Iamiaceae (ID 5 and ID 6) and Microbacteriaceae (ID 16 and ID 17) produced hybridization signals upon hybridization with clone C2 and clone C5, respectively (Fig. 2). The differences in the hybridization signals for expected-positive features between different rRNA gene sequences (e.g., clone 1 vs. clone 3) might be a consequence of different probe hybridization kinetics (Liles et al. 2010).

CK	CK	CK	CK	NC	NC	NC	NC	CK	CK	CK	CK
NC	NC	NC	NC	PC	PC	PC	PC	NC	NC	NC	NC
ID-1	ID-1	ID-1	ID-1	ID-2	ID-2	ID-2	ID-2	ID-3	ID-3	ID-3	ID-3
ID-4	ID-4	ID-4	ID-4	ID-5	ID-5	ID-5	ID-5	ID-6	ID-6	ID-6	ID-6
ID-7	ID-7	ID-7	ID-7	ID-8	ID-8	ID-8	ID-8	ID-9	ID-9	ID-9	ID-9
ID-10	ID-10	ID-10	ID-10	ID-11	ID-11	ID-11	ID-11	ID-12	ID-12	ID-12	ID-12
ID-13	ID-13	ID-13	ID-13	ID-14	ID-14	ID-14	ID-14	ID-15	ID-15	ID-15	ID-15
ID-16	ID-16	ID-16	ID-16	ID-17	ID-17	ID-17	ID-17	ID-18	ID-18	ID-18	ID-18
ID-19	ID-19	ID-19	ID-19	ID-20	ID-20	ID-20	ID-20	ID-21	ID-21	ID-21	ID-21
ID-22	ID-22	ID-22	ID-22	ID-23	ID-23	ID-23	ID-23	ID-24	ID-24	ID-24	ID-24
ID-25	ID-25	ID-25	ID-25	ID-26	ID-26	ID-26	ID-26	ID-27	ID-27	ID-27	ID-27
ID-28	ID-28	ID-28	ID-28	ID-29	ID-29	ID-29	ID-29	ID-30	ID-30	ID-30	ID-30
NC	NC	NC	NC	PC	PC	PC	PC	NC	NC	NC	NC
CK	CK	CK	CK	NC	NC	NC	NC	CK	CK	CK	CK

Fig. 1 Detailed design of a single array with exact positions for each probe (*CK* printing control, *PC* positive hybridization control, *NC* negative hybridization control, *ID-1–ID-30* probes targeting ten families of planktonic actinobacteria)

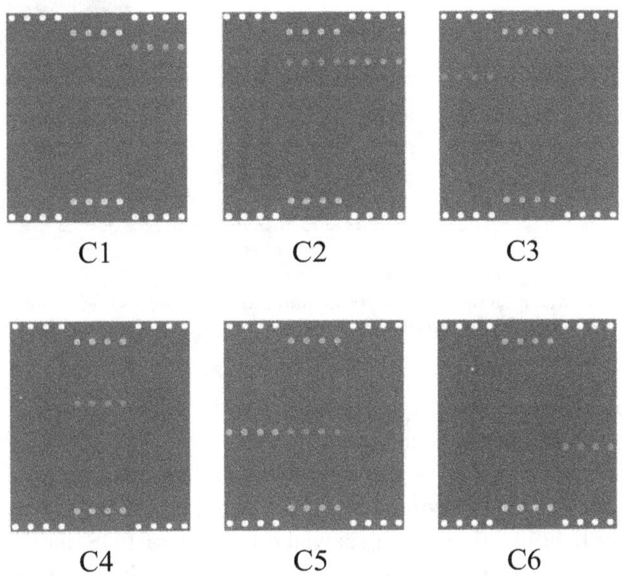

Fig. 2 Hybridization results of six reference clones. **C1**, *Ilumatobacter* sp. (Acidimicrobiaceae); **C2**, Uncultured Iamiaceae (Iamiaceae); **C3**, *Mycobacterium smegmatis* (Mycobacteriaceae); **C4**, Uncultured Cellulomonadaceae (Cellulomonadaceae); **C5**, *Microbacterium* sp. (Microbacteriaceae); **C6**, *Micrococcus* sp. (Micrococcaceae)

Hybridization of water samples

Two water samples from ponds with diseased shrimp populations (PDS) and ponds with healthy shrimp populations (PHS) were assessed by the actinobacterial microarray. Three of thirty probes produced hybridization signals upon hybridization with seawater samples from PHS. There was no hybridization signal produced by other probes. All three probes (ID16, ID17, and ID18) targeting Microbacteriaceae produced strong signals. ID16 (102%) generated the highest signals than others. ID 17 (51%) produced higher signals than ID 18 (8%). The hybridization patterns of the microarray indicated that Microbacteriaceae was present in PHS (Fig. 3a).

Four of the 30 probes generated signals upon hybridization with seawater samples from PDS (Fig. 3b). There was no hybridization signal produced by other probes. Of these four probes, three (ID16, ID17, and ID18) targeted Microbacteriaceae and one (ID7) targeted to Mycobacteriaceae. The signals of ID 7, ID 16, ID 17, and ID 18 were 35, 53, 19, and 19%, respectively. The hybridization patterns of the microarray indicated

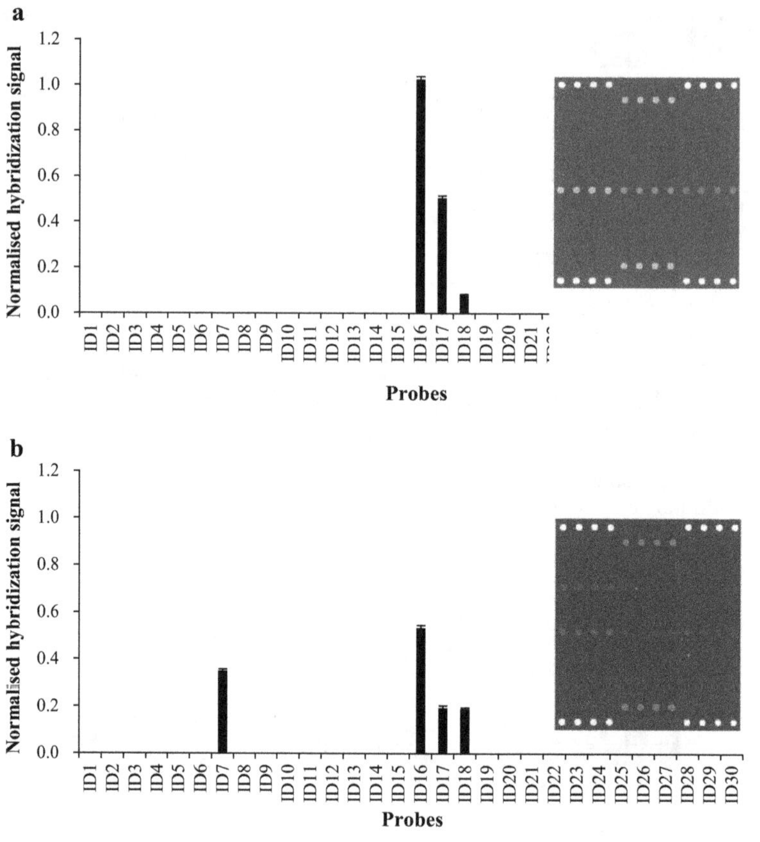

Fig. 3 Microarray analyses of seawater samples. **a** Hybridization of seawater samples from ponds with healthy shrimp populations (PHS). **b** Hybridization of seawater samples from ponds with diseased shrimp populations (PDS)

the presence of Microbacteriaceae and Mycobacteriaceae in PDS. However, the signal intensity of Microbacteriaceae in PDS was much lower than that in PHS.

It was of interest to compare the results of the actinobacterial community in healthy and diseased shrimp ponds, since the composition of bacterial communities in seawater can be altered by shrimps (Sombatjinda et al. 2011). The microarray hybridizations from seawater samples of PDS and PHS revealed some similarities, including the presence of the Microbacteriaceae. Interestingly, there was an apparent decrease in the hybridization signal for Microbacteriaceae and an apparent increase in the hybridization signal for Mycobacteriaceae in PDS compared with PHS. It should be noted that the probe hybridization signals could not be used to characterize the quantitative abundance of any group in an environmental sample due to potential kinetic differences between individual probes (Liles et al. 2010). However, if a comparison was conducted in the same set of probes with different environmental samples, a shift in probe hybridization strength from one sample to another could indicate a real shift in bacterial relative abundance (Liles et al. 2010). Mycobacteriaceae includes pathogens known to cause serious diseases in mammals. For example, *Mycobacterium marinum* was the causative agent of fish tank granuloma in humans (Rahman et al. 2014). However, no relationships between Microbacteriaceae or Mycobacteriaceae and shrimp disease were reported. The dynamics of actinobacterial communities may serve as an indicator of the health status of the shrimp (Berry et al. 2012).

Pyrosequencing results of water samples

The same DNA samples obtained from PHS and PDS were analyzed by pyrosequencing (Fig. 4). The results showed that the relative abundance of actinobacteria was 22.05% and that of Microbacteriaceae was 21.47% in PHS. Acidimicrobiaceae (0.07%), Acidimicrobiales_uncultured (0.04%), Candidatus_Microthrix (0.18%),

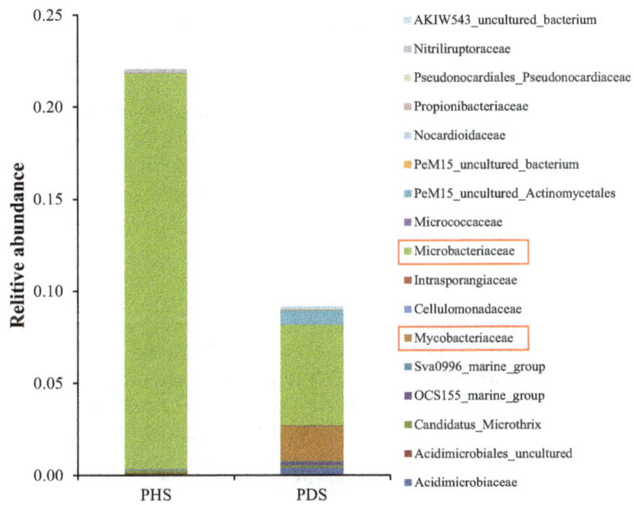

Fig. 4 Relative abundance of actinobacteria detected by Pyrosequencing (*PHS* seawater samples from ponds with healthy shrimp populations, *PDS* seawater samples from ponds with diseased shrimp populations)

Cellulomonadaceae (0.07%), Micrococcaceae (0.04%), Nocardioidaceae (0.02%), Propionibacteriaceae (0.02%), Pseudonocardiaceae (0.04%), and Nitriliruptoraceae (0.09%) had low-relative abundance in PHS. Overall, Microbacteriaceae accounted for 97.37% of actinobacteria and was a dominant actinobacterial family in PHS. Mycobacteriaceae was not detected in PHS.

Pyrosequencing of the PCR amplicons showed that relative abundance of actinobacteria was 9.17%, while that of Microbacteriaceae and Mycobacteriaceae was 5.49% and 1.88% in PDS, respectively. Acidimicrobiaceae (0.43%), Candidatus_Microthrix (0.13%), OCS155_marine_group (0.18%), Sva0996_marine_group (0.05%), Cellulomonadaceae (0.03%), Intrasporangiaceae (0.03%), PeM15_uncultured_Actinomycetales (0.79%), PeM15_uncultured_bacterium (0.03%), Nocardioidaceae (0.05%) and AKIW543_uncultured_bacterium (0.10%) had low-relative abundance in PDS. Microbacteriaceae (59.87% of actinobacteria), and Mycobacteriaceae (20.50% of actinobacteria) were dominant actinobacterial families in PDS.

The pyrosequencing results also indicated that Microbacteriaceae was present at high levels in both samples. The relative abundance of Microbacteriaceae was higher while that of Mycobacteriaceae was lower in PHS than in PDS. Therefore, the power of the microarray for predicting changes in actinobacterial communities in shrimp ponds was supported by the pyrosequencing analysis. Although high-throughput (or 'next-generation') sequencing can detect low-relative-abundance microorganisms in the environment (Andersson et al. 2010; Roesch et al. 2007), its high cost and complex analysis hinder its large-scale application to monitoring bacterial communities (Kircher and Kelso 2010; Seong Woon et al. 2010). The large parallel nature of phylogenetic microarray hybridizations enables rapid identification of bacterial groups and reflection of their changes in the environment (Brodie et al. 2006; Huyghe et al. 2008). Our result showed that actinobacterial microarray can detect high-relative-abundance actinobacteria and reflect their changes in health as well as in shrimp pond conditions. These findings indicated that the actinobacterial microarray has the potential for wide application to the monitoring of changes in actinobacterial communities in aquaculture environments. Future studies will expand beyond actinobacteria to encompass a broader diversity of bacteria. Eventually, this technique may be applied to monitor many microorganisms in the aquaculture environment.

Conclusions

We developed a 16S rRNA gene-based microarray that contained 30 probes targeting 10 dominant families of planktonic actinobacteria in shrimp ponds. The specificity of probes was validated by 16S rRNA genes clones. The availability of the microarray was tested using seawater samples from shrimp ponds, and the results were validated by pyrosequencing. Overall, the results indicated that the actinobacterial microarray might be widely applicable to the monitoring of changes in planktonic actinobacterial communities in aquaculture environments.

Acknowledgements This work was financially supported by the Tianjin science and technology plan Project (15ZCZDSF00620), National High Technology Research and Development Program of China (863 Program, 2012AA092000) and Open Fund of Zhejiang Provincial Top Key Discipline of Aquaculture in Ningbo University (Grant No. xkzsc10). We thank LetPub (www.letpub.com) for its linguistic assistance during the preparation of this manuscript.

References

Andersson AF, Riemann L, Bertilsson S (2010) Pyrosequencing reveals contrasting seasonal dynamics of taxa within Baltic Sea bacterioplankton communities. ISME J 4:171–181. doi:10.1038/ismej.2009.108

Becerril-Espinosa A, Freel KC, Jensen PR, Soria-Mercado IE (2013) Marine Actinobacteria from the Gulf of California: diversity, abundance and secondary metabolite biosynthetic potential. Antonie Van Leeuwenhoek 103:809–819

Berry D, Schwab C, Milinovich G, Reichert J, Ben Mahfoudh K, Decker T, Engel M, Hai B, Hainzl E, Heider S, Kenner L, Muller M, Rauch I, Strobl B, Wagner M, Schleper C, Urich T, Loy A (2012) Phylotype-level 16S rRNA analysis reveals new bacterial indicators of health state in acute murine colitis. ISME J 6:2091–2106. doi:10.1038/ismej.2012.39

Bianca C, Ermanno R, Andrea F, Kaarina S, Pirjo R, Anne R, Maria Angela M, Stefano V, Annick W, Christophe B (2004) Development of a universal microarray based on the ligation detection reaction and 16S rRNA gene polymorphism to target diversity of cyanobacteria. Appl Environ Microbiol 70:7161–7167

Brodie EL, DeSantis TZ, Joyner DC, Baek SM, Larsen JT, Andersen GL, Hazen TC, Richardson PM, Herman DJ, Tokunaga TK, Wan JMM, Firestone MK (2006) Application of a high-density oligonucleotide microarray approach to study bacterial population dynamics during uranium reduction and reoxidation. Appl Environ Microbiol 72:6288–6298. doi:10.1128/aem.00246-06

Bruce KD, Hiorns WD, Hobman JL, Osborn AM, Strike P, Ritchie DA (1992) Amplification of DNA from native populations of soil bacteria by using the polymerase chanin-reaction. Appl Environ Microbiol 58:3413–3416

Caporaso JG, Kuczynski J, Stombaugh J, Bittinger K, Bushman FD, Costello EK, Fierer N, Pena AG, Goodrich JK, Gordon JI, Huttley GA, Kelley ST, Knights D, Koenig JE, Ley RE, Lozupone CA, McDonald D, Muegge BD, Pirrung M, Reeder J, Sevinsky JR, Tumbaugh PJ, Walters WA, Widmann J, Yatsunenko T, Zaneveld J, Knight R (2010) QIIME allows analysis of high-throughput community sequencing data. Nat Methods 7:335–336. doi:10.1038/nmeth.f.303

Desantis TZ, Brodie EL, Moberg JP, Zubieta IX, Piceno YM, Andersen GL (2007) High-density universal 16S rRNA microarray analysis reveals broader diversity than typical clone library when sampling the environment. Microb Ecol 53:371–383

Ferreira NC, Bonetti C, Seiffert WQ (2011) Hydrological and water quality indices as management tools in marine shrimp culture. Aquaculture 318:425–433

He Z, Wu L, Li X, Fields M, Zhou JZ (2004) Empirical establishment of oligonucleotide probe design criteriausing perfect match and mismatch probes and artificial targets. Appl Environ Microbiol 71:3753–3760

Huyghe A, Francois P, Charbonnier Y, Tangomo-Bento M, Bonetti EJ, Paster BJ, Bolivar I, Baratti-Mayer D, Pittet D, Schrenzel J, Geneva Study Grp N (2008) Novel microarray design strategy to study complex bacterial communities. Appl Environ Microbiol 74:1876–1885. doi:10.1128/aem.01722-07

Jörg P, Lau SCK, Jakob P, Rudolf A, Frank Oliver GC (2004) Application and validation of DNA microarrays for the 16S rRNA-based analysis of marine bacterioplankton. Environ Microbiol 6:638–645

Kircher M, Kelso J (2010) High-throughput DNA sequencing-concepts and limitations. BioEssays 32:524–536. doi:10.1002/bies.200900181

Kyselková M, Kopecký J, Felföldi T, Čermák L, Omelka M, Grundmann GL, Moënne-Loccoz Y, Ságová-Marečková M (2008) Development of a 16S rRNA gene-based prototype microarray for the detection of selected actinomycetes genera. Antonie Van Leeuwenhoek 94:439–453

Leung PS, Tran LT (2000) Predicting shrimp disease occurrence: artificial neural networks vs. logistic regression. Aquaculture 187:35–49

Liles MR, Turkmen O, Manske BF, Zhang M, Rouillard JM, George I, Balser T, Billor N, Goodman RM (2010) A phylogenetic microarray targeting 16S rRNA genes from the bacterial division Acidobacteria reveals a lineage-specific distribution in a soil clay fraction. Soil Biol Biochem 42:739–747

Loy A, Lehner A, Lee N, Adamczyk J, Meier H, Ernst J, Schleifer K-H, Wagner M (2002) Oligonucleotide microarray for 16S rRNA gene-based detection of all recognized lineages of sulfate-reducing prokaryotes in the environment. Appl Environ Microbiol 68:5064–5081

Loy A, Schulz C, Lücker S, Schöpfer-Wendels A, Stoecker K, Baranyi C, Lehner A, Wagner M (2005) 16S rRNA gene-based oligonucleotide microarray for environmental monitoring of the betaproteobacterial order "Rhodocyclales". Appl Environ Microbiol 71:1373–1386

Ludwig W, Strunk O, Westram R, Richter L, Meier H, Yadhukumar Buchner A, Lai T, Steppi S, Jobb G, Forster W, Brettske I, Gerber S, Ginhart AW, Gross O, Grumann S, Hermann S, Jost R, Konig A, Liss T, Lussmann R, May M, Nonhoff B, Reichel B, Strehlow R, Stamatakis A, Stuckmann N, Vilbig A, Lenke M, Ludwig T, Bode A, Schleifer KH (2004) ARB: a software environment for sequence data. Nucleic Acids Res 32:1363–1371. doi:10.1093/nar/gkh293

Ma Z, Song X, Wan R, Gao L (2012) A modied water quality index for intensive shrimp ponds of *Litopenaeus vannamei*. Ecol Indic 24:287–293

Ma L, Mao G, Liu J, Gao G, Zou C, Bartlam MG, Wang Y (2016) Spatial-temporal changes of bacterioplankton community along an exhorheic river. Front Microbiol 7:1–12

Rahman SA, Singh Y, Kohli S, Ahmad J, Ehtesham NZ, Tyagi AK, Hasnain SE (2014) Comparative analyses of nonpathogenic, opportunistic, and totally pathogenic Mycobacteria reveal genomic and biochemical variabilities and highlight the survival attributes of *Mycobacterium tuberculosis*. Mbio 5:1–9. doi:10.1128/mBio.02343-14

Rao PSS, Karunasagar I, Otta SK, Karunasagar I (2000) Incidence of bacteria involved in nitrogen and sulphur cycles in tropical shrimp culture ponds. Aquac Int 8:463–472. doi:10.1023/a:1009250004999

Relogio A, Schwager C, Richter A, Ansorge W, Valcarcel J (2002) Optimization of oligonucleotide-based DNA microarrays. Nucleic Acids Res 30:1–10. doi:10.1093/nar/30.11.e51

Roesch LF, Fulthorpe RR, Riva A, Casella G, Hadwin AKM, Kent AD, Daroub SH, Camargo FAO, Farmerie WG, Triplett EW (2007) Pyrosequencing enumerates and contrasts soil microbial diversity. ISME J 1:283–290. doi:10.1038/ismej.2007.53

Sanguin H, Herrera A, Oger-Desfeux C, Dechesne A, Simonet P, Navarro E, Vogel TM, Moenne-Loccoz Y, Nesme X, Grundmann GL (2006) Development and validation of a prototype 16S rRNA-based taxonomic microarray for Alphaproteobacteria. Environ Microbiol 8:289–307. doi:10.1111/j.1462-2920.2005.00895.x

Schenk PM, Kazan K, Wilson I, Anderson JP, Richmond T, Somerville SC, Manners JM (2000) Coordinated plant defense responses in Arabidopsis revealed by microarray analysis. PNAS 97:11655–11660. doi:10.1073/pnas.97.21.11655

Seong Woon R, Abell GCJ, Kyoung-Ho K, Young-Do N, Jin-Woo B (2010) Comparing microarrays and next-generation sequencing technologies for microbial ecology research. Trends Biotechnol 28:291–299

Sombatjinda S, Boonapatcharoen N, Ruengjitchatchawalya M, Wantawin C (2011) Dynamics of microbial communities in an earthen shrimp pond during the shrimp growing period. Environ Nat Resour Res 1:171–180

Subramani R, Aalbersberg W (2012) Marine actinomycetes: an ongoing source of novel bioactive metabolites. Microbiol Res 167:571–580

Tanner MA, Shoskes D, Shahed A, Pace NR (1999) Prevalence of corynebacterial 16S rRNA sequences in patients with bacterial and "nonbacterial" prostatitis. J Clin Microbiol 37:1863–1870

Ward AC, Bora N (2006) Diversity and biogeography of marine actinobacteria. Curr Opin Microbiol 9:279–286

Woese CR (1987) Bacterial evolution. Microbiol Rev 51:221–271

Xiong J, Zhu J, Wang K, Wang X, Ye X, Liu L, Zhao Q, Hou M, Qiuqian L, Zhang D (2014a) The temporal scaling of bacterioplankton composition: high turnover and predictability during shrimp cultivation. Microb Ecol 67:256–264

Xiong J, Zhu J, Zhang D (2014b) The application of bacterial indicator phylotypes to predict shrimp health status. Appl Microbiol Biotechnol 98:8291–8299

Xiong J, Chen H, Hu C, Ye X, Kong D, Zhang D (2015) Evidence of bacterioplankton community adaptation in response to long-term mariculture disturbance. Sci Rep 5:1–11

Zhang Y, Wang C, Wang Z, Ma D, Pan S, Zhu S, Li F, Wang B (2008) Construction of double suicide genes system controlled by MDR1 promoter with targeted expression in drug-resistant glioma cells. J Neuro Oncol 86:3–11. doi:10.1007/s11060-007-9431-2

Zhang R, Wu Q, Piceno YM, Desantis TZ, Saunders FM, Andersen GL, Liu WT (2013) Diversity of bacterioplankton in contrasting Tibetan lakes revealed by high-density microarray and clone library analysis. FEMS Microbiol Ecol 86:277–287

Zhang D, Wang X, Xiong J, Zhu J, Wang Y, Zhao Q, Chen H, Guo A, Wu J, Dai H (2014) Bacterioplankton assemblages as biological indicators of shrimp health status. Ecol Indic 38:218–224

13

Prediction of fillet yield in hatchery populations of Asian sea bass, *Lates calcarifer* (Bloch, 1790) using body weight and measurements

Suthajaree Yenmak · Weerakit Joerakate · Supawadee Poompuang

Abstract The purpose of this study was to develop and validate prediction equations for fillet traits in Asian sea bass. Ninety-day old fish (average weight = 9.48 ± 0.35 g; length = 8.78 ± 0.6 cm) from four hatchery stocks (Chachoengsao, Chon Buri, Chumphon, and Samut Songkram) were raised in earthen ponds. After 300 days, weight and body measurement data of live fish (n = 400) were collected. Mean individual weight was greatest for the Chacheongsao stock (1166.32 ± 23.42 g) and was similar for the other stocks, ranging from 982.96 ± 25.07 to 997.44 ± 24.71 g. Fillet percentage varied slightly from 47.33 to 49.88%. Positive high correlations were observed for weight and body measurements with fillet weight, whereas there were significant but weak correlations for body measurements with fillet yield. Prediction equations developed from body weight using simple linear regression models yielded R^2 values of 0.97–0.98 for fillet weight for each stock. Correlations between values predicted from the body weight models and actual values were 0.98–0.99 for fillet weight. Stepwise regression was performed to develop prediction models for fillet yield from body measurements. The best fillet yield prediction models identified length and an additional 2–3 measurement as potential predictors depending on the population. Prediction biases were close to zero despite low to moderate (r values of 0.20–0.51) degrees of predictive power of the models. However, prediction models for fillet yield should be further developed to increase predictability and be applicable to new data.

Keywords Barramundi · Fillet yield · Stepwise regression · Predictive power · Value-added fish products

Introduction

Asian sea bass, also known as barramundi (*Lates calcarifer*), is the predominant fish species used for coastal aquaculture in Thailand. It is a significant fishery commodity representing 10% of all seafood production. Asian sea bass production increased by 51% from 8004 metric tons in 2001 to 16,501 tons in 2014 as reported by the Thailand Department of Fisheries (DOF 2016). Farmed sea bass is normally sold as whole fish at 500–700 g for local markets, with a retail price of 150 baht/kg (4.50 USD/kg). The consumer preference is increasingly leaning towards value-added and convenience products, such as frozen fish and fish fillets.

S. Yenmak · W. Joerakate · S. Poompuang (✉)
Department of Aquaculture, Faculty of Fisheries, Kasetsart University, Bangkok 10900, Thailand
e-mail: supawadee.p@ku.ac.th

S. Yenmak
Center for Advanced Study in Agriculture and Food, Institute for Advanced Studies, Kasetsart University, Bangkok 10900, Thailand

Therefore, the production of larger fish (1000–2000 g) has increased during the past decade. These types of products are also exported to Taiwan, Singapore, Malaysia and Australia (Food and Agriculture Organization 2017). Value addition is one of the approaches expected to increase profitability of the sea bass farming and processing industries (Department of Fisheries 2016).

Improvement in production traits, including growth performance and yield, is considered the most important breeding goals in fish farming (Gjedrem 1997, 2000; Kause et al. 2002; Thodesen et al. 2012). Several studies have been conducted comparing growth performance and fillet percentage for various fish strains utilizing body weight and volume measurements, as well as body shape, e.g., Nile tilapia *Oreochromis niloticus* (Rutten et al. 2004; Rutten et al. 2005; Peterman and Phelps 2012), striped bass *Morone* spp. (Bosworth et al. 1998), common carp *Cyrpinus carpio* (Cibert et al. 1999) and catfish *Ictalurus* spp. (Argue et al. 2003; Bosworth et al. 2001). Fillet yield, an economically important trait for the fish processing industry, however, is difficult to measure with high accuracy without sacrificing the animal and cannot be used as a criterion during broodstock selection. A more practical approach is through selection of candidates on phenotypic characteristics that are highly correlated with fillet traits. Studies have evaluated the ability of total weight and various body shape measurements to predict fillet traits and found body weight to be an accurate predictor of fillet weight (Rutten et al. 2004; Kause et al. 2007; Nguyen et al. 2010; Gjerde et al. 2012; Thodesen et al. 2012). By contrast, most studies obtained modest to low prediction accuracies for fillet percentage models developed from body measurements, although a few have shown potentially useful prediction power (Bosworth et al. 2001; Sang et al. 2009).

Prediction models for fillet yields are typically developed using relationships between growth metrics and fillet yields. There is a lack of information regarding phenotypic correlations between growth traits and processing characteristics for Asian sea bass stocks in Thailand. Therefore, the objective of this study was to investigate the potential for total weight and body measurements of live fish to predict fillet weights and fillet yields in four different stocks of Asian sea bass and to develop prediction equations for these traits.

Materials and methods

Fish origin and fry production

Sea bass fingerlings at approximately 90 days post-hatch (dph) were obtained from private hatcheries, including Chachoengsao (CHAC), Chon Buri (CHON), Chumphon (CHUM), and Samut Songkram (SAMU). The CHAC and SAMU stocks originated from wild populations of the Bang Pakong and Meklong Rivers, respectively, in the upper Gulf of Thailand. The CHON stock was founded from broodstock obtained from cage culture operations along the east coast of the Gulf of Thailand, mainly in Chon Buri Province. The CHUM stock originated from wild populations of the Chumphon River, which runs through the west coast of the Gulf of Thailand. These hatchery populations represent primary sources of seed stocks for Asian sea bass farming in Thailand (Department of Fisheries 2016; Senanan et al. 2015). In general, the hatcheries produced sea bass fry from group mating with a sex ratio of 1:2 (male:female). About 18–24 brood fish, aged 3–5 years with body weight of 4–5 kg were stocked in each spawning tank ($2 \times 4 \times 2.5$ m^3). They were injected with 1 ml of synthetic hormone (Suprefact) under the dorsal fin, in the afternoon, to induce gamete maturation. Spawning normally took place 24–28 h post-injection (second night). Eggs were collected at 38–40 h (second morning) post-spawning and transferred to nursing tanks ($1.5 \times 2.5 \times 1$ m^3), where larvae were reared until 20 dph. Fry were then transferred to concrete tanks ($2 \times 4 \times 1$ m^3) and size graded at 30 and 60 dph at the hatcheries to minimize cannibalism.

Fish culture

The grow-out experiment was conducted at the Kasetsart University Fisheries Research Station in Samut Songkram Province. Four 1 m deep earthen ponds (20 m $\times$ 40 m or 800 m^2) were each equipped with four, 36 m^2 stationary cages ($6 \times 6 \times 1.2$ m), providing a randomized complete block design (RCBD). In each pond (block), populations were randomly assigned to cages, each with 50 fish (a density of 1.4 fish per m^2). Average size at stocking ranged from 9.48 ± 0.62 to 9.74 ± 0.16 g and 8.13 ± 0.82 to 9.04 ± 0.90 cm total lengths. The cages were covered with netting to exclude predatory birds. Aeration was provided through polyethylene pipes on the pond bottoms below the cages to maintain sufficient dissolved oxygen concentration. Water was supplied from a brackish water canal connected to the Meklong River. Fresh well water was used to dilute brackish water to obtain salinities within 10–25 ppt. There was no water exchange during grow-out, but the ponds were topped off regularly to compensate for evaporation, maintaining the water depth within the cages at 80–90 cm. The cage netting was cleaned regularly to allow water flow through the cages.

Fish were fed to apparent satiation with a 40% protein commercial floating feed (Betagro Feed Mills, Bangkok, Thailand) twice daily at 07:00 and 17:00 h. Feed consumption was closely monitored and uneaten was removed prior to the next feeding. Fish health was observed and mortalities were immediately removed during feedings. Temperature and other water quality parameters were checked twice daily. Dissolved oxygen (DO) was measured using a portable model YSI 550A meter (Yellow Springs Instruments, Inc.; Yellow Springs, OH, USA) and pH using a model YSI EcoSense pH 100A meter (Yellow Springs Instruments, Inc.; Yellow Springs, OH, USA). Salinity was measured weekly using a portable salinity refractometer (Atago, Co., Ltd.; Tokyo, Japan). Aquatic plants and algae that may interfere with fish growth were removed regularly from the ponds.

After 300 days of rearing, fish were harvested and transported to the Taiban Fishery processing plant in Samut Prakarn Province, approximately 70 km from the research station. Fish were weighed (*W* in grams) and tagged to identify stock. They were placed on a light-colored background on their left side and photographed using a Nikon D40X digital camera. A centimeter scale was included in the images to ensure calibration of body measurement to the nearest 0.01 cm. SigmaScan Pro 4.0 software (www.sigmaplot.com) was used to measure 12 distances between body locations as illustrated in Fig. 1. A total of 400 fish (100 from each stock) were filleted by four skilled processors (*P*) at the processing plant, with each processor filleting individuals from all four stocks. The filleting process was as follows: (1) scales were removed manually. (2) Fish were cut from the top of the head, down the side behind the pectoral fins and along the side of the dorsal fins, and (3) the fillet was removed by cutting along the backbone from head to tail. The weight of the two fillets with skin attached (FW) was measured and recorded in grams. Fillet yield (FY) was calculated as $FY = (FW/W) \times 100$.

Statistical analysis

The distribution of data was examined using SAS PROC UNIVARIATE procedure (SAS Inst. Inc., Cary, NC, USA). The output indicated that all the analysis variables were normally distributed. Analysis of variance was performed to examine the effects of pond (block) and genetic factors (stock) on weight, standard length and

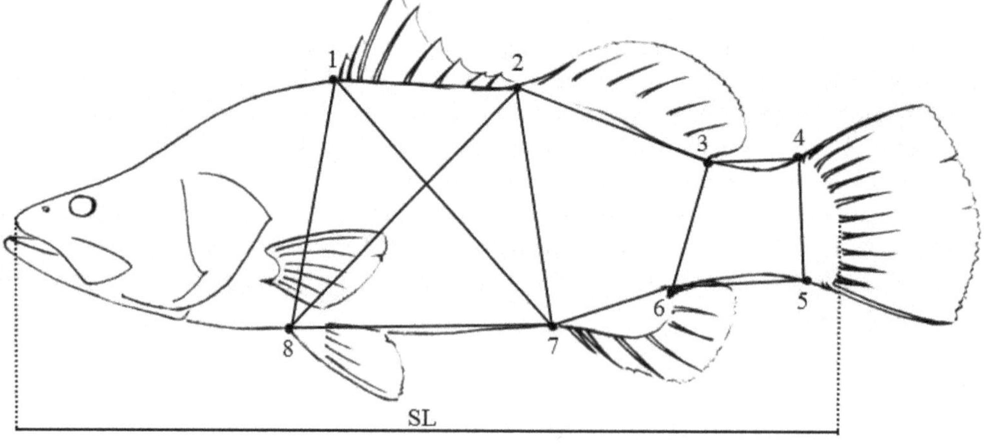

Fig. 1 Body measurement in Asian sea bass: distances from anterior 1st dorsal-anterior 2nd dorsal (*M12*), anterior 1st dorsal-anterior anal (*M17*), anterior 1st dorsal-anterior pelvic (*M18*), anterior 2nd dorsal-posterior dorsal (*M23*), anterior 2nd dorsal-anterior anal (*M27*), anterior 2nd dorsal-anterior pelvic (*M28*), posterior dorsal–dorsal membrane of caudal fin (M34), posterior dorsal-posterior anal (*M36*), dorsal–ventral membrane from caudal fin (*M45*), ventral membrane of caudal fin-posterior anal (*M56*), posterior anal-anterior anal (*M67*), anterior anal-anterior pelvic (*M78*), and *SL* standard length

condition factor of preprocessed fish. There was a significant genetic effect on fish weights ($P < 0.0001$), whereas a block effect was not significant. Consequently, fillet yield was analyzed with total weight fitted as a covariate using the general linear model procedure (PROC GLM: SAS). The fixed effects of processor and stock were added to the model as: $y_{ijk} = \mu + P_i + S_j + \beta * W_{ijk} + e_{ijk}$, where y_{ijk} is an observation of FY, μ is the overall mean, P_i is a fixed effect of processor, S_j is a fixed effect of stock, β is the regression coefficient of body weight, W_{ijk} is a co-variable of total weight and e_{ijk} is a random residual error. In addition, fillet weight was analyzed with the fixed effects of processor and stock as: $y_{ijk} = \mu + P_i + S_j + e_{ijk}$, where y_{ijk} is an observation of FW and other parameters are as described previously.

Scatter plots revealed relationships between total weight and fillet traits (Fig. 2). Data for body weight and body measurements were pooled across stocks and phenotypic correlations among traits were calculated using Pearson correlations with PROC CORR in SAS. Residual correlations were calculated after adjustments for systematic effects (Bosworth et al. 2001; Hung and Nguyen 2014). Collinearity of independent variables was tested using a variance inflation factor (VIF). Due to high correlations among predictors, fillet weight was predicted from total weight and length using simple linear regression models (PROG REG) in SAS. Because there were many independent variables to consider, stepwise regression was used to develop prediction equations for fillet yields. Goodness of fit for the models was assessed using coefficient of determination (R^2). In addition, the best model was determined based on conceptual predictive criterion (Cp) and Akaike's information criterion (AIC). Probability values < 0.05 were considered significant for estimates of parameters (intercept and beta). To assess the ability of these equations to predict new data, fillet records of each stock were evaluated using a fourfold cross-validation procedure. In the cross-validation, the data were randomly split into four subsets. Each iteration of the cross-validation used 75% of the records for training and the remaining 25% for testing. Predictive abilities of the equations were evaluated using the root mean squared error of prediction (RMSEP), calculated as the standard deviation of the residuals and the correlation between the predicted and the actual values (r) calculated for the best model. Prediction bias was calculated as the difference between the actual and the predicted values.

Results

Analyses of variance demonstrated highly significant differences in total weight and standard length among population samples of Asian sea bass ($P < 0.0001$) (Table 1). At 390 days of age, CHAC exhibited the highest weight (1166.32 ± 23.42 g). Least squares means for total weight among the remaining three stocks were similar and ranged from 997.44 ± 24.71 to 982.96 ± 25.07 g. Analysis of processing yields

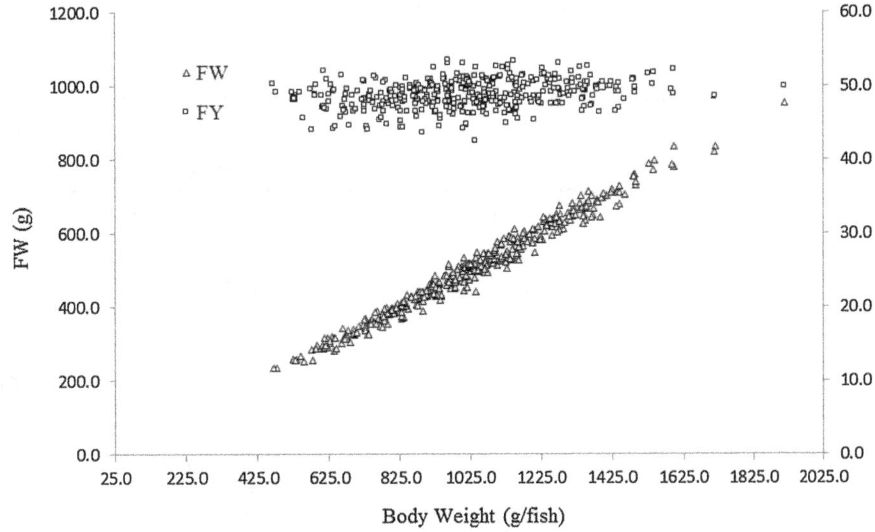

Fig. 2 Scatter plots showing relationships between body weight and fillet traits (*FW* fillet weight, *FY* fillet yield)

demonstrated significant differences among stocks. The CHAC samples exhibited the highest fillet weight of 596.23 ± 12.27 g, whereas weights among the remaining stocks were similar, ranging from 478.85 ± 13.32 to 491.53 ± 12.95 g. In addition, fillet yields differed among stocks with CHON (49.88 ± 0.19%) having the highest yield, followed by SAMU (48.24 ± 0.19%), CHUM (47.80 ± 0.20%), and CHAC (47.33 ± 0.19%).

Phenotypic correlations of total weight with body measurements and fillet traits are presented in Table 2. Positive high correlations were observed for weight, standard length, and body measurements with fillet weight (0.99, 0.93, and 0.40–0.91). Significant but weak correlations were observed for total weight, standard length and body measurements with fillet yield (0.21, 0.24, and 0.11–0.24). A modest relationship (0.34) was observed between fillet weight and fillet yield.

Regression equations for predicting fillet weight from total weight, standard length, and body measurement (*M18*) in each stock are presented in Table 3. Across populations, total weight accounted for 98% of the variation in fillet weight, whereas standard length and *M18* accounted for 86–91 and 61–84% of the variability in fillet weight, respectively. The model predictive power (RMSEP) for fillet weight indicated that the regression models developed from total weight resulted in a prediction error of 14.84–21.86 g. The errors of prediction were higher using standard length (39.84–44.99 g) and *M18* (51.60–71.58 g) as predictors. Correlation coefficients between actual and predicted fillet weights were high for total weight (0.99). Fillet weights were under-predicted by 10.6, 0.38, and 6.03 g for CHAC, CHON and CHUM stocks, and over-predicted by 0.42 g for SAMU stock. Results for fillet yield predictions are presented in Table 4. Table 5 provides validation statistics for model selection. The best models were chosen when the Cp approached the number of parameters and AIC was smallest. The R^2 values were low for all predictors, which explained only 15–28% of the variation in fillet yield. The RMSEP ranged from 1.21 to 1.70%. Correlation coefficients between actual and predicted fillet yields were low for CHUM (0.20) and moderate for CHAC (0.49), CHON (0.51), and SAMU (0.39) stocks. Fillet yields were over-predicted by 0.03% for CHAC and under-predicted by 0.06, 0.04 and 0.06% for CHON, CHUM, and SAMU stocks, respectively.

Discussion

Results of the grow-out experiment indicated significant stock effects for all body size parameters. In particular, fillet weight varied significantly from 478.8 ± 13.32 g to 569.2 ± 12.27 g over the size range (960–1190 g) of fish examined. The CHAC stock exhibited greater total and fillet weights than the other stocks, but had the lowest fillet yield (47.3%). Interestingly, three stocks (CHON, CHUM and SAMU) exhibited similar weight were found to differ in fillet yields, among which CHON had the highest fillet percentage (49.88%). Such differences could result from variations in body shape among populations likely due to genetic composition and effects of selection within stocks. Differences in fillet yields (31–38%) between strains of Nile tilapia have been reported (Rutten et al. 2004; Peterman and Phelps 2012). Fillet

Table 1 Least squares means (± SE) for total weight (g), standard length (cm), fillet weight (g), and fillet yield (%) for Asian sea bass, after 300 days of rearing

Stock	Weight (g)	Length (cm)	Fillet weight (g)	Fillet yield (%)
CHAC	1166.32 ± 23.42^{a}	37.19 ± 0.25^{a}	569.23 ± 12.27^{a}	47.33 ± 0.19^{c}
CHON	982.96 ± 25.07^{b}	35.48 ± 0.27^{b}	486.10 ± 13.13^{b}	49.88 ± 0.19^{a}
CHUM	984.72 ± 25.42^{b}	35.37 ± 0.28^{b}	478.85 ± 13.32^{b}	47.80 ± 0.20^{c}
SAMU	997.44 ± 24.71^{b}	35.83 ± 0.27^{b}	491.53 ± 12.95^{b}	48.24 ± 0.19^{b}
ANOVA				
F (stock/cage)	14.09	10.36	11.36	6.00
P value	< 0.0001	< 0.0001	< 0.0001	0.0005
F (processor)			0.46	0.86
P value			0.7127	0.4634

ANOVA results showing *F* values for differences among populations for preprocessed fish and among populations and processors for fillet weight and fillet yield. Entries in columns with different superscripts differ significantly at $P < 0.05$

Table 2 Pearson's correlation coefficients (above the diagonal) and residual correlations between total weight (*W*), standard length (*L*), body measurements *M12, M17, M18, M23, M27, M28, M34, M36, M45, M56, M67 and M78*, fillet weight (*FW*) and fillet yield (*FY*)

Variables	*W*	*L*	*M12*	*M17*	*M18*	*M23*	*M27*	*M28*	*M34*	*M36*	*M45*	*M56*	*M67*	*M78*	*FW*	*FY*
W		0.93	0.77	0.91	0.87	0.81	0.87	0.90	0.52	0.84	0.76	0.40	0.52	0.84	0.99	0.21
L	0.92		0.76	0.91	0.86	0.76	0.88	0.89	0.53	0.83	0.74	0.41	0.48	0.86	0.93	0.24
M12	0.77	0.76		0.85	0.62	0.56	0.71	0.78	0.40	0.69	0.53	0.34	0.40	0.72	0.76	0.14
M17	0.92	0.91	0.86		0.85	0.79	0.91	0.95	0.47	0.87	0.73	0.34	0.44	0.92	0.91	0.22
M18	0.86	0.77	0.62	0.85		0.79	0.90	0.91	0.53	0.83	0.79	0.38	0.55	0.83	0.86	0.22
M23	0.81	0.77	0.56	0.79	0.78		0.80	0.78	0.38	0.75	0.75	0.22	0.63	0.76	0.80	0.19
M27	0.89	0.89	0.72	0.91	0.90	0.80		0.93	0.49	0.86	0.76	0.31	0.53	0.87	0.87	0.23
M28	0.91	0.90	0.79	0.95	0.91	0.77	0.93		0.52	0.89	0.78	0.34	0.50	0.90	0.89	0.24
M34	0.53	0.53	0.40	0.49	0.55	0.40	0.51	0.55		0.63	0.57	0.71	0.32	0.43	0.51	0.12
M36	0.85	0.83	0.68	0.87	0.84	0.75	0.87	0.88	0.64		0.75	0.42	0.50	0.80	0.83	0.21
M45	0.76	0.73	0.53	0.72	0.77	0.72	0.75	0.78	0.60	0.75		0.44	0.54	0.71	0.76	0.18
M56	0.29	0.39	0.33	0.36	0.39	0.24	0.33	0.38	0.71	0.43	0.49		0.15	0.25	0.40	0.11
M67	0.52	0.47	0.39	0.44	0.54	0.64	0.54	0.50	0.31	0.50	0.53	0.12		0.31	0.51	0.16
M78	0.87	0.88	0.75	0.36	0.39	0.75	0.33	0.39	0.46	0.43	0.70	0.31	0.32		0.84	0.23
FW	0.98	0.92	0.75	0.91	0.86	0.80	0.88	0.91	0.52	0.84	0.75	0.39	0.50	0.80		0.34
FY	0.25	0.27	0.16	0.23	0.25	0.21	0.23	0.25	0.13	0.21	0.20	0.14	0.17	0.22	0.39	

Table 3 Linear regression models showing intercept, regression coefficients, and coefficient of determination (R^2) for predicting fillet weight from total weight (*W*), standard length (*L*), and body measurement *M18* in Asian sea bass stocks

Stock	Intercept	Regression coefficients			Prediction	*P* value	Validation		
		W	*L*	*M18*	R^2		RMSEP	*r*	Bias
CHAC	− 37.70	0.52	–	–	0.97	< 0.0001	21.86	0.98	10.60
	− 1293.06	–	50.08	–	0.86	< 0.0001	44.99		
	− 584.01	–	–	98.84	0.61	< 0.0001	76.58		
CHON	− 31.14	0.53	–	–	0.98	< 0.0001	17.74	0.99	0.38
	− 1155.57	–	46.26	–	0.91	< 0.0001	39.80		
	− 708.32	–	–	107.16	0.84	< 0.0001	52.88		
CHUM	− 13.09	0.50	–	–	0.98	< 0.0001	18.04	0.99	6.03
	− 1036.36	–	42.84	–	0.90	< 0.0001	39.84		
	− 664.29	–	–	106.55	0.78	< 0.0001	59.71		
SAMU	− 26.69	0.52	–	–	0.98	< 0.0001	14.74	0.99	− 0.42
	− 1044.18	–	42.86	–	0.88	< 0.0001	40.29		
	− 726.49	–	–	108.61	0.80	< 0.0001	51.60		

Validation of prediction equations is based on root mean squared error of prediction (RMSEP), correlation coefficient between predicted and observed values (*r*), and prediction bias

percentages of Asian sea bass (47.3–49.8%) were similar to fillet yields observed in hybrid striped bass (49.7%) (Bosworth et al. 1998). This is not surprising as both species are anadromous perciformes fish, having slightly compressed, streamlined body shapes. In other species, fillet percentages varied from 35% for river catfish *Pangasianodon hypophthalmus* (Sang et al. 2009) and gilthead seabream *Sparus auratus* (Navarro et al. 2009) to 32–41% for common carp (Kocour et al. 2007; Bauer and Schlott 2009) and 42% for channel catfish (Argue et al. 2003). Fillet yields were as high as 58% for Coho salmon *Oncorhynchus kisutch* (Neira et al. 2004), 64% for rainbow trout *O. mykiss* (Kause et al. 2002), and 69% for Atlantic salmon *Salmo salar* (Powell et al. 2008). In giant freshwater prawn *Macrobrachium rosenbergii*, a meat yield of 45% was reported (Hung and Nguyen 2014). With regard to phenotypic correlations among growth metrics, fillet weight was closely

Table 4 Stepwise regression equations for predicting fillet yield from body measurements in Asian sea bass stocks. Predictor variables are standard length (L) and body measurements (*M17*, *M18*, *M28*, and *M67*)

Stock	Intercept	Regression coefficients				
		L	*M17*	*M18*	*M28*	*M67*
CHAC	30.88	0.67	− 0.34	− 0.06	–	–
CHON	38.44	0.76	− 0.90	0.24	–	− 0.76
CHUM	52.27	0.80	− 0.35	− 0.07	–	–
SAMU	41.59	0.05	–	− 0.01	0.36	–

Table 5 Validation statistics (conceptual predictive criterion, Cp and Akaike's Information Criterion, AIC), model property (coefficient of determination, R^2, and root mean square of prediction, RMSEP), and relationship between actual and predicted values (r) and prediction bias of the best models for predicting fillet yield in Asian sea bass populations

Population/traits	Cp	AIC	R^2	RMSEP	*r*	Bias	*P* value
CHAC							
L	0.0005	407.55	0.20	1.85			< 0.0001
L + *M18*	1.0026	405.36	0.22	1.84			0.0137
L + *M18* + *M17*	2.0000	403.58	0.28	1.21	0.49	− 0.03	0.0102
CHON							
L	0.0021	349.98	0.15	1.78			< 0.0001
L + *M18*	1.0001	348.05	0.17	1.77			0.0003
L + *M18* + *M17*	2.0021	345.85	0.20	1.73			0.0153
L + *M18* + *M17* + *M67*	3.0001	344.38	0.22	1.70	0.51	0.06	0.0096
CHUM							
L	0.0026	304.71	0.12	1.46			< 0.0001
L + *M18*	1.0032	303.26	0.14	1.45			0.0935
L + *M18* + *M17*	2.0000	302.43	0.15	1.43	0.20	0.04	0.0002
SAMU							
L	0.0034	332.54	0.12	1.53			< 0.0001
L + *M18*	1.0000	330.71	0.15	1.52			0.0296
L + *M18* + *M28*	2.0030	329.06	0.17	1.52	0.39	0.06	0.0002

related to fish weight, as well as standard length and body measurements, while there were weak correlations between fillet yield and fish size. Results were mostly in agreement with other previous studies, in that high correlations between fillet weight and total weight suggest that growing fish to a larger size is an efficient way to increase fillet weight.

Prediction equations developed in this study were based on simple linear regression, as some of the independent variables are highly correlated, e.g., total weight and standard length. In general, the most useful prediction equation is the one that accounts for the largest proportion of the variation (R^2) in the predicted traits. An R^2 of 1 would be a perfect prediction while an R^2 of 0 would indicate no predictive power. The predictions of fillet weight from total weight gave R^2 values of 0.98–0.99 and would be the equations recommended for use in practice for sea bass populations in this study. The proportion of variation explained by the prediction equations in our study using fish weight for fillet weight (0.98) was similar to the 0.95 R^2 reported for tilapia by Rutten et al. (2004). Similarly, Sang et al. (2009) predicted river catfish fillet weight of from body weight, and obtained an R^2 value of 0.86, which was lower than the values obtained in this study. However, a lesser degree of fit was exhibited when standard length (R^2 values of 0.86–0.81) or body measurement (*M18*) (R^2 values of 0.61–0.80) was used as predictors for fillet weight. In addition to R^2 values, RMSEP and correlation coefficients between observed and predicted values (r) of the equations should be considered if predictions are to be made for Asian sea bass stocks other than those in this study. Validation of the model developed from body weight showed the highest correlation coefficients between predicted and observed fillet weights (0.98–0.99) and lowest RMSEP values (14.74–21.86 g), suggesting that these equations would be applicable to other Asian sea bass stocks.

Fillet yield, as a proportional trait consisting of fillet weight and total weight, showed weak relationships to total weight, standard length, and body measurements. Total weight is an important variable, normally included in prediction equations for fish fillet yield. The proportion of variation in fillet yield explained by the regression equation was low to moderate (R^2 = 0.16–0.36) for total weight in tilapia (Rutten et al. 2004), but much higher for the ratio between body volume and body weight in river catfish (0.75) (Sang et al. 2009). However, only body measurement traits not total weight were included in the stepwise-developed regression equations in our study. The first included variable was standard length, which accounted for 12–20% of the variation in fillet yields, depending on the stock. Addition of other variables (*M17*, *M18*, *M28* and *M67*) depending on the stock accounted for an additional 5–8% of the variation in fillet yields. Although the best model for each population gave a low to moderate degree of predictive ability for fillet yield, prediction bias was close to zero. The relationship between fish size and fillet yield in this study generally confirms the principle that larger fish provide higher fillet yield percentages. However, regarding the prediction equations, for example, the regression coefficient of standard length on fillet yield was 0.67 for the CHAC stock. This value represents the expected change in the predicted fillet yield for each unit change in fish length, when the remaining variables are fixed. Therefore, for each 1 cm increase in body length, an individual fillet yield is expected to increase 0.67%. This prediction ability for predict fillet yield is probably too low to be practical for predicting new data for the CHUM stock (r = 0.20). At this stage, it appears that our regression models for predicting fillet yields in three stocks are adequate (r values of 0.39–0.51). For future work, it would be useful to develop similar models for larger fish (1500–2000 g) using a larger data set to increase predictive power. With the use of up to 2700 records and a multiple regression models, Sang et al. (2009) obtained an r value of 0.86 for predicting fillet yield in river catfish.

In conclusion, total weight is the best single measurement for predicting fillet weight in Asian sea bass. Regression models developed in this study are applicable to fish weighing 960–1190 g from all populations sampled. Using separate models yielded similar prediction power and accuracy for fillet weight. This study demonstrates the use of various body measurements in predicting fillet yields. Despite relatively weak correlations between body measurements and fillet yields, we were able to develop useful prediction models for fillet yields in three stocks of Asian sea bass. Although fillet yield is the most important trait for fish processors, increased fillet percentages could be achieved along with increased fillet weights, given the positive correlation between the traits. The application of prediction models would allow for rapid genetic improvement via breeding to increase fillet yields and enable producers to be more competitive in the value-added markets of Asian sea bass.

Acknowledgements We would like to acknowledge the Center for Advanced Studies in Agriculture and Food (CASAF), Institute for Advanced Studies, Kasetsart University (Grant number CASAF 094) for providing a scholarship funding to the first author. We acknowledge the Kasetsart University Research and Development Institute (KURDI) for their support through grant number Kor-Sor (Dor) 34.59.

References

Argue BJ, Liu Z, Dunham RA (2003) Dress-out and fillet yields of channel catfish, *Ictalurus punctatus*, blue catfish, *Ictalurus furcatus*, and their F1, F2 and backcross hybrids. Aquaculture 228:81–90

Bauer C, Schlott G (2009) Fillet yield and fat content in common carp (*Cyprinus carpio*) produced in three Austrian carp farms with different culture methodologies. J Appl Ichthyol 25:591–594

Bosworth BG, Libey GS, Notter DR (1998) Relationships among total weight, body shape, visceral components, and fillet traits in palmetto bass (striped bass female *Morone saxatilis* × white bass male *M. chrysops*) and paradise bass (striped Bass female *M. saxatilis* × yellow bass male *M. mississippiensis*). J World Aquacult Soc 29:40–50

Bosworth BG, Holland M, Brazil BL (2001) Evaluation of ultrasound imagery and body shape to predict carcass and fillet yield in farm-raised catfish. J Anim Sci 79:1483–1490

Cibert C, Fermon Y, Vallod D, Meunier FJ (1999) Morphological screening of carp *Cyprinus carpio*: relationship between morphology and fillet yield. Aquat Liv Resour 12:1–10

Department of Fisheries (2016) Fishery statistics analysis and research group. Information and Communication Technology Center, Department of Fisheries, no. 3/2016

Food and Agriculture Organization (2017) Fishstat Plus Version 2.30. FAO Fisheries and Aquaculture Department, Fishery Information, Data and Statistics Unit. www.fao.org/fishery/statistics/software/fishstat/en

Gjedrem T (1997) Flesh quality improvement in fish through breeding. Aquacult Int 5:197–206

Gjedrem T (2000) Genetic improvement of cold-water species. Aquacult Res 31:25–33

Gjerde B, Mengistu SB, Ødegård J, Johansen H, Altamirano DS (2012) Quantitative genetics of body weight, fillet weight and fillet yield in Nile tilapia (*Oreochromis niloticus*). Aquaculture 342–343:117–124

Hung D, Nguyen HN (2014) Modeling meat yield based on measurements of body traits in genetically improved giant freshwater prawn (GFP). Aquacult Int 22:619–631

Kause A, Ritola O, Paananen T, Mantysaari E, Eskelinen U (2002) Coupling body weight and its composition: a quantitative genetic analysis in rainbow trout. Aquaculture 211:65–79

Kause A, Paananen T, Ritola O, Koskinen H (2007) Direct and indirect selection of visceral lipid weight, fillet weight, and fillet percentage in a rainbow trout breeding program. J Anim Sci 85:3218–3227

Kocour M, Mauger S, Rodina M, Gela D, Linhart O, Vandeputte M (2007) Heritability estimates for processing and quality traits in common carp (*Cyprinus carpio* L.) using a molecular pedigree. Aquaculture 270:43–50

Navarro A, Zamorano MJ, Hildebrandt S, Gines R, Aguilera C, Afonso JM (2009) Estimates of heritabilities and genetic correlations for growth and carcass traits in gilthead seabream (*Sparus auratus* L.), under industrial conditions. Aquaculture 289:225–230

Neira R, Lhorente JP, Araneda C, Diaz N, Bustos E, Alert A (2004) Studies on carcass quality traits in two populations of Coho salmon (*Oncorhynchus kisutch*): phenotypic and genetic parameters. Aquaculture 24:117–131

Nguyen NH, Ponzoni RW, Abu-Bakar KR, Hamzah A, Khaw HL, Yee HY (2010) Correlated response in fillet weight and yield to selection for increased harvest weight in genetically improved farmed tilapia (GIFT strain), *Oreochromis niloticus*. Aquaculture 305:1–5

Peterman MA, Phelps RP (2012) Fillet yields from four strains of Nile Tilapia (*Oreochromis niloticus*) and a Red variety. J Appl Aquacult 24:342–348

Powell J, White I, Guy D, Brotherstone S (2008) Genetic parameters of production traits in Atlantic salmon (*Salmo salar*). Aquaculture e 274:225–231

Rutten MJM, Bovenhuis H, Komen H (2004) Modeling fillet traits based on body measurements in three Nile tilapia strains (*Oreochromis niloticus* L.). Aquaculture 231:113–122

Rutten MJM, Bovenhuis H, Komen H (2005) Genetic parameters for fillet traits and body measurements in Nile tilapia (*Oreochromis niloticus* L.). Aquaculture 246:125–132

Sang NV, Thomassen M, Klemetsdal G, Gjøen HM (2009) Prediction of fillet weight, fillet yield, and fillet fat for live river catfish (*Pangasianodon hypophthalmus*). Aquaculture 288:166–171

Senanan W, Pechsiri J, Sonkaew S, Na-Nakorn U, Sean-In N, Yashiro R (2015) Genetic relatedness and differentiation of hatchery populations of Asian seabass (*Lates calcarifer*) (Bloch, 1790) broodstock in Thailand inferred from microsatellite genetic markers. Aquacult Res 46:2897–2912

Thodesen J, Rye M, Wang YX, Bentsen HB, Gjedrem T (2012) Genetic improvement of tilapias in China: genetic parameters and selection responses in fillet traits of Nile tilapia (*Oreochromis niloticus*) after six generations of multi-trait selection for growth and fillet yield. Aquaculture 366–367:67–75

14

Natural spawning, embryonic and larval development of F2 hybrid grouper, tiger grouper *Epinephelus fuscoguttatus* × giant grouper *E. lanceolatus*

Fui Fui Ching · Norazmi Othman · Azaharie Anuar · Rossita Shapawi · Shigeharu Senoo

Abstract This study aims to reveal the first report of the natural spawning of F1 hybrid grouper (TGGG), a crossbreed between the tiger grouper, *Epinephelus fuscoguttatus* × giant grouper, *E. lanceolatus*, since its first production in 2006. This marks the completion of its full cycle after a 10-year period. In order to establish a seed rearing protocol for a novel F2 hybrid TGGG, natural spawning, embryonic and larval developments were thoroughly observed. Five batches of natural spawning were recorded with an average of 1.50–15.3 kg eggs collected, while fertilization and hatching rates were recorded at 85.3–97.6%, and 63.0–98.3%, respectively. F2 larvae hatched out at 17:50 hours with an average body size of 1.74 ± 0.01 mm, and a yolk sac volume of 0.85 ± 0.197 mm^3. The first feeding was initiated 3 days after hatching, which coincided with the onset of functional feeding apparatus and active swimming behavior. Larval dorsal and pelvic spines were formed at 6 days AH coupled with dynamic feeding activity, as more food was found in the digestive tract. Meanwhile, the F2 hybrid grouper shifted habitat from pelagic to benthic as early as 25 days AH, and entered a juvenile stage at 35 days AH, attaining a skin coloration similar to that of the F1 juvenile. This study concluded that naturally spawned eggs of F2 hybrid TGGG were exceptionally high in quality, although larvae were small and fragile, and performed vigorous feeding activities and cannibalistic behavior. Thus, these findings can serve as primary data to further develop the optimal rearing protocol to enhance the overall rearing performance.

Keywords Hybrid grouper · Natural spawning · Embryonic and larval development · Morphology · Sensory organ · Behavior changes

Introduction

Hybridization is a process that involves the crossbreeding of two animals or plants from the different taxa, to produce a new species (Dunham et al. 1987). In aquaculture, hybridization is a powerful tool that offers the hope of producing aquatic organisms with valuable traits, where the offspring carries the characteristics of hybrid vigor or positive heterosis. Generally, the preferred new offspring may result in shorter grow out and production cycle, higher survival, fast growth, better food conversion, enhanced flesh quality, high disease

Fui Fui Ching (✉) · N. Othman · A. Anuar · R. Shapawi
Borneo Marine Research Institute, Universiti Malaysia Sabah, Jln UMS, 88450 Kota Kinabalu, Sabah, Malaysia
e-mail: cfuifui@ums.edu.my

S. Senoo
Aquaculture Technology and Production Center, Kindai University, 3153 Shirahama, Wakayama 649-2211, Japan

resistance as well as the ability to tolerate a wider range of rearing environments, as similarly reported in various marine and freshwater hybrid fish species (Bunlipatanon and U-taynapun 2017; Bartley et al. 1991).

Hybrid groupers have taken the Asian aquaculture industry by storm since 2006, with the first production of hybrid TGGG, a crossbreed between tiger grouper (*Epinephelus fuscoguttatus*) and giant grouper (*E. lanceolatus*). This novel hybrid grouper has gained immediate popularity from aquaculturists and seafood consumers, owing to its production success and premium organoleptic properties, which has led to high commercial value.

To date, there are over ten hybrid groupers that have been produced, among which are crossbreeding in between orange spotted grouper (*E. coioides*), coral grouper (*E. corallicola*), mouse grouper (*Cromileptes altivelis*), camouflage grouper (*E. polyphekadion*), and red spotted grouper (*E. akaara*), among others (Liufu et al. 2007; Addin and Senoo 2011; Huang et al. 2014; Koh et al. 2010). TGGG are widely cited as the most successful hybrid combination, as it is able to grow quickly (Ch'ng and Senoo 2008), has a higher survival and better feeding performance (Othman et al. 2015), and is able to tolerate a wide range of rearing parameters (Shapawi et al. 2018; De et al. 2016).

Luin et al. (2013) studied the complete sexual maturation and gonad development of hybrid TGGG in captivity, and indicated that hybrid TGGG started to mature as female and male at approximately 9.0 and 11.0 kg, respectively. This initial report has opened up the possibility of using the hybrid progeny to develop F2 generation, and perform backcross breeding. In 2016, fertilized eggs were found in hybrid TGGG broodstock tank, and preserved in a hatchery at Borneo Marine Research Institute, Universiti Malaysia Sabah, Malaysia. These matured broodstock of hybrid TGGG had spawned naturally, and produced approximately 20 kg of fertilized eggs within 6 months. This marks the complete closed cycle of hybrid TGGG, after 10 years being introduced into aquaculture (Fig. 1). However, similar to other groupers, high larval mortality is considered the bottleneck that is hindering the establishment of F2 TGGG production.

In an attempt to establish a seed production technique for a new F2 hybrid TGGG, the authors observed the embryonic and larval development in relation to functional sensory organ and behavior changes of this new species. The findings are significantly useful for exploring the ideal rearing protocol, and later further improving the growth performance and overall production.

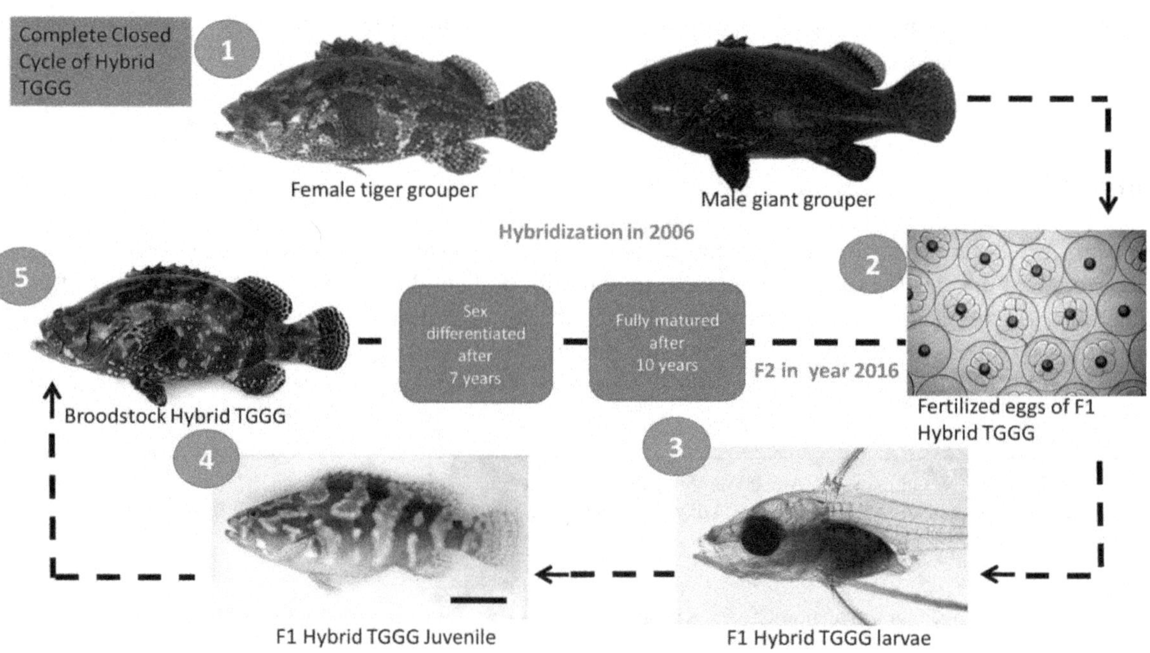

Fig. 1 Complete closed cycle of hybrid tiger grouper, *Epinephelus fuscoguttatus* × giant grouper, *E. lanceolatus* (TGGG)

Materials and methods

Broodstock management

The broodfish of hybrid TGGG used in this study have been reared for a period of 10 years in the hatchery of Borneo Marine Research Institute, Universiti Malaysia Sabah, Malaysia. Broodfish (n = 28) were initially kept in a tank with a capacity of 150 tonnes, and then transferred to one unit of 20 tonnes of cylindrical fiber reinforced plastic (FRP) tank equipped with a recirculation system for experiment purposes. The body weight of the broodfish ranged from 12 to 20 kg. They were fed until satiation daily with enriched prey fish (*Sardinella* sp.) supplemented with pure cod liver oil (Seven seas, Merck Company, United Kingdom), squid oil and vitamin premixes. The water salinity, temperature, dissolved oxygen and pH were maintained at 30–32 ppt, 27.0 ± 1.0 °C, 6.89–7.02 mg L^{-1} and 7.80–8.01, respectively.

Egg collection and larval rearing

Newly fertilized eggs were obtained from broodstock hybrid TGGG that spontaneously spawned between January and April 2016. The eggs were collected around 03:00 hours from a flow-through screen that was initially installed next to a 20 tonne broodstock tank, and were later incubated in a 1000-L FRP tank. Eggs were stocked at about 30 eggs L^{-1}, and gently aerated at 250 mL min^{-1}. The water temperature, pH and dissolved oxygen (DO) in the incubation tank ranged from 29.5 to 30.5 ± 0.5 °C, 6.9–7.2 and 6.8–6.9 mg L^{-1}, respectively.

The first feeding time was commenced at the first hour after the mouth had opened (h AMO), which coincided with the functionality of four feeding-related organs (eyes, mouth, intestine and anus). A 12L:12D (light and dark) light regime was provided by 12 h natural light (640–750 lx) at daytime (06:00–18:00) and 12 h dark (18:00–06:00) at night-time. During the first feeding, rotifer, *Brachionus plicatilis* sp. complex and commercial *Nannochloropsis oculata* (K2, New World Aqua, Korea) were provided to the larvae at a density of 30 individual mL^{-1} and 0.5×10^{-6} cells mL^{-1}, respectively.

From 0 h AMO, larvae (n = 10) were sampled from the larval rearing tank to measure total length (TL), yolk sac volume (YV) and oil globule volume (OGV). Larval TL was measured at 6 hourly intervals up to 78 h after hatching (h AH), and continued on a 5-day interval basis. Meanwhile, larval YV and OGV were measured until both yolk sac and oil globule were completely absorbed.

Daily routine regimes include bottom cleaning and water renewal for rearing maintenance purposes. Bottom cleaning was performed to remove debris, excessive feeds and dead larvae. Filtered seawater was later added (10–20%) in order to restore the amount of removed water, which was siphoned out during the bottom cleaning process. In addition, the feeding densities of both rotifers and *N. oculata* in the rearing water were maintained uniformly for the first 20 days of rearing, followed by *Artemia* and mixed species of copepods (80–450 μm) from day 20 to 40 and artificial feed from day 30 onwards. The water quality was monitored twice daily at 09:00 and 16:00 hours.

Sampling I: egg development

Newly fertilized eggs were collected from the incubation tank using a fine scoop net (60 μm mesh size) and transferred to a 500-mL beaker containing filtered seawater. The eggs were taken out and placed on a clean petri dish under a compound microscope (Nikon, Eclipse E600) for observation purposes. The time taken for the eggs to reach different development stages was recorded, and the size of the ovulated eggs was measured. Images of the eggs were captured by a digital camera (Canon, 60D, Japan). The fertilization rate (%) was recorded by calculating the total number of fertilized eggs/total number of eggs × 100. Meanwhile, the hatching rate (%) was determined by calculating total number hatchlings/total number of fertilized eggs × 100.

Sampling II: larval development

Yolk sac and oil globule absorptions Ten newly hatched larvae (n = 10) were sampled at 0 h after hatching (h AH) from the incubation tank. Larvae were positioned on a slide glass, under a profile projector (Mitutoyo, PJ3000, Japan), for measurements of total length (TL), yolk sac and oil globule length (mm) and height (mm), respectively. The yolk sac volume was measured by the formula for a prolate spheroid = $\pi/6\ lh2$, where l represents the yolk sac length and h represents the yolk sac height. The oil globule volume (OGV) was measured by the formula OGV = $\pi/6\ d^3$, where d represents the diameter of oil globule (Bagarinao 1986).

Indication of first feeding time The same specimen used for yolk sac and oil globule absorption was also used to observe larval first feeding time according to the method of Ching et al. (2014), whereby time of the completion of all feeding-related organs was recorded and defined as onset of first feeding time.

Larval morphology, sensory development and behavioral changes Starting from 0 h AH, ten larvae (n = 10) were randomly sampled from the rearing tank at a 6-h interval for the first 48 h. They were alternated with daily sampling from 3 days after hatching (d AH), and the total length and subsequent larval growth stages were carefully measured under a compound microscope and profile projector, respectively. The formation of sense organ was also recorded during the microscopic observation. As for behavioral observation, ten larvae (n = 10) were sampled daily and transferred to a 500-mL beaker, and their swimming behavior was recorded.

Results

Spawning and egg development

Five batches of natural spawning were recorded from January to April 2016 with each batch lasting for 3–5 days. The number of eggs (diameter 860 ± 30 μm) in each spawning ranged from 1.50 to 15.3 kg, with fertilization and hatching rates recorded at 85.3–97.6% and 63.0–98.3%, respectively (Table 1).

Egg development of F2 hybrid TGGG is shown in Fig. 2. At 1 min (0:01 h) after fertilization (AF), 2-cell stage was observed and first cleavage occurred (Fig. 2a). At 1 h 40 min AF, eggs developed to the 4-cell stage (Fig. 2b); and later at 2 h 15 min AF, the eggs developed to the 6-cell stage (Fig. 2c). At 5 h AF, eggs entered into a morula stage (Fig. 2d) and reached the gastrula stage at 8 h AF (Fig. 2e). Meanwhile, at 10 h 45 min AF, the first appearance of Kupffer's vesicles was observed (Fig. 2f).

Later at 11 h 50 min AF, the larval head and myomere were formed (Fig. 2g). At 12 h 35 min AF, the lens vesicle was noticed (Fig. 2h), and at 13 h 30 min AF, the first movement of the embryo was observed (Fig. 2i). The larval heart beat was first detected at 17 h 10 min AF (Fig. 2j), and the hatching started at 17 h 45 min AF (Fig. 2k). Later, newly hatched larvae were seen at 17 h 46 min AF (Fig. 2 l) with an average size of 1.74 ± 0.01 mm.

Larval development

The changes of the larval yolk sac and oil globule volume are shown in Fig. 3. Yolk sac and oil globule volumes were 0.85 ± 0.197 and 0.27 ± 0.05 mm^2, respectively, at 0 h AH (28.5 ± 0.1 °C). The F2 hybrid

Table 1 Natural spawning occurrence number, duration, total eggs, mean fertilization and hatching rates of F2 hybrid TGGG collected from January to April 2016

Batch	Month	Duration (day)	Total eggs (kg)	Fertilization rate (%)	Hatching rate (%)
1	January	3	1.5	97.67 ± 0.6	98.3 ± 0.6
2	February	3	1.6	85.3 ± 0.6	96.0 ± 1.0
3	February	5	1.4	91.3 ± 1.1	74.6 ± 2.5
4	March	5	15.3	92.7 ± 0.6	74.7 ± 2.1
5	April	5	3.64	91.0 ± 1.0	63.0 ± 2.0

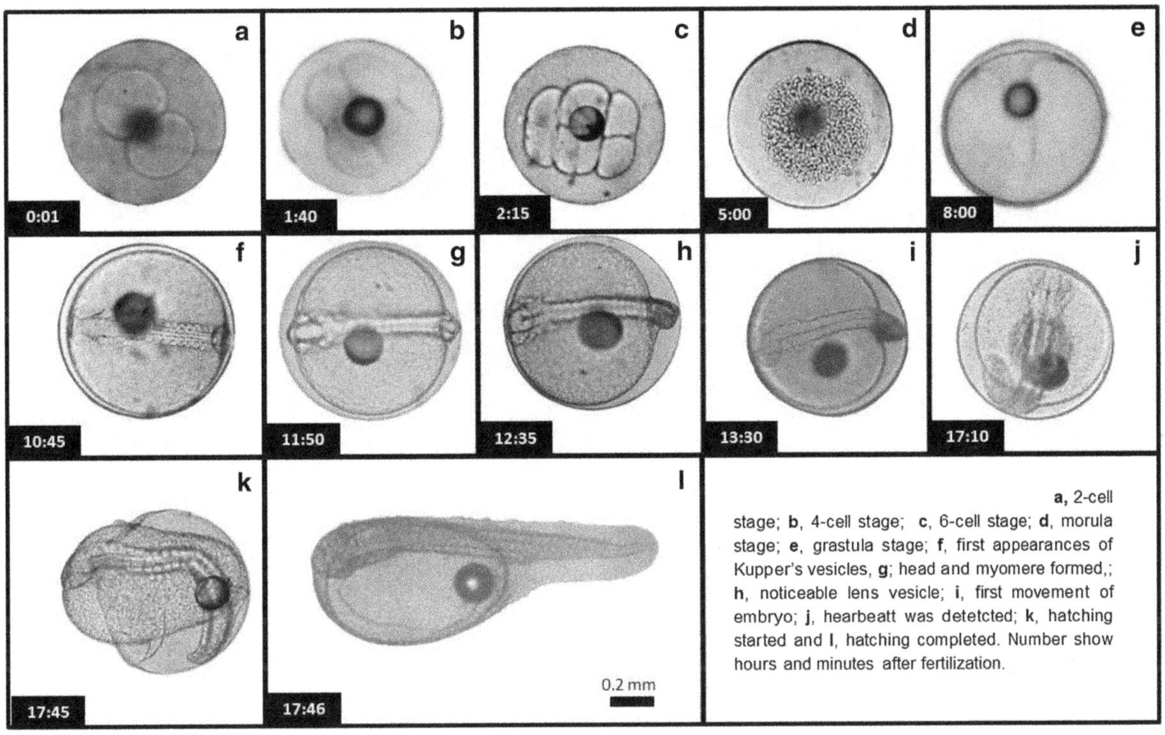

Fig. 2 Embryo development of F2 hybrid TGGG, *E. fuscoguttatus* × *E. lanceolatus*: **a** 2-cell stage, **b** 4-cell stage, **c** 6-cell stage, **d** morula stage, **e** gastrula stage, **f** first appearances of Kupffer's vesicles, **g** head and myomere formed, **h** noticeable lens vesicle, **i** first movement of embryo, **j** heartbeat was detected, **k** hatching started and **i** hatching completed. Numbers show hours and minutes after fertilization

TGGG larvae took approximately 60 and 72 h to absorb both yolk sac and oil globule, respectively. The larval development at an early stage is shown in Fig. 4, and its subsequent growth in Fig. 5. Its correlation to behavioral changes can be seen in Table 2.

Newly hatched larvae were seen floating near the water surface with the head positioned down, and no specific swimming pattern was observed. Larvae also were found with unopened mouth, non-pigmented eyes, closed anus and large yolk sac. The heart beat was observed under a microscope, and the onset of cardiac function in larvae was marked (Fig. 4a). At 6 h AH, the larvae were seen with pigmented eyes (Fig. 4b). At 18 h AH, they were well pigmented with visible pectoral fins, and melanophores can be observed. The anus was seen to be open.

At 24 h AH, the larvae grew with reduced yolk sac (0.08 ± 0.01 mm^3), and started to perform vertical swimming. They were able to avoid any disturbance created surrounding them. Such a reaction is highly associated with the presence of free neuromast and capula. However, during this time, all feeding-related organs were still undeveloped, with no sign of first feeding occurrence.

At 3 days AH, the larvae grew larger and were able to perform both vertical and horizontal swimming patterns, and all four feeding-related organs were well-developed when the larvae have pigmented eyes, an open mouth with a functional lower jaw, peristaltic digestive tract and an open anus (Fig. 4d). Rotifer was first found in the larval digestive tract at this stage. Meanwhile, the yolk sac was absorbed completely, and marked the onset of the larval exogenous feeding phase.

As larvae entered 6 days AH, the dorsal and pelvic spines started to form, and the larvae commenced active swimming behavior coupled with dynamic feeding activities at this stage, as more rotifer were found in the digestive tract. By the age of 25 days AH, both dorsal and pelvic spines were elongated, spiny and covered by melanophore layers. The larvae were seen as more aggressive, particularly in foraging prey and preferred *Artemia* and copepod compared to rotifers. Interestingly, the larvae started to aggregate around the central part of the bottom tank, instead of the other columns of the tank, and eventually switched to the benthic habitat.

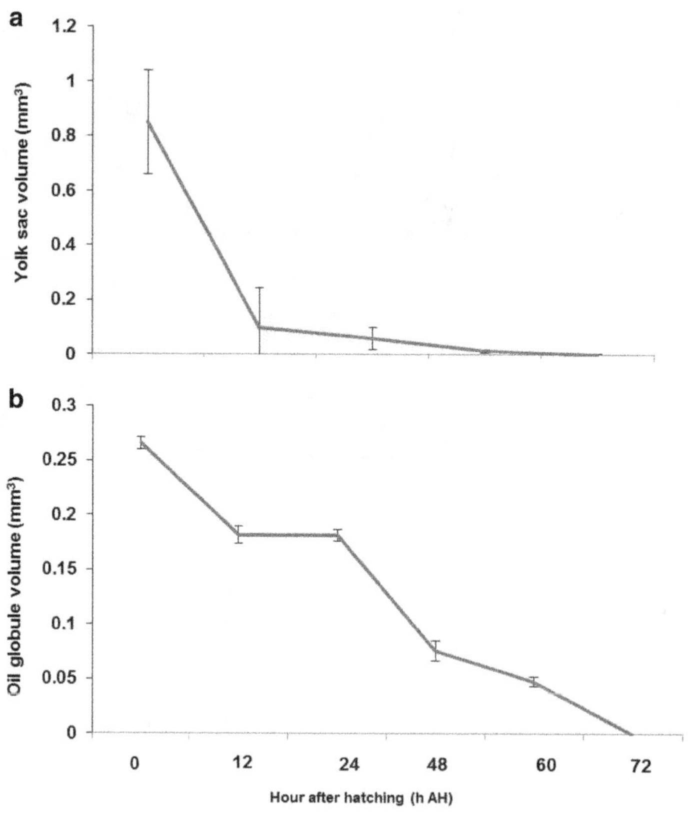

Fig. 3 The changes of yolk sac and oil globule volumes (mm^3) of F2 hybrid TGGG, *E. fuscoguttatus* × *E. lanceolatus*

By the age of 35 days AH, F2 hybrid TGGG entered the juvenile stage, as their morphological characteristics started to resemble those of an adult. The body is no longer transparent with slight pigmentation, and the lateral line can be seen clearly. The stomach was covered with a silver pouch, and was obvious to the naked eyes. At this stage, caudal fins were well-developed, along with the dorsal and pelvic spines. Although juveniles are seen as more aggressive compared to the larval stage, they remain at the bottom, with less swimming activity. However, they performed speedy swimming, particularly when approaching food.

Discussion

Natural spawning

This study revealed the first record of natural spawning in hybrid TGGG in captivity, since its first report on the successful sex differentiation and gonad maturation by Luin et al. (2013). To the best of our knowledge, this is the first report on natural spawning in hybrids among *Epinephelus* sp. and marked the first close cycle of hybrid TGGG. The majority of hybrid fish of various fish species documented were either sterile (Argue and Dunham 1999), low in fertility (Wagner and Oplinger 2013) or functionally fertile (Hodson 1989; Ma and Yamazaki 1986).

Obtaining matured broodstock are among the challenges in aquaculture (Song et al. 2005), and the natural spawning of hybrid TGGG is evidently possible in captivity through sustained efforts on broodstock management, to develop the closed cycle hybrid TGGG. This is beneficial to both the aquaculture and seafood industry, particularly in the Asian region, where hybrid TGGG are prominent (Shapawi et al. 2018).

Spawning triggers are environmental cues that lead the fish to spawn. The sudden changes in the environmental parameters in captivity such as fluctuation of water temperature, salinity, and photoperiod, often trigger several fish to spawn, namely, Yellow perch, *Perca flavescens* (Starzynski and Lauer 2015), striped

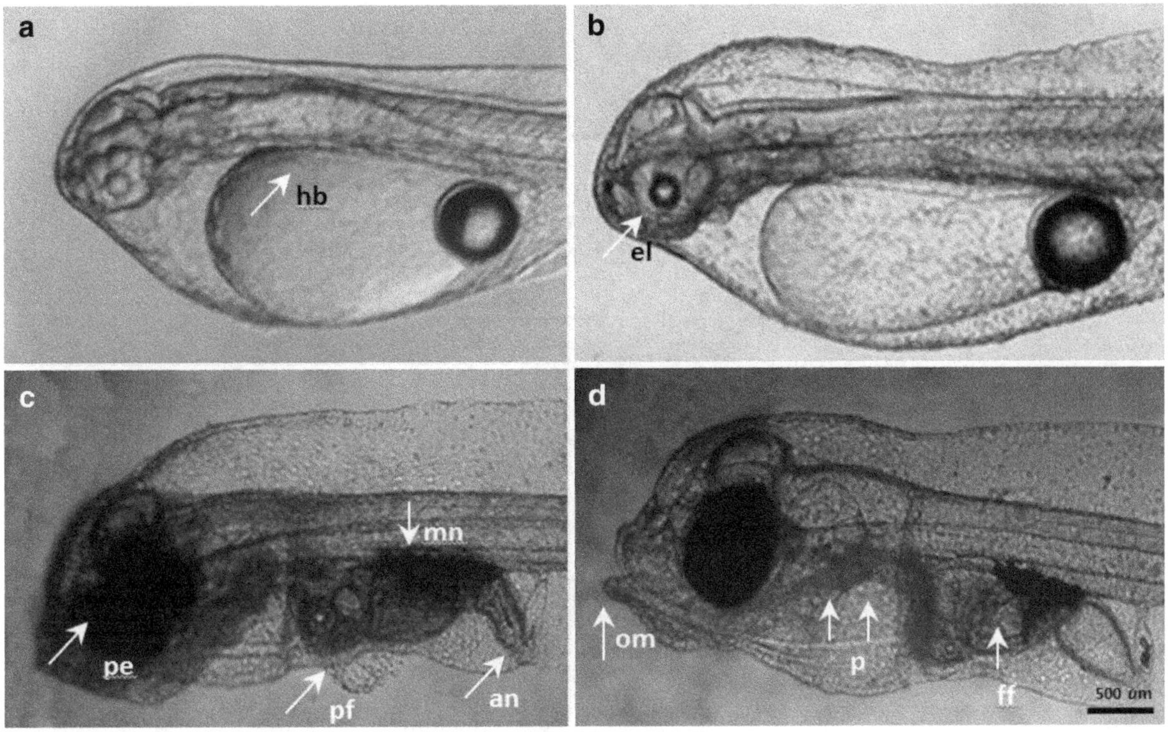

Fig. 4 Early stage larval development of F2 hybrid TGGG, *E. fuscoguttatus* × *E. lanceolatus*. **a** Visible heart beat (hb) detected; **b** onset of pigmentation around eye lens (el); **c** well pigmented eyes (pe), first appearance of pectoral fins (pe), opened anus (an) and visible melanophores; **d** functional and opened mouth (om), peristaltic movement detected in esophagus (p), first feeding detected (ff)

bass, *Morone saxatilis* (Henderson-Arzapalo and Colura 2011), yellowtail, and *Seriola quinqueradiata* (Mushiake et al. 1998), among others. In this study, the natural spawning of F1 hybrid TGGG was triggered by the sudden change of water depth from 3.0 to 1.5 m.

The water level fluctuation is a crucial factor in inducing reproductive adaptations in many fish species, and this study assumes a sudden change of water depth had triggered stress in the matured hybrid TGGG to release hormones, a chemical messenger that travels into the blood and eventually responds in a variety of different ways, including spawning. From an aquaculture point of view, the synchronization of spawning triggered by water depth could allow for some important advantages in hatchery operation. These include less use of costly synthetic hormones, reduced physical injury, and physiological stress of capturing, handling, injecting, and holding broodfish, which can have a greater detrimental effect on spawning success than almost any other factor (Rottmann et al. 1999).

Fertilization and hatching rates

The quality of a fish gamete is often defined as its ability to be fertilized and hatched, and later develops into a normal embryo (Chevassus-au-Louis and Lazard 2009; Bobe and Labbe 2010). Fertilization and hatching rates are powerful indicators that are used to predict the production of high quality fish larvae, larval survivability and economical utilization of overall hatchery operational costs. The fertilization and hatching rate of F2 hybrid TGGG were averaged at 91.5% and 80.7%, respectively. This is considered relatively higher compared to the parental species recorded at 86.8% and 87.2%, respectively, as reported by Ch'ng and Senoo (2008) and other hybrid groupers such as: (1) spotted grouper × tiger grouper (*E. polyphekadion* × *E. fuscoguttatus*) at 51.0% and 38.3%; (2) coral grouper x tiger grouper (*E. corallicola* × *E. fuscoguttatus*) at 75.0% and 35.4%; (3) and orang spotted grouper × giant grouper (*E. coioides* × *E. lanceolatus*) at 91.0 and 33.6%, respectively (Ch'ng and Senoo 2008; Ivan et al. 2008; Addin and Senoo 2011).

The findings reveal the possibility of this new species to be introduced as a suitable candidate of aquaculture species, judging from the high quality of fertilization and hatching rates. The high fertilization and

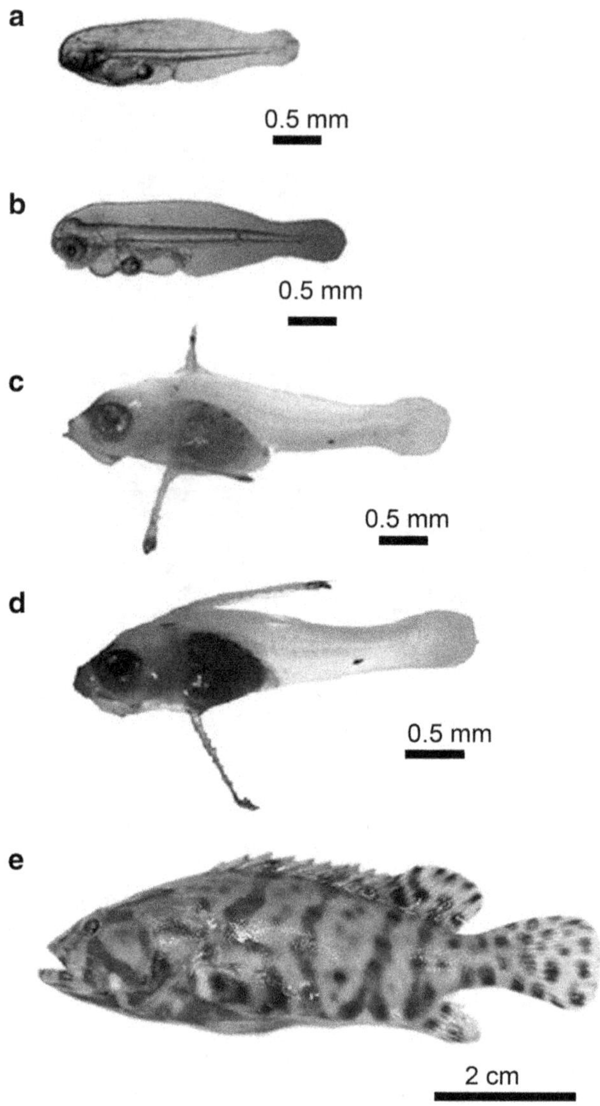

Fig. 5 Morphological changes of F2 hybrid TGGG, *E. fuscoguttatus* × *E. lanceolatus*. **a** 0 h after hatching (h Ah), **b** 12 h AH, **c** 6 days after hatching (d AH), **d** 25 days AH, **e** 35 days AH

hatching rates of F2 hybrid grouper in this study is contributed by the high quality of eggs and milts produced by the fertile hybrid TGGG reared in captivity, where water quality was strictly monitored to the finest level, and the food involved a nutritionally balanced diet.

Broodstock management can be pricy in aquaculture by taking the overall time and cost into consideration. However, given the importance of effective broodstock management in the provision of quality gametes, this critical management should not be neglected, as similarly claimed by Bromage (1998), Izquierdo et al. (2001) and Migaud et al. (2013).

Egg and larval development

F2 hybrid TGGG took 17 h 50 min to hatch, which is slightly shorter compared to the 18-h period taken by F1 hybrid TGGG to hatch, as reported by Ch'ng and Senoo (2008). This is also considered faster than the 19-h period seen in tiger grouper, *E. fuscoguttatus* (Boonlipatanon et al. 2002), the 30-h period seen in the giant grouper, *E. lanceolatus* (Garcia-Otega et al. 2014), the 24-h period seen in the mouse grouper, *C. altivelis* (Senoo et al. 2004), the 20–24 h range seen in some other hybrid groupers (Addin and Senoo 2011; Senoo 2008); and even the 19-h period seen in the backcross hybrid grouper (Gan et al. 2016) (see Table 3).

Table 2 Correlation between morphological and behavioral changes of F2 hybrid TGGG, *E. fuscoguttatus* × *E. lanceolatus*

Morphological features	Days after hatching (d AH)	Behavioral changes
Hatching commenced, unpigmented eyes, heart beat detected, large yolk sac	0	Pelagic behavior, spread all over water column, no swimming activity
Mouth not formed, anus closed, first appearance of eye lens	1	Vertical swimming
Deeply pigmented eyes, peristaltic movement on digestive tract, opened and functional eyes and anus, pectoral fins formed	2	Horizontal swimming, active S-posture, positively phototactic, onset of first feeding
Eye movement detected, yolk sac completely absorbed, air bladder inflated, intestinal tract movement	3	Swimming all water column, actively prey on Brachionus sp.
All fins formed (dorsal and anal), clear pigmentation on head, abdomen and tail	6	Swimming at the center and surface of water.
Spiny spines covered with melanophore, elongated spine.	10	Aggregated around central part of tank, schooling behavior started to form
Slight pigmentation on body, non-transparent, colored abdomen	25	Active and increased swimming speed, prey on artemia and copepode
Well-developed fins, lateral line can be seen, silvered abdomen, enter juvenile stage	35	Benthic habitat. Fed on artificial feed and onset of cannibalism.

Glamuzina et al. (2001) and Sargent et al. (1999) stated that most hybrid fish eggs hatch quicker, and this might be closely related to the adequate supply of all the required nutrients, particularly the good quality of lipid source to maintain energy reserves in eggs for quicker embryogenesis of marine fish. The enriched prey fish diet given to the F1 hybrid TGGG broodstock prior to spawning seems to be responsible for their positive effect on shorter hatching of F2 hybrid TGGG. This is in agreement with a study of Sargent et al. (1989), which stated that prey fish such as *Sardinella* sp. used in this study, store fat or lipid in their muscle mostly as triglycerides, which are an important source of energy to support fish maturation and reproduction.

Similar to other groupers, the larvae of F2 hybrid TGGG are small and fragile upon hatching, and depend exclusively on the nutritive reserves of the yolk sac and oil globules as energy and food sources. A larger yolk sac volume of F2 hybrid TGGG compared to other hybrid groupers reported by Addin and Senoo (2011) seems to be advantageous, as they would have a lengthier preparatory time to gain feeding experience and improve feeding ability, before being entirely reliant on external food sources (Ching et al. 2012). These findings also reveal that F2 hybrid TGGG have 40-h-long nutritional transition period, and this feature makes them stronger to withstand mortality related to starvation in the early stages. Long yolk sac and oil globule absorption duration is widely cited by many as advantageous features to avoid a high larval mortality, due to starvation at early stages in various marine fish larvae species (Pena and Dumas 2005; Yoseda et al. 2006; Ching et al. 2016).

Table 3 Hatching hours (h) of different grouper species

Species	Hatching hour (h)	References
Tiger grouper (TG)	24	Ching et al. (2012)
Mouse grouper (MG)	24	Senoo et al. (2004)
Orange spotted grouper (OG)	20	Kawahara et al. (1997)
Giant grouper (GG)	30	Garcia-Otega et al. (2014)
TG ×GG	23	Ch'ng and Senoo (2008)
OG ×TG	23	Ivan et al. (2008)
OG ×GG	20	Koh et al. (2010)
MG ×TG	23	Senoo (2008)
MG ×GG	23	Senoo (2008)
CG ×TG	19	Addin and Senoo (2011)
CG ×GG	20	Addin and Senoo (2011)
SG ×TG	20	Addin and Senoo (2011)
CG ×TG	20	Addin and Senoo (2011)
TG ×SG	20	Addin and Senoo (2011)
SG ×GG	19	Addin and Senoo (2011)
OGGG ×GG	19	Gan et al. (2016)

Water temperature was recorded at 28–30 °C

F2 hybrid TGGG is morphologically smaller compared to other groupers, and is thus probably difficult to be cultured and potentially display poor growth performances, judging from many cited reports that smaller body size larvae are highly sensitive to various abiotic and biotic factors and vulnerable to mortality (Sakakura et al. 2007; Kohno 1998). Therefore, this study provides substantial information on larval morphological changes in relation to its behavior, as they aged to juveniles to ease F2 hybrid TGGG rearing procedures. The system used can be manipulated to suit its developmental stages to further enhance their overall growth performance.

Morphological development of F2 hybrid grouper is almost similar to that reported in F1 hybrid TGGG by Ch'ng and Senoo (2008), and other groupers (Hussain and Higuchi 1980; Glamuzina et al. 2001). At an early larval stage, the presence of cupula and free neuromast observed in F2 hybrid TGGG as early as 6 h AH indicates their remarkable ability to sense stimuli in water columns. This information is particularly important to develop a handling stress-free rearing technique to culture F2 hybrid TGGG, as both cupula and free neuromast are easily broken by mechanical contact (Mukai and Kobayashi 1992).

In the later stage, the F2 hybrid TGGG, similar to other serranidae fish, goes through metamorphosis processes that represent significant transformation in the post-larvae stage (Park et al. 2014). In this study, the first spinous ray of the ventral fin was observed as early as 10 days after hatching (d AH), indicating their strong swimming capabilities at early stages. This feature is considered beneficial, particularly to coordinate active swimming movement that later contributes to improved feeding performance. However, extra precaution should be given in handling F2 hybrid TGGG at this stage, as larvae start to aggregate and the spine might tangled or broken, causing them to die.

The onset of the first feeding was observed at 3 days AH, as similarly reported in other groupers (Ching et al. 2012; Ch'ng and Senoo 2008; Yoseda et al. 2006). Although F2 hybrid TGGG larvae is characterized by a small mouth size upon hatching, they were able to attain successful initiation of first feeding when *Nannochloropsis* sp. were co-introduced with rotifer into the rearing tank. A similar study reported the use of microalgae to compensate for delays in successful initiation of first feeding for small mouth size marine fish larvae, such as sea bream, *Sparus aurata* (Muller-Feuga et al. 2003), turbot *Scophthalmus maximus* and halibut, *Hippoglossus hippoglossus* (Reitan et al. 1997). In addition, additional microalgae in the tanks created the "green water" rearing condition, thus offering a positive effect on the survival rates of F2 hybrid TGGG at an early stage.

Major morphological deformities including jaw, skeletal, operculum, fin, scale disorientation, and abnormal swimming behavior, among others, were not found throughout the larval to juvenile stages of F2 hybrid TGGG. In general, aquaculture is often threatened by the present susceptibility of cultured fish to disease, low tolerance in a wide range of water qualities, and rearing systems which eventually produce a weak fish with low growth performance. In comparison to the F1 hybrid TGGG, the following generation is characterized by greater hybrid vigor or positive heterosis characteristics (Shapawi et al. 2018). Despite the excellent egg quality and comparable larval growth observed in F2 hybrid TGGG, cannibalism remains the critical factor responsible for the low survival, as similarly observed in other fish species. In the future, further studies on the ideal rearing system for the enhancement of growth performance is investigated to demonstrate the advantages of F2 hybrid TGGG.

Acknowledgements This project was funded by Universiti Malaysia Sabah through UMSGreat Research Grant GUG0042-STWN-M-1/2016. We thank all hatchery staffs at the Borneo Marine Research Institute and Aquaculture Technology and Production Center, Kindai University, for their assistance in sample collection.

References

Addin MA, Senoo S (2011) Production of hybrid groupers: spotted grouper, *Epinephelus polyphekadion* × tiger grouper, *E. fuscoguttatus* and coral grouper, *E. corallicola* × tiger grouper. In: 2011 International symposium on grouper culture—technological innovation and industrial development, Taiwan

Argue BJ, Dunham RA (1999) Hybrid fertility, introgression and backcrossing in fish. Rev Fish Sci 7(3–4):137–195

Bagarinao T (1986) Yolk resorption, onset of feeding and survival potential of larvae of three tropical marine fish species reared in the hatchery. Mar Biol 91:449–459

Bartley DM, Rana K, Immink AJ (1991) The use of inter-specific hybrids in aquaculture and fisheries. Rev Fish Biol Fish 10:325–337

Bobe J, Labbe C (2010) Egg and sperm quality in fish. Gen Compar Endocrinol J 165:535–548

Boonlipatanon P, Detsathit S, Singhabun A (2002) Report of natural spawning and larviculture of tiger grouper, *Epinephelus fuscoguttatus* (Forsskal, 1775), at Krabi Coastal Aquaculture Station, Thailand. Marine Finfish Aquaculture Newsletter. Network of Aquaculture Centers in Asia-Pacific Publication, pp 1–3

Bromage N (1998) Broodstock management and the optimisation of seed supplies. Aquac Sci 46:395–401

Bunlipatanon P, U-taynapun K (2017) Growth performance and disease resistance against Vibrio vulnificus infection of novel hybrid grouper (*Epinephelus lanceolatus* × *Epinephelus fuscoguttatus*). Aquacult Res 48(4):1711–1723

Ch'ng CL, Senoo S (2008) Egg and larval development of a new hybrid grouper, tiger *grouper Epinephelus fuscoguttatus* × giant grouper *E. lanceolatus*. Aquacult Sci 56(4):505–512

Chevassus-au-Louis B, Lazard J (2009) Current situation and prospects for international fish farming: consumption and production. Cahiers Agricult 18:82–90

Ching FF, Nakagwa Y, Kato K, Murata O, Miyashita S (2012) Effects of delayed first feeding on the survival and growth of tiger grouper, *Epinephelus fuscoguttatus* (Forsskål, 1775), larvae. Aquac Res 43(2):303–310

Ching FF, Miura A, Nakagawa Y, Kato K, Senoo S, Sakamoto W, Takii K, Miyashita S (2014) Flow field control via aeration adjustment for the enhancement of larval survival of the kelp grouper *Epinephelu sbruneus* (Perciformes:Serranidae). Aquac Res 45(5):874–881

Ching FF, Miura A, Nakagawa Y, Kato K, Sakamoto W, Takii K, Miyashita S, Senoo S (2016) Aeration rate adjustment at night to prevent sinking syndrome-related death in the tiger grouper *Epinephelus fuscoguttatus* (Perciformes:Serranidae) larvae. Aquac Res 47(1):165–175

De M, Ghaffar MA, Bakar Y, Das SM (2016) Effect of temperature and diet on growth and gastric emptying time of the hybrid, *Epinephelus fuscoguttatus*♀ × *E. lanceolatus*♂. Aquac Rep 4:118–124

Dunham RA, Smitherman RO, Goodman RK (1987) Comparison of mass selection, crossbreeding, and hybridization for improving growth of channel catfish. Progress Fish Cult 49(4):293–296

Gan HL, Luin M, Shapawi R, Ching FF, Senoo S (2016) Egg development of backcrossed hybrid grouper between OGGG (*Epinephelus coioides* x *Epinephelus lanceolatus*) and giant grouper (*Epinephelus lanceolatus*). Int J Aquatic Sci 7(1):13–18

Garcia-Otega A, Daw A, Hopkins K (2014) Feeding hatchery-produced larvae of the giant grouper *Epinephelus lanceolatus*. In: Conference: hatchery technology for high quality juvenile production: proceedings of the 40th U.S.–Japan aquaculture panel symposium, Honolulu, Hawaii

Glamuzina B, Glavi N, Skaramuca B, Koul V, Tutman P (2001) Early development of the hybrid *Epinephelus costae* ♀ × *E. marginatus* ♂. Aquaculture 198:55–61

Henderson-Arzapalo A, Colura RL (2011) Laboratory maturation and induced spawning of striped bass. Progress Fish Cult 49(1):60–63

Hodson RG (1989) Hybrid striped bass biology and life history, vol 300. Southern Regional Aquaculture Center Publication

Huang W, Liu Q, Xie JF, Wang WM, Xiao J, Li SS, Zhang HF, Zhang Y, Liu SJ, Lin HR (2014) Characterization of triploid hybrid groupers from interspecies hybridization (*Epinephelus coioides* ♀ × *Epinephelus lanceolatus* ♂). Aquac Res 47(7):2195–2204

Hussain NA, Higuchi M (1980) Larval rearing and development of the brown spotted grouper, *Epinephelus tauvina* (Forskål). Aquaculture 19:339–350

Ivan KCC, Muhd-Shaleh SR, Senoo S (2008) Egg and larval development of a new hybrid orange-spotted grouper *Epinephelus coioides* × tiger grouper *E fuscoguttatus*. Aquac Sci 56(3):441–451

Izquierdo MS, Fernandez-Palacios H, Tacon AGJ (2001) Effect of broodstock nutrition on reproductive performance of fish. Aquaculture 197(1–4):25–42

Kawahara S, Shams AJ, Al-Bosta AA, Mansor MH, Al-Baqqal AA (1997) Effects of incubation and spawning water temperature and salinity on egg development of the orange-spotted grouper (*Epinephelus coioides*, Serranidae). Asian Fish Soc 9(1997):239–250

Koh ICC, Muhd-Shaleh SR, Akazawa N, Ota Y, Senoo S (2010) Egg and larval development of a new hybrid orange spotted grouper *Epinephelus coioides* × giant grouper *E.lanceolatus*. Aquac Sci 58(1):1–10

Kohno H (1998) Early life history features influencing larval survival of cultivated tropical finfish. In: De Silva SS (ed) Tropical mariculture. Academic Press, San Diego, pp 72–110

Liufu YZ, Liu HZ, Lin XC, Huang HR (2007) Preliminary study on the hybrid red-spotted grouper (*Epinephelus akaara*) × orange-spotted grouper (*Epinephelus coioides*). Acta Sci Nat Univ Sun 46(3):72–75

Luin M, Ching FF, Senoo S (2013) Sexual maturation and gonad development in tiger grouper (*Epinephelus fuscoguttatus*) × giant grouper (*E. lanceolatus*) hybrid. J Aquc Res Dev 5(2):1–5

Ma HF, Yamazaki F (1986) Fertility of hybrids between female masu salmon, *Oncorhynchus masou* and male pink salmon, *O. gorbuscha*. Bull Faculty Fish Hokkaido Univ 37(4):295–302

Migaud H, Bell G, Cabrita E, McAndrew B, Davie A, Bobe J, Herráez MP, Carrillo M (2013) Gamete quality and broodstock management in temperate fish. Rev Aquac 5(1):5194–5223

Mukai Y, Kobayashi H (1992) Development of free neuromasts in larvae of cyprinid fish. Mem Fac Agric Kinki Univ 27:1–14

Muller-Feuga A, Robert R, Cahu C, Robin J, Divanach P (2003) Uses of microalgae in aquaculture. In: Strottrup JG, McEvoy A (eds) Live feeds in marine aquaculture. Oxford, London, pp 253–299

Mushiake K, Kawano K, Kobayashi T, Yamasaki T (1998) Advanced spawning in yellowtail, *Seriola quinqueradiata*, by manipulations of the photoperiod and water temperature. Fish Sci 64(5):727–731

Othman AR, Kawamura G, Senoo S, Ching FF (2015) Effects of different salinities on growth, feeding performance and plasma cortisol level in hybrid TGGG (tiger grouper, *Epinephelus fuscoguttatus* × giant grouper, *Epinephelus lanceolatus*) juveniles. Int Res J Biol Sci 4(3):15–20

Park JY, Han KH, Cho JK, Myeong JI, Park JM (2014) Early osteological development of larvae and juveniles in red spotted grouper, *Epinephelus akaara* (Pisces:Serranidae). Dev Reproduct 20(2):87–707

Pena R, Dumas S (2005) Effect of delayed first feeding on development and feeding ability of *Paralabrax maculatofasciatus* larvae. J Fish Biol 67(3):640–651

Reitan KI, Rainuzzo JR, Oie G, Olsen Y (1997) A review of the nutritional effects of algae in marine fish larvae. Aquaculture 155:207–221

Rottmann RW, Shireman JV, Chapman FA (1999) Capturing, handling, transporting, injecting and holding brood fish for induced spawning, vol 422. Southern Regional Aquaculture Center Publication, pp 1–2

Sakakura Y, Shiotani S, Chuda H, Hagiwara A (2007) Flow field control for larviculture of the seven-band grouper *Epinephelus septemfasciatus*. Aquaculture 268:209–215

Sargent J, Henderson RJ, Tocher DR (1989) The lipids. In: Halver JE (ed) Fish nutrition. Academic Press, Cambridge, pp 153–218

Sargent JR, McEvoy L, Estevez A, Bell G, Bell M, Henderson J, Tocher DR (1999) Lipid nutrition of marine fish during early development: current status and future directions. Aquaculture 179(1999):217–229

Senoo S (2008) Consideration of artificial egg collection technique on fish—IV (Fish Culture in Southeast Asia 80). Aquanet Magn 204:64–67

Senoo S, Shapawi R, Abdul Rahman R (2004) Induced spawning technique of mouse grouper, *Cromileptes altivelis*. Universiti Malaysia Sabah Publication, pp 1–23

Shapawi R, Ching FF, Senoo S, Mustafa S (2018) Nutrition, growth and resilience of tiger grouper (*Epinephelus fuscoguttatus*) × giant Grouper (*Epinephelus lanceolatus*) hybrid—a review. Reviews in Aquaculture, pp 1–12

Song YB, Oh SR, Seo JP, Ji BG, Lim BS, Lee YD (2005) Larval development and rearing of longtooth grouper *Epinephelus bruneus* in Jeju Island, Korea. J World Aquac Soc 36(2):209–216

Starzynski D, Lauer TE (2015) How temperature affects timing and duration of yellow perch spawning in the Indiana waters of Lake Michigan. J Freshw Ecol 30(3):445–453

Wagner EJ, Oplinger RW (2013) Toxicity of copper sulfate to *Flavobacterium psychrophilum* and rainbow trout eggs. J Aquat Anim Health 25:125–130

Yoseda K, Teruya K, Sugaya T, Sekiya S (2006) Effects of delayed initial feeding on larval feeding, early survival, and the growth of red spotted grouper *Epinephelus akaara* larvae. Suisan Gakkaishi 72:702–709

15

Preliminary use of cashew kernel oil in *Clarias gariepinus* fingerlings diet: comparison with fish oil and palm oil

Morgane Paul Magouana Anvo · Rokyatou Sissao · Benié Rose Danielle Aboua · Chantal Yvette Zoungrana-Kaboré · Athanase Kraidy Otchoumou · Essetchi Paul Kouamelan · Aboubacar Toguyéni

Abstract The present study was performed to investigate the effect of fish oil (FO), palm oil (PO), and cashew kernel oil (CKO) as the dietary lipid sources on the growth performance, tissues fatty acids composition and economical profitability in *Clarias gariepinus* fingerlings. Three isonitrogenous (49.32%), isolipidic (15.18%), and isoenergetic (21 kJ/g) diets were formulated, differed only to the added lipid source and coded accordingly as D1 (fish oil), D2 (palm oil), and D3 (cashew kernel oil). Catfish fingerlings (initial body weight = 2.43 ± 0.04 g) were stocked at 1 fish/L and fed to trial diets at 5.6%. At the end of 8 weeks, *C. gariepinus* fingerlings fed with D1 and D3 had significantly identical growth and feed utilization efficiency. Those fed with D2 exhibited significantly and numerically low growth performance compared, respectively, to D1 and D3. Contrary to whole-body composition of fish, the muscle lipids fatty acids composition was influenced by the fatty acids profile of the dietary lipid sources. High net profit value (NPV) of 24.59$, high investment cost analysis (ICA) of 10.14$ and low benefit: cost ratio (BCR) of 2.42 were observed in fish fed with D1 while those fed with D3 allowed to have high NPV (23.15$), low ICA (7.86$) and high BCR (2.95). Like others vegetable oils, cashew kernel oil can replace expensive fish oil in *C. gariepinus* fingerlings diet without negatively affecting its growth performances.

Keywords *Clarias gariepinus* · Vegetable oils · Fish oil · Growth performance

Introduction

The supply of fish products of Burkina Faso in 2009 was 56,400 tons, including 12,000 tons from domestic production with only 400 tons produced by aquaculture, and 44,400 tons come from import (Kabré et al. 2014). Given the importance of a river system, fish farming is a promising alternative to reduce the

---

M. P. M. Anvo (✉) · B. R. D. Aboua · E. P. Kouamelan
Laboratoire d'Hydrobiologie et d'Eco-technologie des eaux, UFR. Biosciences, Université Félix Houphouët-Boigny, 22 BP 582, Abidjan 22, Côte d'Ivoire
e-mail: morgane1er@yahoo.fr

M. P. M. Anvo · R. Sissao · C. Y. Zoungrana-Kaboré · A. Toguyéni
Unité de Recherche en Aquaculture et Biodiversité Aquatique/Laboratoire d'Etude et de Recherche sur les Ressources Naturelles et Sciences de l'Environnement (UR-ABAQ/LERNSE), Université Polytechnique de Bobo-Dioulasso, 01 BP 1091, Bobo-Dioulasso 01, Burkina Faso

A. K. Otchoumou
Laboratoire de biotechnologie, UFR Biosciences, Université Félix Houphouët-Boigny, 22 BP 582, Abidjan 22, Côte d'Ivoire

dependence of this country on the importation of fish. Especially since the per capita fish consumption has increased regularly from 1.5 kg/year in 2003 (Kabré et al. 2014) to more than 5 kg/year in 2013 (FAO 2016).

Aquaculture production in the developing countries is greatly constrained by undersupply, scarcity and high cost of conventional fish feed (Fagbenro and Arowosegbe 1991). Moreover, fish feed alone account for between 60 and 75% of the total cost of fish production (Babalola 2010). Thus, aquaculture production can be profitable only if the operators have access to balanced food at competitive prices (FAO 2016) by using locally available inputs.

Fish oil is a main dietary lipid source used in commercial aquafeeds. Other than providing a source of energy and essential fatty acids, it is commonly used to coat the extruded pellets in order to improve the palatability and appearance of the feed (Ng and Chong 2004). According to Riediger et al. (2009), fish oils are particularly rich in the longer chain omega-3 fatty acids such as eicosapentaenoic acid (EPA) and docosahexaenoic acid (DHA) which are highly desirable in fish and human nutrition, although they are absent in most of vegetable oils (Montero et al. 2007).

However, the stagnation in global fish oil production coupled with increased demand for its use in aquaculture feeds (Barlow 2000) caused its price inflation. This situation led to several researches directed towards the evaluation of less expensive lipid sources to replace partially or totally fish oil. Vegetable oils have been found as potential substitutes for fish oil in fish diets. Numerous studies have shown that vegetable oils can substitute fish oil in fish diets without affecting fish growth. Indeed, palm oil (Ng et al. 2003; Yilmaz et al. 2004; Nyina-wamwiza et al. 2007; Sotolu 2010; Effiong and Akpan 2016), sunflower oil (Ng et al. 2003; Nyina-wamwiza et al. 2007; Effiong and Akpan 2016), soya oil (Sotolu 2010), coconut oil (Aderolu and Akinremi 2009; Hammed et al. 2014; Effiong and Akpan 2016), groundnut oil (Sotolu 2010), shea butter oil (Aderolu et al. 2011), and olive and sesame oil (Effiong and Akpan 2016) have successfully been used in *Clarias gariepinus* diet. Nevertheless, they are not always available and become some time expensive. For example, palm oil which is the second most used oil in aquafeeds is costly in Burkina Faso where the palm trees do not grow. Thus, there is the need to investigate other lipid sources in fish diets.

Anacardium occidentale (cashew) belongs to Anacardiaceae family and is native to north eastern of Brazil. The first plantations were introduced around 1960 by the Tropical Forest Technical Center in Burkina Faso (Bila et al. 2010). This country has an annual production of about 75,000 metric tons of raw cashew nuts (Rabany et al. 2015). Two oils are obtained from its nuts. The first found between the seed coat or pericarp and the nuts, called the cashew nut shell liquid (CNSL), is not suitable for consumption. The second type of oil found in the kernel of the cashew nut, called cashew kernel oil (CKO) is comparable to others vegetable oils (Idah et al. 2014) and is cheaper.

The African catfish *C. gariepinus* is a very popular aquaculture species in a number of African countries. It has been introduced in many countries of Europe and Asia due to its tolerance to drastic rearing conditions (wide range of temperatures, low oxygen and high salinity levels), high nutritive value, good taste, and few bones of this species. In addition, it has high fecundity and growth rate (Hecht et al. 1996). It is regarded as an opportunistic, omnivorous predator since *C. gariepinus* has the ability to efficiently utilize and/or switch between alternative foods sources such as plants and detritus when animals prey become scarce (Potts et al. 2008). Thus, it can feed on wide variety of ingredients of animal and plant origin.

Considering the less expansive price and high availability of cashew kernel oil in Burkina Faso, this oil could be used as an alternative dietary lipid source in *C. gariepinus* feeding in order to reduce the fish production cost in this country where fisheries produce only one-third of the population fish consumption. Thus, the aim of this study was to evaluate the nutritive potential of cashew kernel oil in *C. gariepinus* fingerlings diet in comparison to palm oil and conventional fish oil.

Methods

Fish and system

Clarias gariepinus fingerlings were obtained from the Study of Natural's Resources and Environment Sciences Laboratory, Unit of Aquaculture Research and Aquatic's Biodiversity of Polytechnic University of Bobo-Dioulasso, Burkina Faso. *C. gariepinus* fingerlings (initial weight: 2.43 ± 0.04 g) were separated into three lots according to experimental diets with three replicates. These fingerlings from the same batch were randomly distributed in each aquarium. Fish were stocked at 1 fingerling/L in 50 L aquarium. This density was maintained constant until the end of the experiment. Aquaria were connected to a recirculation system with mechanical filter. Its capacity was 500 L. Sponge, ceramic ring and foam was used as filtration materials. Water flow through each aquarium was maintained constant at 0.5 L/min. Water parameters such as dissolved oxygen, pH, conductivity and temperature were measured each day at 8 AM with multi-meter (HACH, HQ40D). Average values for water quality parameter monitored in the aquarium during the experiment were: pH, 6.77 ± 0.28; temperature, 29.01 ± 1.22 °C; dissolved oxygen, 4.22 ± 0.32 mg/L; and conductivity, 17.03 ± 1.41 μS/cm. To quantify the exact feed intake, the left-over of feed was siphoned out immediately, dried, and weighed. Each morning before feeding, feces were collected (siphoning). Fish sample were taken at the beginning and at the end of the experiment from each aquarium, sacrificed for whole-body composition and muscle lipids fatty acids analysis. At the end of the trial period, livers and viscera were removed from three other fish in each aquarium and weighed for the estimation of the hepatosomatic index (HSI) and the viscerosomatic index (VSI).

Experimental diets and feeding

Three isonitrogenous (49.32%), isolipidic (15.08%) and isoenergetic (21 kJ/g) experimental diets, differing only in the added lipid source and coded D1 (fish oil), D2 (palm oil) and D3 (cashew nut oil) were formulated (Table 1). Each diet was manufactured by mixing all ingredients (including corresponding oil) with 20% of water. Floating pellets (1.8 mm) were obtained from the mixture using fish feed extruder (BEDO machine, BD-GP70), dried in an oven at 40 °C for 72 h, and stored at 18 °C until their use in fish feeding. Fish were fed

Table 1 Formulation, cost and proximate composition of the experimental diets (g/100 g dry matter)

Ingredients (%)	D1	D2	D3	Ingredient cost ($/kg)
Fish meal	60.00	60.00	60.00	5.54
Soybean meal	16.00	16.00	16.00	0.57
Maize meal	13.00	13.00	13.00	0.24
Fish oil (Menhaden)	6.00	0.00	0.00	16.30
Palm oil	0.00	6.00	0.00	1.95
Cashew kernel oil	0.00	0.00	6.00	0.81
Lysine	1.00	1.00	1.00	4.08
Methionine	1.00	1.00	1.00	7.99
Vitamins mixture[a]	1.00	1.00	1.00	2.45
Minerals mixture[b]	1.00	1.00	1.00	1.31
Betaine	1.00	1.00	1.00	8.15
Diet cost ($/kg)	4.67	3.79	3.73	
Proximate composition				
Moisture	9.19	8.98	9.26	
Crude protein	48.98	49.37	49.62	
Crude lipid	15.09	15.24	14.93	
Ash	12.35	11.98	12.18	
Crude fiber	3.45	3.79	3.73	
NFE	20.13	19.62	19.54	
Energy (kJ/g)	21.03	21.09	21.01	

NFE: nitrogen-free extract = 100 − (crude protein + crude lipid + crude fiber + ash)

[a] Vitamins mixture: vit A: 2,500,000 IU; vit D3: 500,000 IU; vit E: 30,000 mg; vit K3: 2000 mg; vit B1: 2000 mg; vit B2: 5000 mg; panthotenic acid: 10,000 mg; niacin 5000 mg; vit B6: 4000 mg; folic acid: 2000 mg; vit B12: 80 mg; vit C: 20,000 mg; biotin: 200 mg and inositol: 80,000 mg

[b] Minerals mixture: calcium = 23 g, Phosphore = 18 g, magnesium = 0.21 g, copper sulfate = 0.8 g, cobalt sulfate = 0.02 g, manganese sulfate = 0.6 g, zinc sulfate = 8.15 g, selenium sulfate = 0.04 g, ferrous sulfate = 0.9 g

to 5.6% of the total fish body weight as recommended by Hogendoorn (1983) during 56 days. All the fish of each tank were counted and weighed every 2 weeks and the feed amount was adjusted accordingly.

Chemical analysis

Proximate composition of feeds ingredient, diets and fish whole-body composition before and after the experiments were analyzed for crude protein, crude lipid, crude fiber, ash, moisture, and nitrogen-free extracts according to (AOAC 1990). Gross energy was calculated on the basis of 23.65 kJ/g for protein, 40 kJ/g for lipid and 16.8 kJ/g for carbohydrate (Jobling 1983). For fatty acid analysis, total lipid was extracted from samples according to Folch et al. (1957). Fatty acid analysis was performed on each lipid sources, experimental diets and on muscles lipids of fish samples from each aquarium. Fatty acids profiles were analyzed according to Legrand et al. (2010) using a gas chromatograph (WATERS Agilent, model HP 6890N) coupled by mass spectrometer (Quattro micro GC, MICROMASS).

Growth performance, nutrient efficiency, biological and economical parameters

The growth performance, nutrient efficiency and biological indices were calculated using the following formulae:

$$\text{Mean weight gain (MWG)} = \frac{\text{Final body weight} - \text{initial body weight}}{\text{Number of fish}}$$

$$\text{Survival rate (SR, \%)} = 100 \times \frac{\text{Final number}}{\text{Initial number}}$$

$$\text{Specific growth rate (SGR, \%/day)} = 100 \times \frac{\ln[\text{final body weight (g)}] - \ln[\text{initial body weight (g)}]}{\text{Number of experimental days}}$$

$$\text{Feed conversion ratio (FCR)} = \frac{\text{Feed intake (g)}}{\text{Fish weight gain (g)}}$$

$$\text{Protein efficiency ratio (PER)} = \frac{\text{Fish weight gain (g)}}{\text{Protein intake (g)}}$$

$$\text{Hepatosomatic index (HSI, \%)} = 100 \times \frac{\text{Liver weight (g)}}{\text{Final weight (g)}}$$

$$\text{Viscerosomatic index (VSI, \%)} = 100 \times \frac{\text{Visceral weight (g)}}{\text{Final weight (g)}}.$$

Economic evaluations in terms of gross profit (GP), net profit value (NPV), investment cost analysis (ICA), and benefit cost ratio of using tree oils in *C. gariepinus* juveniles production were determined according to Faturoti and Lawal (1986) and Mazid et al. (1997).

$$\text{Net profit value (NPV, \$)} = \text{Final mean weight of fish cropped} \times \text{Total number of the survival} \times \text{Cost per kg,,}$$

where 1 kg of fish costs 16.3$,

$$\text{Investment cost analysis (ICA, \$)} = \text{Cost of feeding (\$)} + \text{cost of fingerlings stocked,}$$

where 1 g of fingerling costs 0.016$,

$$\text{Gross profit (GP, \$)} = \text{Net profit value (\$)} - \text{Investment cost analysis (\$)}$$

$$\text{Benefit : cost ratio (BCR)} = \frac{\text{Total cost of fish cropped (\$)}}{\text{Investment cost analysis (\$)}}.$$

Statistical analyses

Data were expressed as mean ± SD ($n = 3$). The effects of diet were tested with one way analysis of variance (ANOVA), followed by Tukey's test. Differences were considered significant when $P < 0.05$. Statistical analyses were performed using STATISTICA 7.1 software.

Results

Fatty acids of lipids sources and experimental diets

The analyzed proximate composition of the trial diets (Table 1) was consistent with formulated values and was approximately similar in all treatments. The fatty acid composition of the experimental diets reflected the fatty acid composition of the lipid sources (Table 2). The dominant saturated fatty acid (SFA) in the three lipid sources and their corresponding diets was palmitic acid. This one was higher in D2 (28.62%) and D3 presented the lowest (16.84%). Oleic acid was the predominant monounsaturated fatty acid (MUFA) in trial diets with highest concentration observed in D3 (32.91), whereas the lowest in D1. Concerning polyunsaturated fatty acids (PUFA), linoleic acid was the highest n-6 fatty acid, while eicosapentaenoic acid (EPA) and

Table 2 Fatty acid composition (% total fatty acids) of the different lipid sources and the experimental diets

Fatty acids	Lipid sources			Diets		
	FO	PO	CKO	D1	D2	D3
14:00	10.35	0.46	0.36	8.19	4.37	4.36
16:00	25.64	42.16	11.45	22.17	28.62	16.84
18:00	4.37	4.12	7.94	4.53	4.43	5.91
20:00	0	0.26	0.19	0.04	0.14	0.12
ΣSFA	40.36	43	19.94	34.93	37.2	27.23
16:1n-7	14.47	0.06	0.24	8.96	3.43	3.5
18:1n-9	10.96	40.61	60.6	13.86	25.24	32.91
18:1n-7	2.92	0.97	0.13	2.02	1.27	0.94
20:1n-9	0.84	0.06	0	0.77	0.47	0.45
ΣMUFA	29.19	47.7	60.97	25.61	30.41	37.8
18:2n-6	1.34	9.06	16.94	10.73	13.69	16.72
18:3n-6	2.37	0	0	2.92	2.01	2.01
20:4n-6	1.2	0	0.5	1.16	0.84	1.03
Σn-6	4.91	9.06	17.44	14.91	16.54	19.76
18:3n-3	1.6	0	0.2	1.48	0.78	0.95
20:5n-3 (EPA)	12.5	0	0	9.45	4.65	4.66
22:5n-3	1.27	0	0	1.74	1.26	1.25
22:6n-3 (DHA)	8.15	0.02	0.03	6.67	3.54	3.55
Σn-3	23.52	0.02	0.23	19.34	10.23	10.41
ΣPUFA	28.43	9.08	17.97	34.25	26.86	30.17
Σn-3/Σn-6	4.79	0.00	0.01	1.29	0.62	0.52

FO fish oil, *PO* palm oil, *CKO* cashew kernel oil, *SFA* saturated fatty acids, *MUFA* monounsaturated fatty acids, *PUFA* polyunsaturated fatty acids, *EPA* eicosapentaenoic acid, *DPA* docosapentaenoic acid, *DHA* docosahexaenoic acid

docosahexanoic acid (DHA) were the predominant n-3 fatty acid. Although palm oil (PO) and cashew kernel oil (CKO) contained few EPA and DHA, their corresponding diet contained high concentration of these fatty acids.

D3 contained the highest concentrations of n-6 PUFA with 16.72% in the form of linoleic acid. D1 contained the highest amounts of n-3 PUFA with 9.45% in the form of EPA and 6.67% as DHA. The concentration of EPA and DHA were identical in D2 and D3.

Growth performance, nutrient efficiency and biological indices

At the end of trial period, *C. gariepinus* fingerlings fed with D1 and D3 had significantly identical FMW, MWG SGR, feed intake, and PER. These parameters were higher in fish fed with D3 than those fed with D2 even though the difference was not significant (Table 3). FCR of fish fed in D1 was significantly identical with that of fish fed in D3 while was significantly higher than that of D2 fed fish. This nutrient utilization index was numerically higher in D3 fed fish than fish fed with D2. No significant survival rate difference was observed between fish batches fed with experimental diets. This parameter was around 91% for all treatments. Biological indices (HIS and VSI) did not show significant differences among fish fed with the trial diets (Table 3).

Economical parameters

The analyzed economical parameters are presented in Table 4. D1 was the most expensive diet while D3 was cheaper. ICA was significantly higher in fish fed with D1. NPV was higher with D1, followed by D3, and the lowest was obtained with D2. No significant difference was observed regarding GP for all treatments, but this parameter was numerically higher in fish fed with D3. BCR was significantly higher with D3 and D2, however, fish fed with D3 shown larger BCR than D2.

Whole-body proximate composition and muscle fatty acids of fish

The whole-body moisture, protein, lipid, and ash content of fish were not significantly affected by the dietary lipid sources (Table 5). However, whole-body proximate compositions of the African catfish of all treatments were higher than the initial body composition for all the parameters measured except the moisture. The fatty acid composition of the muscle lipids generally reflected those of the corresponding diets with a few exceptions (Table 6). However, a few minor differences between diets and muscle tissue of fish have been noticed. Contrary to PUFA and SFA, MUFA concentrations were generally high in the muscle lipids regardless of dietary treatments. Oleic acid (18:1n-9) was higher in muscle lipids compared to their corresponding diet. Linoleic acid (18:2n-6) and γ-Linolenic acid (18:3n-6) were higher in dietary lipids than in muscle tissues.

Table 3 Growth performance, nutrient efficiency and biological parameters of *Clarias gariepinus* fingerlings fed trial diets

Indices	Diets		
	D1	D2	D3
IMW (g)	2.44 ± 0.13	2.46 ± 0.35	2.39 ± 0.29
FMW (g)	33.21 ± 0.53^{a}	28.60 ± 1.19^{b}	31.04 ± 1.17^{ab}
MWG (g)	30.77 ± 0.52^{a}	26.14 ± 1.25^{b}	$28{,}65 \pm 1.19^{ab}$
SGR (%/day)	4.66 ± 0.03^{a}	4.38 ± 0.11^{b}	4.57 ± 0.08^{ab}
Feed intake (mg)	39.98 ± 0.72^{a}	35.31 ± 1.92^{b}	37.78 ± 1.27^{ab}
FCR	1.29 ± 0.00^{a}	1.35 ± 0.03^{b}	1.31 ± 0.0^{ab}
PER	1.57 ± 0.00^{a}	1.51 ± 0.03^{b}	1.55 ± 0.01^{ab}
SR (%)	90.66 ± 2.3	92.00 ± 3.46	91.33 ± 3.05
HSI	1.08 ± 0.18	0.97 ± 0.22	1.03 ± 0.14
VSI	2.09 ± 0.26	2.03 ± 0.16	2.11 ± 0.13

Data are mean values ± SD ($n = 3$); means in the same row with the same superscript were not significantly different ($P > 0.05$)

Table 4 Economical parameters of *Clarias gariepinus* fingerlings fed with trial diets

Indices	Diets		
	D1	D2	D3
Cost of lipid source ($/L)	16.30	1.95	0.81
Cost of feed ($/kg)	4.67	3.79	3.73
ICA ($)	10.14 ± 0.16^{a}	7.50 ± 0.36^{b}	7.86 ± 0.23^{b}
NPV ($)	24.59 ± 0.41^{a}	21.51 ± 1.67^{b}	23.15 ± 1.11^{ab}
GP ($)	14.45 ± 0.49	14.01 ± 1.31	15.29 ± 0.95
BCR	2.42 ± 0.06^{a}	2.86 ± 0.08^{b}	2.95 ± 0.09^{b}

ICA, NPV, GP and BCR are expressed as mean values ± SD (n = 3); means in the same row with the same superscript were not significantly different ($P > 0.05$). Prices in $ (USD); 1$ (USD) = 613.20 FCFA (UEMOA), based on conversion rate of November 16, 2016c

Table 5 Proximate whole-body composition (% wet weight) of *Clarias gariepinus* fingerlings fed trial diets for 8 weeks

	Initial	D1	D2	D3
Moisture (%)	73.56 ± 0.22	71.32 ± 1.15	72.00 ± 0.77	70.88 ± 1.59
Protein (%)	15.32 ± 0.04	17.43 ± 0.84	16.89 ± 0.56	17.45 ± 1.09
Lipid (%)	4.98 ± 0.25	7.27 ± 0.25	6.90 ± 0.73	7.61 ± 0.69
Ash (%)	3.53 ± 0.36	3.98 ± 0.41	4.21 ± 0.34	4.06 ± 0.19

Data are mean values ± SD (n = 3); means in the same row with the same superscript were not significantly different ($P > 0.05$)

Table 6 Muscle fatty acids composition (% total fatty acids) of *Clarias gariepinus* fingerlings fed on trial diets for 8 weeks

Fatty acids	Initial	D1	D2	D3
14:00	3.92	6.94 ± 1.11a	3.75 ± 0.21b	3.91 ± 0.44b
16:00	22	21.34 ± 1.31^{a}	25.17 ± 1.74^{a}	14.68 ± 1.78^{b}
18:00	7.27	6.09 ± 0.45	6.16 ± 0.24	6.96 ± 0.31
ΣSFA	33.19	34.37	35.08	25.55
16:1n-7	3.64	7.71 ± 0.36^{a}	3.20 ± 0.14^{b}	3.46 ± 0.25^{b}
18:1n-9	25.04	20.55 ± 2.21^{a}	35.13 ± 3.62^{b}	43.85 ± 2.33^{c}
18:1n-7	0.31	0.21 ± 0.04	0.11 ± 0.04	0.11 ± 0.08
20:1n-9	1.89	0.12 ± 0.09	0.15 ± 0.06	0.05 ± 0.05
ΣMUFA	30.88	28.59	38.59	47.47
18:2n-6	11.71	11.65 ± 1.18^{a}	13.29 ± 0.69^{a}	15.57 ± 0.59^{b}
18:3n-6	0.46	0.41 ± 0.22	0.2 ± 0.08	0.27 ± 0.10
20:2n-6	0.18	0.74 ± 0.53	0.25 ± 0.15	0.21 ± 0.09
20:3n-6	0.68	1.49 ± 0.74	0.97 ± 0.23	1.03 ± 0.65
20:4n-6	0.81	0.87 ± 0.23	0.74 ± 0.19	0.79 ± 0.26
Σn-6	13.84	15.16	15.45	17.87
18:3n-3	1.98	1.92 ± 0.61	1.09 ± 0.39	1.13 ± 0.32
20:5n-3 (EPA)	2.19	5.54 ± 0.69	2.27 ± 0.41^{b}	2.04 ± 0.66^{b}
22:5n-3	0.99	1.37 ± 51^{a}	0.88 ± 0.8^{b}	0.94 ± 0.11^{b}
22:6n-3 (DHA)	3.58	5.86 ± 1.94^{a}	3.24 ± 1.01^{b}	2.84 ± 0.88^{b}
Σn-3	8.74	16.69	7.48	6.95
ΣPUFA	22.5	31.85	22.93	24.82

Data are mean values ± SD (n=3); means in the same row with the same superscript were not significantly different ($P > 0.05$). SFA saturated acids, *MUFA* monounsaturated acids, *PUFA* polyunsaturated acids, *EPA* eicosapentaenoic acid, *DPA* docosapentaenoic acid, *DHA* docosahexaenoic acid

Dihomo-γ-linolenic acid (20:3n-6) and eicosadienoic acid (20:2n-6) were absent in diets, while they were present in whole-body lipids. EPA concentrations were generally lower in muscle lipids than in the dietary fats.

Discussion

The results of this present study show that fingerlings of *C. gariepinus* responded well to all the experimental diets, since fish of all groups displayed high survival rate (around to 91%) and rapid growth with SGR ranged from 4.38 to 4.66%/day. Nevertheless, African catfish fed with diets containing CKO (high in 18:2n-6) and FO (high in EPA and DHA) exhibited higher weight gain and feed utilization efficiency compared to fish fed with diets containing PO. Besides, fish fed with CKO and FO did not show significant difference in growth. Similar results have been obtained by previous studies showing significant similar growth performance in African catfish fed with vegetable oils and fish oil diets (Legendre et al. 1995; Lim et al. 2001; Sotolu 2010). However, Ng et al. (2003) obtained lower growth performance in African catfish fed with diets containing fish oil as compared to fish fed with diets supplemented with various vegetable oils (sunflower oil, crude palm kernel oil, crude palm oil). This difference is probably due to the fatty acids contents of the FO used. Indeed, Ng et al. (2003) used cod liver oil (very low in n-6 PUFA and high n-3/n-6), while menhaden oil (low n-6 PUFA and low n-3/n-6) has been used in the present study. A minimum level of 0.5–1% dietary n-3 fatty acids has been recommended for *Heterobranchus longifilis* fry (Kerdchuen 1992). In this study trial, diets n-3 fatty acids which ranged from 10.23 to 19.34% of total fatty acids (1.62–2.92% of dry feed) met the reference values recommended for African catfish *Heterobranchus longifilis.* Contrary to Ng et al. (2003) who observed growth decreasing in *C. gariepinus* fed high concentration of n-3 fatty acids, D1 containing the highest concentration of n-3 fatty acids allowed the best growth performance. However, Satoh et al. (1989) showed that n-3 highly unsaturated fatty acids appear to be responsible for the enhanced growth of Channel Catfish (*Ictalurus punctatus*). Besides, n-3 lowering in trout diets reduced the animals' resistance to the IHN virus (Kiron et al. 1995). Thus, growth decreasing of fish fed with D2 might not be associated to the dietary n-3 concentration, since the concentrations of n-3 fatty acids were identical in D2 and D3. However, the n-3 fatty acids of these diets were around 10% of total fatty acids or 1.5% of diets. This value met the *C. gariepinus* n-3 fatty acid requirement which was ranged from 0.5 to 0.75% of the ration (Corraze 1999). The relative high concentration of n-3 including EPA and DHA in PO and CKO diets was originated to the residual oil found in fishmeal. Fish eating a diet containing CKO displayed a higher growth than those fed with PO diet despite those diets contained the same amount of DHA and EPA. The decreasing of weight gain and feed utilization efficiency in fish fed with D2 could be due the low concentration of n-6 fatty acids in this diet (16.54%) compared to D3 (19.76%). This finding is consistent with previous study showing best growth of African catfish fed with sunflower oil (high in 18:2n-6) diet (Ng et al. 2003). In the other hand, D1 and D3 contained higher concentration of linolenic (18:3n-3) acid than D2. Linoleic (18:2n-6) and linolenic acids (18:3n-3) are the main essential fatty acids for freshwater fish (Corraze 1999). Fish are unable to synthesize these fatty acids de novo. Therefore, fish must obtain these fatty acids via feed (Takeuchi et al. 1980). Besides, the higher amount of oleic acid (18:1n-9) in D3 than in D2 might be another raison that fish fed with D3 performed better than those fed with D2. This was in agreement with Stickney and Andrews (1972) who reported that channel catfish fed diets containing beef tallow (high in 18:ln-9 and low in 18:2n-6 and 18:3n-3) grew better.

Organ indices have been used as indicators of change in nutritional and energy status of fish (Adams et al. 1996). In the present study, fish biological indices (HIS and VSI) were not significantly affected by dietary lipid source. The same result was previously found by Mourente et al. (2005) in European sea bass fed with various vegetable oils. Our results were not in the same line with Ochang (2011) who reported higher HSI and lower VSI in *Oreochromis niloticus* fed with soybean oil than those fed with cod liver oil. This result suggests that lipid was deposited in the liver and perivisceral adipose tissue of fish of all treatments at a relatively uniform concentration, since these organs are the main fat-stocking sites in fish (Corraze 1999). Beyond, the identical organism indices in all treatments may indicate that fish health was not affected by lipid source, since these indices were used for assessing fish health (Goede and Barton 1990).

The whole-body proximate composition was not affected by the dietary lipid sources since trial diets had identical protein (49.32 ± 0.32%), lipid (15.08 ± 0.15%), NFE (19.76 ± 0.32%), and ash (12.17 ± 0.18%)

contents. The fatty acids profile is important to assess the nutritional quality of the fish since this profile must generate maximum desirable fatty acids such as n-3 fatty acid, especially EPA and DHA which have important roles in human health. In the present study, muscle lipid fatty acids composition of fish was influenced by the fatty acid profile of the dietary lipid sources. This observation had also been reported in previous studies (Yingst and Stickney 1979; Bell et al. 2002; Ng et al. 2000, 2003). PUFA decreasing in muscle lipids was consecutive to EPA, DHA, linoleic, and γ-Linolenic acids lowering. Specially, the lowering of EPA in fish fed with diets D2 and D3 indicates the decreasing of this fatty acid retention when the fish were fed with plant oil. This is on the same line with Ng et al. (2003) who reported that feeding fish with vegetable oils decreased the muscle content of EPA, DHA and ARA. The MUFA increasing in muscle tissue resulting in oleic acid raising may suggest that *C. gariepinus* have high activity of Δ9-desaturases enzymes which allows the bioconversion of SFA (16:0 and 18:0) to oleic acid. Moreover, stearic and palmitic acids can be biosynthesized de novo by all known organisms, including fish, by the conventional pathway (Sargent et al. 2002). 18:2n-6 and 18:3n-3 are essential fatty acid for many animals including many species of freshwater fish. These essentials fatty acid can be desaturated further and elongated to form the physiologically essential C_{20} (20:4n-6; 20:5n-3) and C_{22} (22:6n-3) PUFA. Desaturation and elongation dependent on the relative activities of fatty acid elongases and desaturases such as Δ5 and Δ6 in animals (Tocher 2003). An absence of either of these fatty acids in the diet has been showed leads to the desaturation and chain elongation of oleic acid, 18:1n-9, to 20:3n-9, which is characteristic of an EFA deficiency in many terrestrial animals (NRC 1993). Thus, when EFA are deficient, increased concentrations of 20:3*n*-9 are incorporated into tissue lipids in place of 20:4n-6, 20:5n-3, or 22:6n-3. The absence of 20:3n-9 and the relative high concentrations of 20:4n-6, 20:5n-3 and 22:6n-3 in the tissue of *C. gariepinus* observed in this study showed that the dietary EFA requirements are met. The appearance of 20:3n-6 and 20:2n-6 might, respectively, be due to the bioconversion of 18:3n-6 and 18:2n-6 (Bell et al. 1986). The lowering of linolenic acid in fish muscle lipids compared to initial might be due to the low amounts of this fatty acid in trial diets. Indeed, the fatty acids of muscle lipids in African catfish reflected the fatty acid profile of the dietary oil that was fed (Ng et al. 2003; Babalola et al. 2011).

The best growth performance allowed to obtain the highest NPV in fish fed with D1, while the high price (16.30$/L) of fish oil used in this diet caused the decreasing of the GP observed in this group. However, highest GP and BCR resulting of the cheaper cost of cashew kernel oil (0.81$/L) were recorded with D3. As regards, palm oil, GP, and BCR were numerically lower than fish fed with cashew kernel oil due to its relative high cost (1.95$/L).

This study showed that expensive fish oil can be replaced by cashew kernel oil in *C. gariepinus* diet without negatively affecting growth performance. So, this lipid source might be recommended in order to reduce fish production cost. However, inclusion of cashew kernel oil in the diet of *C. gariepinus* produced tissue with low concentrations of long chain n-3 PUFA. Nevertheless, it would be interesting to analyze the effects of incorporating this oil in *C. gariepinus* practical diets formulated with unconventional protein source.

Acknowledgements This work was funded by the project "Tilapia's seed improvement-UEMOA-UPB" and by the project HAAGRIM through its mobility scholarships (First cohort). The authors are grateful to Mr. Dibi Koffi Sylvain for biochemical analyses, to Silga Touwendsida Eric and Kagambega Florence for their technical assistances during the experiment.

References

Adams SM, Ham KD, Greeley MS, LeHew RF, Hinton DE, Saylor CF (1996) Downstream gradients in bioindicator responses: point source contaminant effects on fish health. Can J Fish Aquat Sci 53:2177–2187

Aderolu AZ, Akinremi OA (2009) Dietary effects of coconut oil and peanut oil in improving biochemical characteristics of *Clarias gariepinus* juvenile. Turk J Fish Aquat Sci 9:105–110

Aderolu AZ, Bello R, Aarode OO, Sanni RO (2011) Utilization of two dietary plant oil sources on growth, haematology, histometry and carcass analysis of juvenile *Clarias gariepinus*. PAT 7(1):117–130

AOAC (1990) Official methods of analysis, 15th edn. Association of Official Analytical Chemists, Arlington, p 1298

Babalola TOO (2010) Utilization of dietary animal fats and vegetable oils by African catfish (*Heterobranchus longifilis*) fingerlings. Ph.D. Thesis University of Ilorin, Nigeria

Babalola TO, Apata DF, Omotosho JS, Adebayo MA (2011) Differential effects of dietary lipids on growth performance, digestibility, fatty acid composition and histology of African catfish (*heterobranchus longifilis*) fingerlings. FNS 2:11–21

Barlow S (2000) Fishmeal and fish oil: sustainable feed ingredients for aquafeeds. Glob Aquac Advocate 4:85–88

Bell MV, Henderson RJ, Sargent JR (1986) The role of polyunsaturated fatty acids in fish. Comp Biochem Physiol 83B:711–719

Bell JG, Henderson RJ, Tocher DR, McGhee F, Dick JR, Porter A, Smullen RP, Sargent JR (2002) Substituting fish oil with crude palm oil in the diet of Atlantic salmon (*Salmo salar*) affects muscle fatty acid composition and hepatic fatty acid metabolism. J Nutr 132:222–230

Bila KB, Djibo OPC, Sanon B (2010) Analysis of the cashew value chain in Burkina Faso. Deutsche Gesellschaft für Internationale Zusammenarbeit (GIZ)-African Cashew initiative (ACi), Burkina Faso, pp 40

Corraze G (1999) Nutrition lipidique. In: Guillaume J, Kaushik S, Bergot P, Métailler R (eds) Nutrition et alimentation des poissons et crustacés. INRA, Paris, pp 147–170

Effiong MU, Akpan AW (2016) Evaluation of the biochemical responses of catfish (clarias gariepinus burchell, 1822) after replacing fish oil with plant-based oils in their diets. J Appl Sci Environ Manage 20(4):1069–1073

Fagbenro OA, Arowosegbe IA (1991) Utilization of agricultural wastes and byproducts in fish feeds production in Nigeria. In: Proceedings of 6th Annual FISON Conference, Lagos, pp 121–130

FAO (2016) La situation mondiale des pêches et de l'aquaculture 2016. Contribuer à la sécurité alimentaire et à la nutrition de tous, Rome, p 224

Faturoti EO, Lawal LA (1986) Performance of supplementary feeding and organic manuaring on the production of *Oreochromis niloticus*. J West Afr Fish 1:25–32

Folch J, Lees M, Stanley GHS (1957) A simple method for the isolation and purification of total lipids from animal tissues. J Biol Chem 226:497–509

Goede RW, Barton BA (1990) Organismic indices and an autopsy-based assessment as indicators of health and condition of fish. Am Fish Soc Symp 8:93–108

Hammed AM, Amosu AO, Fashina-Bombata HA, Ajepe RG, Olufowobi MT (2014) The use of coconut (*Cocos nucifera*) oil as an alternative source of energy in the diets of African Mudcatfish (*Clarias gariepinus*—Burchell, 1822) juveniles. IJSE R 5(8):431–436

Hecht T, Oellermann L, Verheust L (1996) Perspectives on clariid catfish culture in Africa. Aquat Living Resour 9:197–206

Hogendoorn H (1983) Growth and production of the African catfish, Clarias lazera (C. & V.): III. Bioenergetic relations of body weight and feeding level. Aquaculture 35:1–17

Idah P, Simeon M, Mohammed M (2014) Extraction and characterization of cashew nut (*Anacardium occidentale*) oil and cashew shell liquid oil. Acad Res Int 5(3):50–54

Jobling M (1983) A short review and critique of methodology used in fish growth and nutrition studies. J Fish Biol 23:686–703

Kabré TA, Sanogo S, Naba M (2014) Aquaculture en zone sahélo–soudanienne: diagnostic des réalisations et analyse d'impact socioéconomique de la filière poisson au Burkina Faso. J Appl Biosci 76:6368–6385

Kerdchuen N (1992) L'alimentation artificielle d'un silure africain, *Heterobranchus longifilis* (Teleostei. Clariidae): Incidence du mode d'alimentation et première estimation des besoins nutritionnels. Thesis, University of Paris–VI, p 182

Kiron V, Fukuda H, Takeuchi T, Watanabe T (1995) Essential fatty acid nutrition and defense mechanism in rainbow trout *Oncorrynhus mykiss*. Comp Biochem Physiol 111A:361–367

Legendre M, Kerdchuan N, Corraze G, Bergot P (1995) Larval rearing of an African catfish *Heterobranchus longifilis* (Teleostei, Clariidae): effect of dietary lipids on growth, survival and fatty acid composition of fry. Aquat Living Resour 8:355–363

Legrand P, Beauchamp E, Catheline D, Pédrono F, Rioux V (2010) Short chain saturated fatty acids decrease circulating cholesterol and increase tissue PUFA content in the rat. Lipids 45:975–986

Lim PK, Boey PL, Ng WK (2001) Dietary palm oil level affects growth performance, protein retention and tissue vitamin E concentration of African Catfish, *Clarias gariepinus*. Aquaculture 202(1–2):101–112

Mazid MA, Zaher M, Begum NN, Aliu MZ, Nahar F (1997) Formulation of cost-effective feeds from locally available ingredients for carp polyculture system for increase production. Aquaculture 151:71–78

Montero D, Grasso V, Izquierdo MS, Ganga R, Real F, Tort L, Caballero MJ, Acosta F (2007) Total substitution of fish oil by vegetable oils in gilthead sea bream (*Sparus aurata*) diets: effects on hepatic Mx expression and some immune parameters. Fish Shellfish Immunol 24:147–155

Mourente G, Good JE, Bell JG (2005) Partial substitution of fish oil with rapeseed, linseed and olive oils in diets for European sea bass (*Dicentrarchus labrax* L.): effects on flesh fatty acid composition, plasma prostaglandins E2 and F2a, immune function and effectiveness of a fish oil finishing diet. Aquac Nutr 11:25–40

Ng WK, Chong CY (2004) An overview of lipid nutrition with emphasis on alternative lipid sources in tilapia feeds. In: Bolivar RG, Mair GC, Fitzsimmons K (eds) Proceedings of the Sixth International Symposium on Tilapia in Aquaculture. Bureau of Fisheries and Aquatic Resources, Manila, pp 241–248

Ng WK, Tee MC, Boey PL (2000) Evaluation of crude palm oil and refined palm olein as dietary lipids in pelleted feeds for a tropical bagrid catfish *Mystus nemurus* (Cuvier and Valenciennes). Aquac Res 31:337–347

Ng WK, Lim PK, Boey PL (2003) Dietary lipid and palm oil source affects growth, fatty acid composition and muscle α-tocopherol concentration of African catfish, *Clarias gariepinus*. Aquaculture 215:229–243

NRC (1993) Nutrient requirements of fish. National Academy Press, Washington, DC, p 105

Nyina-wamwiza L, Wathelet B, Kestemont P (2007) Potential of local agricultural by-products for the rearing of African catfish, *Clarias gariepinus* in Rwanda: effects on growth, feed utilization and body composition. Aquac Res 38:206–214

Ochang SN (2011) Effect of replacing cod liver oil with soybean oil as dietary lipid on carcass composition, haematology and sensory properties of the Nile tilapia *Oreochromis niloticus*. Int Aquat Res 3:71–77

Potts WM, Hecht T, Andrew TG (2008) Does reservoir trophic status influence the feeding and growth of the sharptooth catfish, *Clarias gariepinus* (Teleostei: Clariidae)? Afr J Aquat Sci 33:149–156

Rabany C, Rullier N, Ricau P (2015) Analysis of cashew production, processing and trade in Africa. RONGEAD/ICA p 37

Riediger ND, Othman RA, Suh M, Moghadasian MH (2009) A systemic review of the roles of n-3 fatty acids in health and disease. J Am Diet Assoc 109(4):668–679

Sargent JR, Tocher DR, Bell JG (2002) The lipids. In: Halver JE, Hardy RW (eds) Fish nutrition, 3rd edn. Elsevier, USA, pp 181–257

Satoh S, Poe WE, Wilson RP (1989) Studies on the essential fatty acid requirement of channel catfish, *Ictalurus punctatus*. Aquaculture 79:121–128

Sotolu AO (2010) Feed utilization and biochemical characteristics of *Clarias gariepinus* (Burchell, 1822) Fingerlings fed diets containing fish oil and vegetable oils as total replacements. World J Fish Mar Sci 2(2):93–98

Stickney RR, Andrews JW (1972) Effects of dietary lipids on growth, food conversion, lipid and fatty acid composition of channel catfish. J Nutr 102:249–258

Takeuchi T, Arai S, Watanabe T, Shimma Y (1980) Requirement of eel *anguilla japonica* for essential fatty acids. Bull Jpn Soc Sci Fish 46(3):345–353

Tocher DR (2003) Metabolism and functions of lipids and fatty acids in teleost fish. Rev Fish Sci 11(2):107–184

Yilmaz E, Naz M, Akyurt I (2004) Effect of dietary olive pomace oil and Lcarnitine on growth and chemical composition of African catfish, (Burchell, 1822). Isr J Aquac Bamidgeh 56(1):14–21

Yingst WL, Stickney RR (1979) Effects of dietary lipids on fatty acid composition on channel catfish fry. Trans Am Fish Soc 108:620–625

16

Evaluation of different dietary additives based on growth performance, innate immunity and disease resistance in juvenile Amur catfish, *Silurus asotus*

Yaa Tiwaah Amoah · Mohammad Moniruzzaman · Seunghan Lee · Jinho Bae · Seonghun Won · Minji Seong · Sungchul C. Bai

Abstract Three different dietary additives were evaluated based on growth performance, innate immune responses and disease resistance in juvenile Amur catfish *Silurus asotus*. Four diets were prepared by supplementing 0.4% Song-gang® stone (SG), 0.05% Yucca meal (YM), 0.05% β-glucan (BG) in combination with 0.04% SG and 0.05% BG (SG + BG) to a basal commercial diet (control; CONT). Triplicate groups of fish averaging 4.95 ± 0.05 g were fed one of the test diets between 3 and 5% body weight for 8 weeks. At the end of the feeding trial, weight gain, specific growth rate, feed efficiency and protein efficiency ratio of fish fed SG and SG + BG were significantly higher than those of fish fed the BG and CONT diets ($P < 0.05$). Lysozyme activity in fish fed the SG, YM, and SG + BG diets was significantly higher than those of fish fed the CONT diet. Super oxide dismutase (SOD) activity was significantly higher among fish fed the SG, YM, BG and SG + BG diets than fish fed the CONT diet. After 14 days challenge test with *Edwardsiella tarda*, cumulative survival rates of fish fed the SG and SG + BG diets were significantly higher than those of fish fed the BG and CONT diets. Results revealed that dietary SG or SG + BG diets had positive responses over other additives in juvenile Amur catfish. However, Song-gang® stone is proposed to be a cheaper dietary additive contributing 0.04 cents/kg of feed based on economic point of view.

Keywords Amur catfish · Song-gang® stone · Yucca meal · β-Glucan · Growth · Immune responses

Introduction

Aquaculture has continuously intensified due to decreased wild capture and increased demand for the food fish. Production of the food fish from aquaculture reached 70.2 million tons in 2013 (FAO 2015). The intensification of aquaculture exposes cultured animals to stressful conditions, pathogens, and leads to deterioration of the environment, which leads to growth suppression, disease outbreaks, products of low market value and eventually significant economic losses (Sakai 1999; Rufchaie and Hoseinifar 2014). To reduce the economic losses caused by poor growth and disease outbreaks in aquaculture, the identification, development and application of natural, practical and cost-effective dietary additives can modulate immune responses as well as can promote growth of farmed fish (Halver and Hardy 2002; Harikrishnan et al. 2011; NRC 2011).

Y. T. Amoah · M. Moniruzzaman · S. Lee · J. Bae · S. Won · M. Seong · S. C. Bai (✉)
Department of Marine Bio-materials and Aquaculture/Feeds and Foods Nutrition Research Center, Pukyong National University, Busan 608-737, Republic of Korea
e-mail: scbai@pknu.ac.kr

These approaches have recently gained attention among fish nutritionists, aquaculture researchers and farmers globally. Feed additives generally used in aquaculture are growth promoters, immune enhancers, antimicrobial agents, antioxidants, binding agents, enzymes, organic acids, feeding/palatability enhancers, probiotics and prebiotics and hormones (NRC 2011). They are added in the aqua feeds to increase the physical or chemical properties of the feed in terms of enhancing fish performance or the quality of resulting product. Nonetheless, during the selection of feed additives their contribution to feed costs should also be considered, since feed costs depending on the source and quality of the protein account for about 40–50% of the total production costs in aquaculture (Craig and Helfrich 2009).

Immune stimulants are naturally occurring compounds that modulate the immune system by increasing the host's resistance against diseases caused by pathogens (Bricknell and Dalmo 2005). They stimulate non-specific immune responses when given alone or the specific immune responses given with an antigen. Dosages of immune stimulants that have been given to aquatic organisms play an important role in the stimulation of immune response (Meena et al. 2013). Immune stimulants include the extracts from naturally occurring minerals, plants, microbes and animal infractions (NRC 2011). Song-gang® stone (SG) is a naturally sourced and purified mineral material from the Republic of Korea comprising predominantly of iron oxide (Fe_2O_3), silicon dioxide (SiO_2), aluminum oxide (Al_2O_3), potassium oxide (K_2O), sodium oxide (Na_2O), and other trace compounds (Lee et al. 2015a; Won et al. 2017). It is considered as one of the important naturally sourced mineral materials that enhances the health status of fish (Choi et al. 2004; Lee et al. 2015b; Won et al. 2017). Fish farmers in the Republic of Korea use Song-gang stone® (SG) at 0.5% as dietary inclusion which gives a positive response in olive flounder aquaculture (Choi et al. 2004). Lee et al. (2015a) suggested that 0.4% (dry matter basis) is the optimum dietary inclusion level for yellow loess (YL) in the diet of rainbow trout. Likewise, Won et al. (2017) found that 0.4% of SG or YL supplemented diet could replace oxytetracycline (OTC) in juvenile and subadult rainbow trout. The *Yucca schidigera* (Agavaceae) has been recognized as an immune stimulant with anti-inflammatory effects (Cheeke et al. 2006). It contains active phytochemicals and a rich source of steroidal saponins (Piacente et al. 2005). Yucca polyphenols also act as antioxidants (Oakenfull and Sidhu 2005; Piacente et al. 2005). Yucca extracts have been found to promote growth, improve immune system, antimicrobial ability and stimulate appetite and anti-stress characteristics. Inclusion levels of yucca extract in aqua feeds include: 0.075% (Gaber 2006) and 0.1–0.58% for Nile tilapia (Njagi et al. 2017), 0.1% for striped catfish (Güroy et al. 2014). The β-glucan (BG) is an inactivated natural microbial product that is also recognized for its immunomodulatory roles in aquatic organisms (NRC 2011). Positive effects of the β-glucan on growth, survival, and immune response of fish species against infectious pathogens has been demonstrated either alone or in combination with other immune stimulants or probiotics (Meena et al. 2013). It influences non-specific or innate immune responses such as increased lysozyme production and superoxide anion production in activated macrophages, thereby, reducing fish mortality when exposed to pathogens (NRC 2011). Low doses of β-glucan have been suggested to enhance growth and innate immunity in different cultured fish species (Ai et al. 2007; Yoo et al. 2007; Das et al. 2013; Sirimanapong et al. 2015).

The Amur catfish, *Silurus asotus*, belongs to the family Siluridae (sheatfishes). It is distributed in Japan, Republic of Korea, China and Taiwan. It is the second freshwater farmed fish species with a production of 4607 tons in the Republic of Korea in 2014 (FAO 2016). Several dietary additives have been identified to promote growth and immune responses in the aquaculture industry (Choi et al. 2004; Ai et al. 2007; Yoo et al. 2007; NRC 2011; Güroy et al. 2014; Lee et al. 2015b; Shahkar et al. 2015); however, each dietary additive could produce different levels of responses in individual fish species and contribute differently to feed cost. In this regard, the effects of dietary additives from mineral sources have not yet been investigated in Amur catfish. Consequently, this study was designed to evaluate the effects of dietary SG, YM, BG and a combination of SG and BG on growth performance, innate immune responses and disease resistance in juvenile Amur catfish.

Methods

Experimental design and diets

A basal commercial catfish diet (Catfish premium; Woosung Feed Co. Ltd, Republic of Korea) (Table 1) without additive inclusion was used as the control diet and four other diets were prepared by supplementing

Table 1 Proximate composition (% dry matter basis) of the basal diet

Proximate content	% Composition
Moisture	6.7
Crude protein	47.4
Crude lipid	9.5
Crude ash	11.7
Nitrogen free extract[a]	24.7

Woosung Fish Feed Company, Daejeon, Republic of Korea

[a] Nitrogen free extract = 100—(moisture + protein + lipid + ash) (Castell and Tiews 1980)

Table 2 Experimental diets and the inclusion rates of feed additives used to feed the juvenile Amur catfish for 8 weeks

Additives (%)	Diets				
	CONT	SG	YM	BG	SG + BG
Song-gang® stone	0	0.4	0	0	0.4
Yucca meal	0	0	0.05	0	0
β-glucan	0	0	0	0.05	0.05

CONT (control)—basal diet (Woosung Fish Feed Company, Daejeon, Republic of Korea), *SG* 0.4% Song-gang® stone (Davistone Co. Ltd., Busan, South Korea) + basal diet, *YM* – 0.05% Yucca meal (as De Odorase® Alltech Korea, Seoul, Rep. of Korea) + basal diet, *BG* 0.05% β-glucan (as CitriStim®, Flavour Cooperation of Korea, Seoul, Republic of Korea) + basal diet, *SG* + *BG* 0.4% Song-gang® stone + 0.05% β-glucan + basal diet

0.4% Song-gang® stone (SG) which contains 73.4% silicon dioxide (SiO_2) and 15.5% aluminum oxide (Al_2O_3), 0.05% Yucca meal (YM) which contains 30% *Yucca schidigera* extract and 2.39% saponin, 0.05% β-glucan (BG) which comes from yeast cell wall and a combination of 0.4% SG and 0.05% BG (SG + BG) (Table 2). The supplementations of the additives are based on our previous research findings (Yoo et al. 2007; Njagi et al. 2017; Won et al. 2017). The unit price of SG, YM and BG is US$ 0.89/kg, US$ 15.21/kg, US$ 26.84/kg, respectively; and the price contribution for the unit kg of feed of SG, YM and BG is US 0.04 cents (¢), US 0.08 cents (¢), 1.34 cents (¢) and US 1.7 cents (¢), respectively. The commercial feed was grounded into powder with a laboratory grinding machine and sieved before addition of the feed additives. Required levels of the feed additives were weighed with an electronic balance (± 0.01 g) and solubilised with filtered tap water before its addition to the powdered feed and mixed in a laboratory electric feed mixer. Water was added at 50 mL/100 g feed. The resultant mixture was pelleted with a laboratory pelleting machine without heating using a 2-mm diameter module (Baokyong Commercial Co., Busan, Republic of Korea) and air dried for 3 days. The dried feeds were sieved and packaged in zipper bags, labeled and stored at – 20 °C before the feeding trial.

Fish and feeding trial

The study was conducted at the Feeds and Foods Nutrition Research Centre (FFNRC), Pukyong National University, Busan, Republic of Korea. A total of 400 juveniles of Amur catfish were obtained from the Namsangju fish farm (Namsangju, Republic of Korea). Prior to the start of the feeding trial, all fish were fed the basal diet twice a day for 2 weeks to acclimate with the experimental condition. The feeding trial was conducted in an indoor semi-recirculating system with 15 aquaria (50 L) receiving a constant water flow of filtered freshwater at the rate of 2 L/min from the central tank. Half of the water was exchanged daily by pumping filtered fresh water to the central tank. Supplemental aeration was provided to maintain dissolved oxygen level near saturation and water temperature was maintained at 25.0 ± 0.1 °C using water heaters throughout the experimental period. Three hundred fish averaging 4.92 ± 0.05 g [mean ± standard deviation (SD)] were weighed and randomly distributed into the 15 indoor aquaria (20 fish/tank). Each aquarium was then randomly assigned one of three replicates of the five dietary treatments. Fish were fed one of the above mentioned diets twice daily (0900 and 1800 hours) ad libitum for 8 weeks. Experimental tanks were siphoned

2 h after feeding to remove fecal matter and uneaten feed. Mortality was checked daily; dead fish were removed immediately and weighed, and the amount of feeds for the tanks adjusted to the proper percentage of the remaining fish weight in the tanks. Dead fish were not replaced during the experiment. Total fish weight in each aquarium was determined at the end of the fourth week as the mid-term sampling and the amount of diet fed to the fish was adjusted accordingly. Finally, the total fish weights in the aquaria were measured at the end of the eighth week. Fish were starved 24 h before each measurement to avoid inclusion of ingested feed in the weight measurements as well as to minimize stress.

Sample collection and analysis

At the end of the feeding trial, all the fish in each aquarium were counted and weighed to calculate weight gain (WG), specific growth rate (SGR), feed efficiency (FE), protein efficiency ratio (PER), and survival rate (SR). Three fish per aquarium were randomly selected, weighed, and measured to calculate condition factor (CF); and then the fish were dissected to obtain liver and visceral masses for the calculation of hepatosomatic (HSI) and viscerosomatic (VSI) indices, respectively.

Prior to the collection of liver and visceral masses for the HSI and VSI calculations, respectively, the same fish were also used to analyze serological characteristics. For the blood collection, fish were first euthanized with ethylene glycol phenyl ether (200 mg/L). Blood samples were obtained from the caudal vein of the fish with 1 ml syringes. Serum samples were obtained from the blood on clotting by centrifugation at $5000\times g$ for 10 min and stored at − 70 °C for the analysis of lysozyme and superoxide dismutase (SOD) activities. A turbidometric assay was used to analyze serum lysozyme activity (Sigma-Aldrich, St. Louis, MO, USA). The bacterium, *Micrococcus lysodeikticus* (0.75 mg/mL) was suspended in a sodium phosphate buffer (0.1 M, pH 6.24), 200 µl of suspension was placed in each well of 96-well plates and 20 µL of test serum was added subsequently. The reduction in absorbance of the samples was recorded at 450 nm after incubation at room temperature (25 °C) for 0 and 60 min in a microplate reader (UVM 340, Biochrom, Cambridge, UK). A reduction in absorbance of 0.001/min was regarded as one unit of lysozyme activity. The lysozyme activity was calculated using the following equation:

$$\text{Units/ml} = (\Delta A_{450}/\text{minutes} \times 1000)/\text{mL enzyme in reaction mixture.}$$

Superoxide dismutase (SOD) activity was measured by the percentage of reaction inhibition rate of the enzyme with WST-1 (Water Soluble Tetrazolium dye) substrate and xanthine oxidase using a SOD Assay Kit (Enzo ADI-900-157, Enzo Life Sciences Inc., Farmingdale, NY, USA) in accordance with the manufacturer's protocol. Each endpoint assay was monitored by absorbance at 450 nm (the absorbance-wavelength for the colored product of WST-1 reaction with superoxide) after 20 min of reaction time at 37 °C. The percentage of inhibition was normalized by mg protein and presented as SOD activity units. Portions of serum were also used for analyses of biochemical parameters including serum total protein, glutamic oxaloacetic transaminase (GOT), glutamic pyruvic transaminase (GPT) and glucose using the commercial clinical kits with an analyzer (Fuji DRI-CHEM 3500i, Fuji Photo Film Ltd, Tokyo, Japan).

Three fish per aquarium were taken and pooled together according to the type of the diet fed and then homogenized and freeze-dried at − 80 °C for analysis. Proximate compositions of basal diet and fish were determined according to the standard procedures of AOAC (2000) in duplicate. Samples were dried to a constant weight at 105 °C to determine the moisture content. Ash content was determined by incineration in a muffle furnace at 550 °C for 3.5 h, crude lipid by ether extraction using the Soxtec system 1046 (Tecator AB, Hoganas, Sweden) and crude protein by the Kjeldahl method ($N \times 6.25$) after acid digestion.

The lyophilized pathogenic bacterium, *Edwardsiella tarda* (ATCC 15947, American Type Culture Collection) was obtained from the Department of Biotechnology, Pukyong National University, Busan, Republic of Korea. The *E. tarda* culture and identification process was followed by Park et al. (2016) with slight modifications. Briefly, thawed and fresh *E. tarda* was cultured in tryptic soy broth (TSB; Sigma-Aldrich) at 26 °C for 24 h and stored at − 80 °C for further use. Bacterial growth was measured at OD_{600} by spectrometer (Lambda 35; Perkin-Elmer, Waltham, MA, USA) followed by plate counting of colonies grown on tryptic soy agar (TSA; Sigma-Aldrich). The bacterial colony was identified by the API 20 E pathogenic bacteria identification kit (BioMérieux, Durham, NC, USA). Five fish were selected from each aquarium and acclimatized for 24 h after the feeding trial. Each fish was injected intraperitoneally with 0.1 mL of fresh bacterial suspension at a concentration of 2×10^8 colony-forming units (CFU)/mL based on a preliminary virulence test (50% lethal dose, LD_{50}) in groups of fish using the pathogenic bacterium *E. tarda* at different concentrations ($10^6 - 10^{10}$ CFU/mL) for 7 days (data not shown). The cumulative survival rate of fish was monitored and fish mortality was recorded daily up to 14 days. For the confirmation of virulence of the bacterium, *E. tarda*, swabs from skin, gill, liver and kidney of dead fish were checked on modified selective agar at 22 °C for 72 h in incubation and we confirmed that the mortality was due to the *E. tarda* by 16S rRNA gene sequencing.

Statistical analysis

All data were subjected to one-way analysis of variance (ANOVA) using SAS Program Version 9.1(SAS Institute, Cary, NC, USA) to test the effects of the different dietary treatments. When a significant effect was observed, Tukey's post hoc test was used to compare the means. Treatment effects were considered significant at confidence level of $P < 0.05$. Survival analysis in challenge test was done based on Kaplan and Meier (1958) to determine the differences among survival curves (K–M curves). All data are presented as mean ± SD.

Results and discussion

The present study demonstrated that the dietary additives produced different response levels in terms of growth, feed utilization, immune responses and disease resistance in juvenile Amur catfish. The results revealed a more positive response of juvenile Amur catfish to Song-gang® stone (SG) as well as combination of SG and β-glucan (SG + BG) compared with Yucca meal (YM), β-glucan (BG) and control (CONT) diets. The additives, β-glucan and Yucca meal have already shown their immunostimulating effects on fish (Yoo et al. 2007; Njagi et al. 2017). In this study, the results demonstrated the potential effect of a mineral-based additive such as Song-gang® stone in the juvenile Amur catfish.

Growth

Growth performance of juvenile Amur catfish fed the experimental diets for 8 weeks is summarized in Table 3. At the end of the feeding trial, weight gain (WG) and specific growth rate (SGR) of fish fed Song-gang® stone (SG) and a combination of SG and β-glucan (SG + BG) containing diets were significantly higher than those of fish fed the BG and control (CONT) diets ($P < 0.05$). However, there were no significant differences among those of fish fed the SG, SG + BG and YM diets, between those of fish fed the YM and BG diets as well as between those of fish fed the BG and CONT diets. Likewise, feed efficiency (FE) and protein efficiency ratio (PER) of fish fed the SG and SG + BG diets were significantly higher than those of fish fed the BG and CONT diets ($P < 0.05$). However, there were no significant differences in FE and PER among fish fed the SG, SG + BG and YM diets ($P > 0.05$). Moreover, fish fed the YM, BG and CONT diets did not show significant differences in terms of FE and PER. The observations with the SG diet is comparable to the findings by Lee et al. (2015b) and Won et al. (2017) with Song-gang® stone and yellow loess, respectively, in rainbow trout *Oncorhynchus mykiss* as well as Shahkar et al. (2015) with Macsumsuk® in Nile tilapia *Oreochromis niloticus*. Growth and feed efficiency improved in olive flounder *Paralichthys olivaceus* fed a combination of β-glucan and BAISM-contained diets (Yoo et al. 2007). Measurement and analysis of condition factor (CF), hepatosomatic index (HSI) and viscerosomatic index (VSI) are important in assessing food value in an animal, its health status and general well-being as well as the degree of pollutants in the environment where the animal lives (Ighwela et al. 2014). There were no significant differences in HSI, VSI and CF among fish fed the experimental diets which was an indication of good nutritional and health status of fish during the feeding trial. In this study, the highest survival rates (100%) recorded among the groups of experimental fish during the study could be attributed to the high acceptability of the experimental diets by the catfish.

Table 3 Growth performance of juvenile Amur catfish fed the experimental diets for 8 weeks

Parameters	Diets[A]				
	CONT	SG	YM	BG	SG + BG
WG[B]	475 ± 6.8^{a}	493 ± 5.2^{c}	486 ± 6.3^{bc}	481 ± 8.3^{ab}	492 ± 2.6^{c}
SGR[C]	3.89 ± 0.03^{a}	3.96 ± 0.02^{c}	3.93 ± 0.02^{bc}	3.91 ± 0.03^{ab}	3.95 ± 0.01^{c}
FE[D]	127 ± 2.3^{a}	138 ± 0.6^{b}	133 ± 0.7^{ab}	132 ± 4.9^{a}	138 ± 4.8^{b}
PER[E]	2.69 ± 0.05^{a}	2.91 ± 0.01^{b}	2.80 ± 0.02^{ab}	2.78 ± 0.1^{a}	2.92 ± 0.1^{b}
HSI[F]	1.11 ± 0.1	1.18 ± 0.1	1.18 ± 0.1	1.17 ± 0.1	1.23 ± 0.2
VSI[G]	3.28 ± 0.6	3.21 ± 0.4	3.05 ± 0.4	3.15 ± 0.6	3.07 ± 0.5
CF[H]	0.61 ± 0.04	0.58 ± 0.1	0.55 ± 0.05	0.59 ± 0.03	0.61 ± 0.1
Survival[I]	100 ± 0.0	100 ± 0.0	100 ± 0.0	100 ± 0.0	100 ± 0.0

Values are mean ± SD for the triplicate groups of fish where values in each row with different superscripts are significantly different ($P < 0.05$)

[A]Footnote: Refer to Table 2

[B]Weight gain (WG) = (final weight initial weight) × 100/initial weight

[C]Specific growth rate (SGR; %) = (ln final weight - ln initial weight) × 100/days

[D]Feed efficiency (FE; %) = wet WG (g) × 100/dry feed intake (g)

[E]Protein efficiency ratio (PER) = wet weight gain/protein intake

[F]Hepatosomatic index (HSI; %) = (liver weight/body weight) × 100

[G]Viscerosomatic index (VSI; %) = (visceral weight/body weight) × 100

[H]Condition factor (CF) = [fish weight (g)/fish length (cm)^3] × 100

[I]Survival (%) = (final number of fish/initial number of fish) × 100

Whole body proximate analysis

Whole body proximate compositions of juvenile Amur catfish fed the experimental diets are summarized in Table 4. There were no significant differences in the whole body moisture and ash content among fish fed the experimental diets at the end of the feeding trial. Crude protein content in fish fed the YM diet was significantly higher than those of fish fed the SG, BG, SG + BG and CONT diets ($P < 0.05$). However, there were no significant differences in the whole body crude protein contents among fish fed the SG, BG, SG + BG and CONT diets. In contrast, crude lipid in fish fed the YM diet was significantly lower than those of fish fed the SG, BG, SG + BG and CONT diets ($P < 0.05$). Fish fed the SG, BG, SG + BG and CONT diets showed no significant differences in case of the whole body crude lipid content. Gaber (2006) reported the higher body protein content in Nile tilapia fed the Yucca-supplemented diet compared with the control diet and significantly lower body lipid content than that of the control diet. Similarly, Yilmaz et al. (2012) also reported lower crude lipid content in sea bass *Dicentrarchus labrax* fed the herbal extracts compared with the control diet. It can be speculated that plant extracts could reduce the crude lipid content in cultured fish.

Table 4 Whole body proximate composition of Amur juvenile catfish fed the experimental diets for 8 weeks (% dry matter basis)

	Diets[A]				
	CONT	SG	YM	BG	SG + BG
Moisture	77.43 ± 0.8	77.63 ± 0.2	77.32 ± 0.2	77.68 ± 0.9	78.81 ± 0.2
Crude protein	67.01 ± 0.4^{a}	66.99 ± 0.6^{a}	69.36 ± 0.2^{b}	66.93 ± 0.01^{a}	67.20 ± 0.1^{a}
Crude lipid	19.56 ± 0.5^{a}	19.32 ± 0.5^{a}	17.76 ± 0.8^{b}	19.51 ± 0.2^{a}	19.50 ± 0.4^{a}
Ash	11.47 ± 0.2	11.78 ± 0.02	11.53 ± 0.5	12.46 ± 0.1	13.12 ± 0.1

Values are mean ± SD from the triplicate groups of fish where values in each row with different superscripts are significantly different ($P < 0.05$)

[A]Footnote: Refer to Table 2

Serum parameters

Serum parameters of juvenile Amur catfish fed the experimental diets are summarized in Table 5. Hematological and serum parameters are good indicators of the health status of an organism and these can vary with season, temperature and nutritional status (Blaxhall 1972). Lee et al. (2016) suggested that any unhealthy condition caused by dietary treatment could affect the hematological characteristics of fish. In this study, there were no significant differences ($P < 0.05$) in terms of serum glucose, total protein (TP), glutamic oxaloacetic transaminase (GOT) and glutamic pyruvic transaminase (GPT) among fish fed the experimental diets which is in agreement with Lee et al. (2015a). Overall, in this study, the results of the serum parameters suggested that the dietary additives had no adverse effects on health status of fish (Blaxhall 1972).

Table 5 Serum parameters in juvenile Amur catfish fed the experimental diets for 8 weeks

	Diets[A]				
	CONT	SG	YM	BG	SG + BG
Glucose (mg/dl)	67.7 ± 29.0	65.3 ± 14.5	61.7 ± 18.9	54.7 ± 5.03	53.3 ± 10.2
Total protein (g/dl)	2.8 ± 0.0	2.9 ± 0.2	2.7 ± 0.1	2.9 ± 0.1	2.9 ± 0.1
GOT (U/l)[B]	153.7 ± 16.2	95.3 ± 37.8	109.7 ± 3.7	140.0 ± 58.6	94.3 ± 26.5
GPT (U/l)[C]	6.3 ± 0.5	5.3 ± 0.5	5.7 ± 0.5	5.7 ± 0.6	6.3 ± 0.5

Values are mean ± SD from the triplicate groups of fish where values in each row without superscripts are insignificantly different ($P > 0.05$)

[A]Footnote: Refer to Table 2

[B]GOT = Glutamic oxaloacetic transaminase

[C]GPT = Glutamic pyruvic transaminase

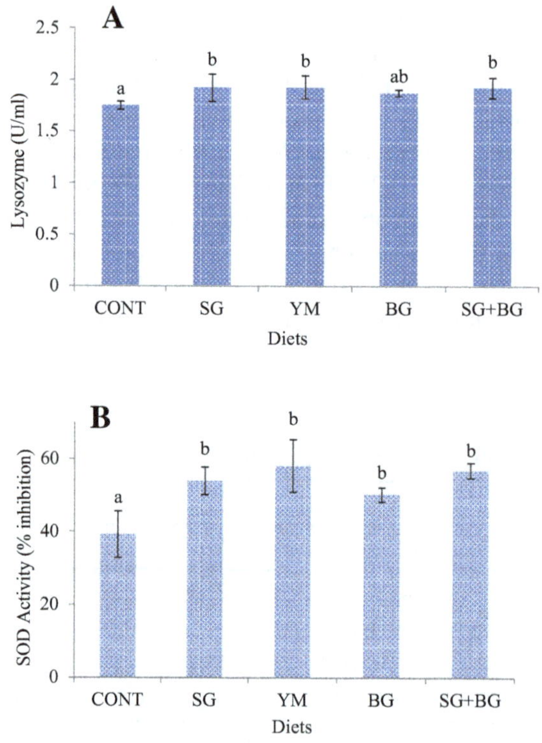

Fig. 1 Lysozyme (**a**) and superoxide dismutase, SOD (**b**) activities of juvenile Amur catfish fed the different dietary additives for 8 weeks

Innate immune responses

Results on the innate immune responses of juvenile Amur catfish fed the experimental diets are presented in Fig. 1 (a, b). Lysozyme activity was significantly higher among fish fed the SG, YM and SG + BG diets than in fish fed the CONT diet ($P < 0.05$). However, there were no significant differences in lysozyme activities between fish fed the BG and CONT diets ($P > 0.05$). Moreover, fish fed the SG, YM, BG and SG + BG diets did not show significant differences in terms of lysozyme activities. Superoxide dismutase (SOD) activity was significantly higher among fish fed the SG, BG, YM and SG + BG diets than in fish fed the CONT diet ($P < 0.05$). However, there were no significant differences in SOD activities among fish fed the SG, YM, BG, and SG + BG diets ($P > 0.05$). The results of the innate immune responses may attribute to the increased immunity of fish fed the additive supplemented diets compared to that of control diet. Innate immune response is the fundamental defense mechanism in fish and also plays a key role in the acquired immune response and homeostasis through a system of receptor proteins (Halver and Hardy 2002; Uribe et al. 2011; Reverter et al. 2014; Srivastava and Pandey 2015). Lysozyme is produced mainly by macrophages in response to microbial components and many other immune stimulants (Siwicki and Anderson Siwicki and Anderson 1993; Ringø et al. 2012) and it is a preferred marker of the immune response due to its close association with leucocytes (Kiron 2012). On the other hand, SOD are metalloenzymes that catalyze the dismutation of the superoxide radical (O^{2-}) into hydrogen peroxide (H_2O_2) and molecular oxygen (O_2) (Shao et al. 2010), which provides an important defense against oxidative damage. In line with the present study, improved immune responses have been recorded in cultured fish with several feed additives such as the combination of β-glucan and BAISM (Yoo et al. 2007), β-glucan (Ai et al. 2007), herbal extracts (Chelladurai et al. 2014), yellow loess (Lee et al. 2015b), Song-gang® stone (Lee et al. 2015b) and Yucca meal (Njagi et al. 2017).

Challenge test

The cumulative survival rate of juvenile Amur catfish fed the experimental diets after 14 days of intraperitoneal injection with *E. tarda* is shown in Fig. 2. Mortality was recorded in fish fed the CONT diet from the third day after the injection, which may indicate the first sign of pathogenicity of injected bacteria on visual observation during the challenge test. At the end of 14 days, the cumulative survival rates of fish fed the SG and SG + BG diets were significantly higher than those of fish fed the BG and control diets ($P < 0.05$). However, there were no significant differences in cumulative survival rates among fish fed the SG, SG + BG and YM diets ($P > 0.05$). Moreover, fish fed the YM, BG, and CONT diets did not show significant differences in cumulative survival rates at the end of the challenge test. The bacterium, *E. tarda* is a gram-negative bacterium associated with the freshwater ecosystem. It can colonize in a wide variety of amphibians, reptiles and fish which may ultimately cause diseases in these animals that evidenced in emphysematous putrefactive disease of the catfish (Meyer and Bullock 1973). In this study, results of the challenge test

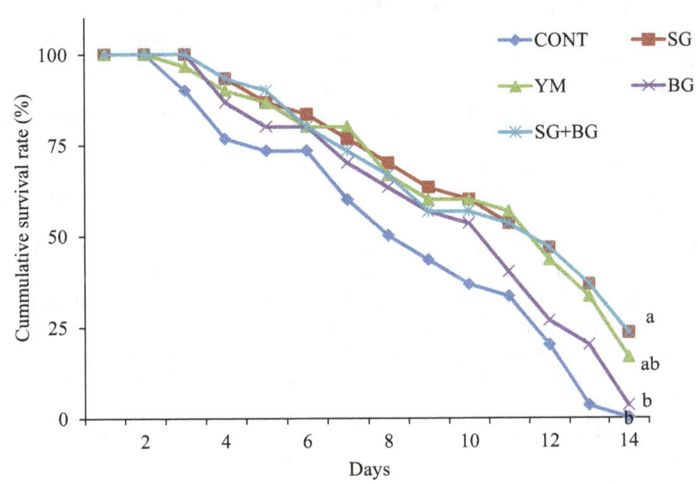

Fig. 2 Cumulative survival of juvenile Amur catfish after intraperitoneal injection with *Edwardsiella tarda* for 14 days

corroborated to the results of the WG, SGR, FE and PER in juvenile Amur catfish. These results with SG are similar to those of Lee et al. (2015a) and Won et al. (2017) reported with yellow loess and SG, respectively, in rainbow trout (*Oncorhynchus mykiss*) and Shakhar et al. (2015) with Macsumsuk® in Nile tilapia (*Oreochromis niloticus*) which were natural sources of minerals.

Conclusion

In conclusion, based on the WG, SGR, FE, PER, lysozyme, SOD and resistance against bacteria, dietary Song-gang® stone (SG) and a combination of Song-gang® stone and β-glucan (SG + BG) could have more positive responses compared to the YM or BG diet alone in terms of improved growth performance, innate immune responses and disease resistance in juvenile Amur catfish. Moreover, as the price contribution for the unit kg of feed of Song-gang® stone is only 0.04 cents (¢) kg^{-1} of feed, it could be the more cost-effective feed additive for catfish culture among the additives that we had tested.

Acknowledgements This research work was supported by the Pukyong National University Research Fund in 2016 (C-D-2016-0237). The authors would like to appreciate the help and cooperation of students and staff of Feeds and Foods Nutrition Research Center (FFNRC) at PKNU during the feeding trial and the analyses of the samples.

References

Ai Q, Mai K, Zhang L, Tan B, Zhang W, Xu W, Li H (2007) Effects of dietary β-1, 3 glucan on innate immune response of large yellow croaker, *Pseudosciaena crocea*. Fish Shellfish Immunol 22:394–402

Blaxhall PC (1972) The haematological assessment of the health of freshwater fish. J Fish Biol 4:593–604

Bricknell I, Dalmo RA (2005) The use of immunostimulants in fish larval aquaculture. Fish Shellfish Immunol 19:457–472

Castell JD, Tiews K (1980) Report on the EIFAC, IUNS and ICES working group on the standardization of methodology in fish nutrition research. EIFAC Technical Paper 36. Hamburg, Federal Republic of Germany, 21-23 March

Cheeke PR, Piacente S, Oleszek W (2006) Anti-inflammatory and anti-arthritic effects of *Yucca schidigera*: a review. J Inflamm 3:6

Chelladurai G, Veni T, Mohanraj J, Nagarajan R (2014) Effect of Herbal Extracts Supplemented diets on Non Specific Immunity and Resistance to *Aeromonas hydrophila* in Indian cat fish (*Mystus montanus*). J Zool Bios Res 1:10–14

Choi S, Ko S, Park G, Lim S, Yoo G, Lee J, Bai SC (2004) Utilisation of song-gang stone as dietary additive in juvenile olive flounder *Paralichthys olivaceus*. J Fish Mar Sci Edu 17:39–45

Craig S, Helfrich LA (2009) Understanding Fish Nutrition, Feeds, and Feeding. Virginia Cooperative Extension. Publication no. 420-256

Das BK, Pattnaik P, Debnath C, Swain DK, Pradhan J (2013) Effect of β-glucan on the immune response of early stage of *Anabas testudineus* (Bloch) challenged with fungus *Saprolegnia parasitica*. SpringerPlus 2:197

FAO (2015) FAO Global Aquaculture Production database updated to 2013—summary information

FAO (2016) Fisheries and Aquaculture Information and Statistics Branch http://www.fao.org/figis/servlet/SQServlet?file=/work/FIGIS/prod/webapps/figis/temp/hqp_7287501540920455907.xml&outtype=html-15/07/2016

Gaber MM (2006) The effects of plant-protein-based diets supplemented with yucca on growth, digestibility, and chemical composition of Nile Tilapia (*Oreochromis niloticus,* L) Fingerlings. J World Aqua Soc 37:74–81

Güroy B, Mantoglu S, Kayali S, Sahin I (2014) Effect of dietary *Yucca schidigera* extract on growth, total ammonia–nitrogen excretion and haematological parameters of juvenile striped catfish *Pangasianodon hypophthalmus*. Aquac Res 45:647–654

Halver JE, Hardy RW (2002) Fish Nutrition. Academic Press, California, pp 529–531

Harikrishnan R, Balasundaram C, Heo MS (2011) Impact of plant products on innate and adaptive immune system of cultured finfish and shellfish. Aquaculture 317:1–15

Ighwela KA, Ahmad AB, Abol-Munafi AB (2014) The selection of viscerosomatic and hepatosomatic indices for the measurement and analysis of *Oreochromis niloticus* condition fed with varying dietary maltose levels. Int J Fauna Biol Stud 1:18–20

Kaplan EL, Meier P (1958) Nonparametric estimation from incomplete observations. J Am Stat Assoc 53:457–481

Kiron V (2012) Fish immune system and its nutritional modulation for preventive health care. Anim Feed Sci Tech 173:111–133

Lee S, Taddese F, Won S, Yun H, Kim S, Park J, Bai SC (2015a) Natural feed additives enhance rainbow trout performance. Global Aquaculture Alliance, Innovation, Missouri, pp 73–75

Lee YK, Katya K, Yun H, Yoon MY, Park K, Sung JS, Shin HS, Bai SC (2015b) Evaluation of dietary yellow loess as an antibiotic replacer on growth, immune responses, serological characteristics and disease resistance in rainbow trout Oncorhynchus mykiss. Aquac Nutri. https://doi.org/10.1111/anu.12348

Lee S, Moniruzzaman M, Yun H, Park Y, Mann J, Bai SC (2016) Comparative studies on effects of extruded pellets and dough type diets on growth, body composition, hematology and gut histology of juvenile Japanese Eel, *Anguilla japonica* (Temminck et Schlegel). J Fish Aqua Sci 11:378–384

Meena DK, Das P, Kumar S, Mandal SC, Prusty AK, Singh SK, Akhtar MS, Behera BK, Kumar K, Pal AK, Mukherjee SC (2013) Beta-glucan: an ideal immunostimulant in aquaculture (a review). Fish Physiol Biochem 39:431–457

Meyer FP, Bullock GL (1973) *Edwardsiella tarda,* a new pathogen of channel catfish, *Ictalurus punctatus.* J Appl Microb 25:155–156

Njagi GW, Lee S, Won S, Hong JW, Hamidoghli A, Bai SC (2017) Effects of dietary Yucca meal on growth, haematology, non-specific immune responses and disease resistance of juvenile Nile tilapia, *Oreochromis niloticus* (Linnaeus, 1758). Aquac Res pp 1–10 (in press) doi:10.1111/are.13264

NRC (2011) Nutrient requirements of fish. National Academy Press, Washington, pp 221–227

Oakenfull D, Sidhu GS (2005) Saponins. In Toxicants of Plant Origin Volume 2. Edited by: Cheeke P.R. Boca Raton, Fla: CRC Press; 1989, pp 97–141

Park Y, Moniruzzaman M, Lee S, Hong J, Won S, Lee JM, Yun H, Kim KW, Ko D, Bai SC (2016) Comparison of the effects of dietary single and multi-probiotics on growth, non-specific immune responses and disease resistance in starry flounder, *Platichthys stellatus.* Fish Shellfish Immunol 59:351–357

Piacente S, Pizza C, Oleszek W (2005) Saponins and Phenolics of *Yucca schidigera* Roezl: chemistry and Bioactivity. Phytochem Rev 4:177–190

Reverter M, Bontemps N, Lecchini D, Banaigs B, Sasal P (2014) Use of plant extracts in fish aquaculture as an alternative to chemotherapy: current status and future perspectives. Aquaculture 433:50–61

Ringø E, Olsen RE, Vecino JLG, Wadsworth S, Song SK (2012) Use of immunostimulants and nucleotides in aquaculture: a review. J Mar Sci Res Dev 1:104–126

Rufchaie R, Hoseinifar SH (2014) Effects of dietary commercial yeast glucan on innate immune response, hematological parameters, intestinal microbiota and growth performance of white fish (*Rutilus frisii kutum*) fry. Croatian J Fish 72:156–163

Sakai M (1999) Current research status of fish immunostimulants. Aquaculture 172:63–92

Shahkar E, Park G, Lee DM, Kwak S, Bai SC (2015) Effects of dietary Macsumsuk® supplementation on growth performance, haematological parameters, disease resistance and body composition of juvenile Nile tilapia, *Oreochromis niloticus* L. J Appl Anim Res 43:125–130

Shao XP, Liu WB, Xu WN, Lu KL, Xia W, Jiang YY (2010) Effects dietary copper sources and levels on performance, copper status, plasma antioxidant activities and relative copper bioavailability in *Carassius auratus* gibelio. Aquaculture 308:60–65

Sirimanapong W, Adams A, Ooi EL, Green DM, Nguyen DK, Browdy CL, Collet B, Thompson KD (2015) The effects of feeding immunostimulant β-glucan on the immune response of *Pangasianodon hypophthalmus.* Fish Shellfish Immunol 45:357–366

Siwicki AK, Anderson DP (1993) Non-specific defence mechanisms assay in fish: II. Potential killing activity of neutrophils and macrophages, lysozyme activity in serum and organs. In: Siwicki AK, Anderson DP, Waluga J (eds) Disease diagnosis and prevention methods, FAO-project GCP/INT/526/JPN. IFI, Olsztyn, Poland, pp 105–111

Srivastava PK, Pandey AK (2015) Role of immunostimulants in immune responses of fish and shellfish. Biochem Cell Arch 15:47–73

Uribe C, Folch H, Enriquez R, Moran G (2011) Innate and adaptive immunity in teleost fish: a Review. Vet Medic 56:486–503

Won S, Moniruzzaman M, Lee S, Hong J, Park JK, Kim S, Bai SC (2017) Evaluation of dietary natural mineral materials as an antibiotic replacer on growth performance, non-specific immune responses and disease resistance in rainbow trout, *Oncorhynchus mykiss.* Aquac Res pp 1–13 (in press) doi:10.1111/are.13295

Yılmaz S, Ergün S, Çelik EŞ (2012) Effects of herbal supplements on growth performance of sea bass (*Dicentrarchus labrax*): change in body composition and some blood parameters. J BioSci Biotech 1:217–222

Yoo G, Lee S, Kim YC, Okorie OE, Park GJ, Han YO, Choi SM, Kang JC, Sun M, Bai SC (2007) Effects of Dietary β-1, 3 Glucan and feed stimulants in juvenile olive flounder, *Paralichthys olivaceus.* J World Aquac Soc 38:138–145

17

A sensory and nutritional validation of open ocean mussels (*Mytilus galloprovincialis* Lmk.) cultured in SE Bay of Biscay (Basque Country) compared to their commercial counterparts from Galician Rías (Spain)

K. Azpeitia · Y. Ríos · I. Garcia · J. Pagaldai · D. Mendiola

Abstract Mussel shell biometry, nutritional quality as well as consumer sensory evaluation of experimental open ocean cultured mussel *Mytilus galloprovincialis* (Lamarck 1819) were analyzed and compared to that of commercial mussels from Galician Rías available in the local market. Both mussel products were of the same commercial size. In this study, open ocean mussels were significantly higher and wider than those of Galician Rías. In addition, with the exception of ash content, both mussel products showed similar biochemical composition. Regarding fatty acid profiles, however, statistical differences were detected. These differences were not fully reflected in the sensory assessment. In terms of consumer acceptability, both mussel products were considered equally satisfactory.

Keywords Fatty acids · Mussel · *Mytilus galloprovincialis* · Open ocean aquaculture · Proximate composition · Sensory analysis

Introduction

Aquaculture is a seafood supply that has increased gradually over the last few decades (Langan and Horton 2003). This growth is due to progress in technology, changes in policy, and an increased societal awareness of sustainability (Bushek et al. 2004; Coen et al. 2007; Costa-Pierce 2008; Forrest et al. 2009; Bostock et al. 2010). In particular, mollusk aquaculture represents more than 75% (13.9 million tons) of the world's aquaculture, with mussel production representing around 13% (1.8 million tons) of its annual production (FAO 2014). Spain, France, and the Netherlands produce one-third of the global mussel aquaculture production (Buck et al. 2010). Specifically, Spain, with an average production near 200,000 tons per year, is the second largest producer worldwide and the first in Europe (FAO 2014). A large part (95%) of its annual production is produced in the Galician Rías from the SW Bay of Biscay (Galician region). Since 2008, mussel products from that area have had a Protected Designation of Origin certifying quality and traceability within the standards of the EU seafood policy.

Shellfish species are a valuable food source in the socioeconomic context of the Basque Country (SE Bay of Biscay) (Gracia 1996), yet mussel aquaculture is limited. Indeed, over the last few decades, every

K. Azpeitia · D. Mendiola (✉)
Marine Research Division, AZTI-Tecnalia, Herrera Kaia; Portualdea, s/n, 20110 Pasaia, Spain
e-mail: dmendiola@azti.es

Y. Ríos · I. Garcia · J. Pagaldai
AZTI-Tecnalia, Astondo Bidea, Edificio 609, Parque Tecnológico de Bizkaia, 48160 Derio, Bizkaia, Spain

mollusk (especially bivalves) consumed in this area has been imported. The lack of development of an aquaculture or shellfish sector in the region of the Bay of Biscay is mainly linked to high wave-energetic scenarios (Galparsoro et al. 2012) and previous issues related to estuarine water quality (Chust et al. 2009). A further restriction is current legislation, which aims to prevent conflict between users of marine habitats. These conflict-prevention methods hinder the development of near-shore aquaculture as the limited inshore spaces are protected marine habitats (Galparsoro et al. 2010; Borja et al. 2011; Pascual et al. 2011).

In addition to these local limitations, the main obstacles in mussel aquaculture in general are those regarding infectious diseases and parasites (Bower et al. 1994), marine phycotoxins (Svensson 2003; Moroño et al. 2003) as well as carrying capacity and competition for the use of marine space (Martínez et al. 2007; Costa-Pierce 2008; Byron et al. 2011). Limited marine space coupled with commercial objectives often results in non-sustainable practices for commercial mussel cultivation (Smaal 2002; Waite et al. 2005; Duarte et al. 2008). It is encouraging, however, that ecosystem-based approach-management strategies have been recommended to promote the sustainable development of mussel cultivation (Soto et al. 2008).

One such strategy is open ocean aquaculture which represents a promising alternative for the sustainable cultivation of bivalve species (Polk 1996; Hesley 1997; Stickney 1998, Bridger and Costa-Pierce 2003). However, this activity is still in development stages (Hickman 1992), and main mussel culture techniques (i.e., rafts, pole racks and/or longline systems) have been established almost exclusively in protected near-shore waters (Pérez-Camacho et al. 1991; Myrand et al. 2009) or in estuarine habitats (Penney et al. 2002; LeBlanc et al. 2005; Drapeau et al. 2006). Therefore, few studies have reported successful biological and/or engineering results on open ocean mussel aquaculture to date (Langan 2001; Langan and Horton 2003; Buck 2007).

Open ocean aquaculture is not only a financially attractive but also an environmentally sound one with a highly nutritious end product for the consumer. Indeed, there is an increasing demand for products sourced from open ocean aquaculture methods, resulting in increased revenue in international markets (Orban et al. 2002; Fuentes et al. 2009; Pogoda et al. 2013). Further, developing open ocean aquaculture methods will be environmentally beneficial (Shumway et al. 2003), as mussels and other bivalves are filter feeders, capturing and eating microscopic particles, plants, and nutrients by filtering, and thus cleaning ocean water (Shpigel et al. 1993). For the consumer, as most farmed mussels have fast growth, they provide year-round nutritious food, much in demand (Seed and Suchanek 1992; Gosling 2003). Mussels are appreciated by humans for their organoleptic properties (Grienke et al. 2014) as well as for their health benefits. They provide high-quality protein, minerals, essential trace elements, fat-soluble vitamins (i.e., vitamin D), and essential fatty acids EFA, especially long-chained polyunsaturated fatty acids: omega-3 (ω3 LC-PUFAs) (Stankovic et al. 2011; James 2013).

Regarding consumption, mussel shape and appearance are decisive factors for consumers' purchase (Fuentes et al. 2009), whereas meat quality (and hence, apparent health benefits) is regulated by biochemical composition (Orban et al. 2002; Filgueira et al. 2006; Fuentes et al. 2009). Mussel meat quality has been described by consumers as the result of not only chemical and biological characteristics, but also organoleptic properties, such as the appearance of the muscle, the intrinsic flavour, and the absence of undesirable components (Vernocchi et al. 2007). As reported by Fuentes et al. (2009) mussel organoleptic characteristics can be origin dependent. This means that different production sites, with different conditions and culture technologies, are expected to promote changes in growth performance and biochemical composition of mussel individuals, which could in turn reflect different product qualities at consumer level (Oliveira et al. 2015).

To date, no studies have investigated organoleptic properties on open ocean mussel products from the SE Bay of Biscay. Both biochemical composition and organoleptic properties are of special importance to anticipate consumer acceptance and market feasibility (Gökoglu 2002; Oliveira et al. 2015) before considering any implementation of mussel activity in the region.

The specific goals of this study were to characterize and compare the biometric parameters, nutritional content, and sensory aspects between experimental mussel product from the SE Bay of Biscay (open ocean) and commercial mussel product from the SW Bay of Biscay (Galician Rías).

Materials and methods

Location and sampling

An experimental suspension longline culture system (Langan 2000) was developed to cultivate shellfish in the open waters of the south eastern corner of the Bay of Biscay (Basque region; Spain). This region extends over 150 km and it is oriented E–W. The climate is temperate, oceanic, with moderate winters and warm summers. According to Köppen's classification, the area is associated with a marine temperate climate (Cfb) (Collins and Borja 2004). Seeds (17.40 ± 5.07 mm) were naturally collected during June 2013 and cultivated at the experimental site (43°21.39′N; 02°26.90′W) over a 1-year period. The open ocean site is already classified as a Class A production area (according to EU regulation (EC) 854/2004). Therefore, those mussels are not subjected to depuration requirements. The site is located at 2 nautical miles off the coast without any protection from land masses (Ryan 2004). Galparsoro et al. (2012) described strong hydrodynamic conditions on Basque Country's coast. A detailed description of current and waves in the experimental site can be found in Azpeitia et al. (2016).

Mussel individuals were cultured within 17 m water column depth, on mussel-growing ropes of 12 m in length. Mussel ropes hung from 5 m headline depth, with a distance between ropes of 0.5 m and an initial stocking density of 681 ± 37 individuals per lineal meter. After 13 months of cultivation, in July 2014, 200 mussel specimens of "commercial size" (Peteiro et al. 2006, i.e., 70 mm) were carefully detached from the ropes by cutting the byssal threads and cleaned of encrusting organisms. Selected mussel individuals were transported in plastic self-draining boxes covered by flaked ice. Later in the laboratory, mussels were conserved in plastic boxes, stored at 4 °C and covered with flaked ice for 24 h previous to analysis to preserve them. Similarly, 200 mussel individuals from Ría de Arosa (Galician Region; Spain) with a comparative commercial size (69.65 ± 6.11 mm) were obtained at early hours on the morning of the day of the sensory analysis. Commercial mussels were bought from a local seafood retailer to ensure a comparable size product with the same conservation procedure (plastic box and stored at 4 °C). These mussels were also stored at 4 °C and covered with flaked ice for some hours previous to analysis. Subsequently, two stocks of 50 mussel individuals from Galicia and from the experimental open ocean site were frozen at −80 °C to await morphometric and biochemical analysis.

Morphometry

Twenty mussel individuals from each origin (experimental open ocean and commercial) were carefully detached by cutting the byssal threads, cleaned of encrusting organisms and measured for morphometric analysis. The individuals were measured for their shell length (L; maximum anterior–posterior axis), shell height (H; maximum dorso-ventral axis), and shell width (W; maximum lateral axis) using a digital caliper to the nearest 0.1 mm. The biometric terminology used in this study is that described by Seed (1968). Mussel shape was analyzed by linear allometric relation between log-transformed mussel shell dimensions [length (L), height (H) and width (W)].

Biochemical composition and fatty acids

Three replicate samples of 10 pooled mussel individuals from each origin were considered for biochemical characterization (including proximate and glycogen analysis) and lipids determination (including total lipids and fatty acids), respectively.

The fresh individuals were first minced in a food processor (IKA® M 20 universal mill, IKA 1603601, Germany). Moisture and ash were determined according to the gravimetric determination. Briefly, minced tissue pools were placed in pre-weighed porcelain trays for drying at 80 °C for 24 h and then weighed to the nearest 0.001 g. Subsequently, dry mussels' tissue pools were ashed at 450 °C for 4 h in a muffle furnace, cooled in a desiccator and weighed again to determine the ash-free dry weight content.

Total protein was determined according to the Kjeldahl's method (AOAC 1975). Samples were digested in a digestor BUCHI B-435, distilled using a Rotavapor® R-210 (BÜCHI Labortechnik AG, Switzerland) and automatic volumetric titration developed through a Mettler Toledo T- 50 (Mettler Toledo, United States).

Glycogen concentration (mg·g^{-1} mussels' tissue) was calculated based on a mussel glycogen standard (Sigma, Saint Louis, MO) following the methodology reported by Gallardi et al. (2014). Briefly, 30% KOH was added to 0.5 g of homogenized mussels' tissue (ratio 2:1). The samples were then heated in a shaking-water bath at 100 °C for 20 min, vortexed for 30 s and subsequently chilled on ice for 5 min. After cooling, 200 μL of each sample was transferred to a 4 mL glass vial followed by 200 μL of 95% ethanol. The solution was vortexed again briefly and then placed in a boiling-water bath for 15 min followed by the addition of 1.2 mL of lukewarm water. The samples were again briefly vortexed and then allowed to stand at room temperature for 5 min before measuring the glycogen content. Glycogen content was measured by colorimetric reaction using 25 μL of prepared sample. 10 μL of 80% aqueous phenol and 200 μL of sulphuric acid were added to the samples on 96-well plates. Absorbance was measured using a multi- detection microplate reader (Synergy HT, BIO-TEK) at 490 nm. The concentration of glycogen in the samples was calculated based on a mussel glycogen standard (Sigma, Saint Louis, MO).

Total lipids were determined based on Blight and Dyer method (Hanson and Olley 1963). In this way, total lipids from a known weight of tissue was extracted and purified by homogenization in a suitable excess of chloroform/methanol/water (2:2:1.8 v/v/v). Following this, total lipids in chloroform solution were determined by weighting the eluate after removal of the solvent by rotary evaporation. All lipid extracts were then stored in chloroform at −30 °C to await analysis.

Fatty acid methyl esters (FAME) were separated and quantified using an Agilent 7890 Gas Chromatographer equipped with auto-injector (Agilent Technologies, United States). The gas chromatograph is equipped with a flame-ionization detector (290 °C) and a DB-WAS 122-7032 Agilent Technologies capillary column (30 m _ 0.25 mm I.D., film 0.25 μm). Helium was used as a carrier gas and initial oven temperature was 50 °C, followed by an increase at a rate of 10 °C·min^{-1} to a final temperature of 240 °C for 9 min. Individual FAME were identified by comparison of retention times with four commercial fatty acid standards (PUFA 1, PUFA 3, BAME and 37-component) supplied by Supelco (Bellefonte, PA) for area percent normalization. At each sample relative quantities were expressed as weight percent of total fatty acids.

Sensory evaluation/consumer testing

Thirty regular domestic consumers of mussel were randomly recruited by email from AZTI's database. They were segmented by gender (50% men) and age range (18–70 years). To remove debris and any remaining byssal thread, 150 mussels were washed and scrubbed for the sensory analysis. Samples were steamed at 100 ± 1 °C for 1 min and then served at 65 ± 1 °C. Mussels were presented to consumers on white plastic plates labeled with 3-digit random numbers, containing two mussels from each origin (experimental from open ocean and commercial from Galician Rías). Consumers tasted one mussel from each origin. The consumers received oral and written instructions and were provided with water and unsalted-crisp bread for palate cleansing after each tasting. Each consumer tasted the samples in randomized order and evaluated five attributes (appearance, odor, taste, texture, and global acceptance) with a structured 9-point scale ranging from 1 (equivalent to I dislike it very much) to 9 (equivalent to I like it very much). The 9-point hedonic scale (Peryam and Girardot 1952; Peryam and Pilgrim 1957) is the most commonly used scale for testing consumer preference and acceptability of foods (Schutz and Cardelo 2001). This kind of scales produce central tendency effects on panelist, which reduces the options to a 7-point scale (Villanueva et al. 2005). Despite this possible reduction, a 9-point scale with an arithmetic adjustment can always be straightforward re-scaled to a 7-point scale (Epler et al. 1998). This enables the comparisons with different scale formats with less response options (Dawes 2012), which are common in current scientific literature. Consumers' purchase intention was also assessed with a structured 5-point scale ranging from 1 (equivalent to I would not buy it) to 5 (equivalent to I will buy it). The same test was repeated three times with each panelist. Consumers tasting was performed in the sensory laboratory of AZTI- Tecnalia specifically designed according to ISO 8587:2006.

Statistical analysis

Relative shell morphology was investigated through the general linear function, where y and x are size-related measures and *a* and *b* are constants (i.e., intercept and slope; which represents the coefficient of relative shape)

(Pauly 1983). Similarly, allometric equations were generated separately for mussels from each origin, using linear regression and logarithmic transformation. The differences between allometric coefficients (*b*) from each origin were evaluated using ANCOVA (Rohlf and Sokal 1981).

Means and standard deviations (mean ± SD) of data on morphometrics, proximate composition (humidity, ash, protein, glycogen, and lipid content), as well as fatty acid profiles, of mussels were plotted using Sigmaplot (11.0) statistical and graphical software (Systat software). Statistical analyses were carried out using a statistical package (SPSS 20.0 for Windows®). Sensory attributes and evaluation data were analyzed using the R-project software (R Development Core Team 2008). A significance level of 0.05 was used. Normality was tested by Shapiro–Wilk test and the homoscedasticity was checked using Leven's test. The differences between mussel products type (experimental and commercial products) were compared using independent sample *t* test. Arcsine transformation (Zar 1984) was applied to the data expressed as percentage.

Results

Morphometry

The biometric results of the two mussel products (experimental open ocean and commercial from Galician Rías) tested within the present study are shown in Table 1. No statistical differences between the shell lengths (L) of both mussel products were detected. Conversely, significant differences were detected for the measurements of shell height (H) and shell width (W) between both mussel products. As observed, the open ocean mussel individuals from the present study displayed significantly higher morphometric values than those from the commercial samples.

The allometric relationships between shell length (mm) and shell width (mm), shell length and shell height (mm), or between shell width (mm) and shell height (mm) of both mussel products are shown in Table 2. Based on ANCOVA analysis (Table 3), the factor of product type significantly affected both the linear relations between log-transformed shell length (mm) and log-transformed shell height (mm), as well as the linear relation between log-transformed shell length (mm) and log-transformed shell width (mm). Higher values were found in the experimental mussels. Finally, no ANCOVA analysis was performed to compare the linear relation between log-transformed shell height (mm) and log-transformed shell width (mm) as significant interaction between the factor (product type) and the covariable [log-transformed shell width (mm)] was observed. Hence, no direct relation between the change in shell width (mm) and the change in shell height (mm) among mussel products could be assessed.

Table 1 *M. galloprovincialis*. Biometric characteristic of the mussel shell and proximate composition of mussel tissue pools from experimental open ocean (SE the Bay of Biscay; Basque region) and sheltered Rías (SW Bay of Biscay; Galician region) conditions

	Experimental	Commercial	One-way ANOVA	
	Open ocean	Galician Rías	*F*	*P* value
L (mm)	70.29 ± 7.96	69.65 ± 6.11	0.037	0.849
H (mm)	37.56 ± 4.59*	33.58 ± 2.48	10.471	0.003
W (mm)	24.58 ± 4.04*	21.18 ± 4.59	5.538	0.024
Moisture (%)	78.77 ± 1.54	76.57 ± 1.77	2.661	0.178
Protein (dry weight %)	49.52 ± 4.29	44.79 ± 4.02	1.962	0.234
Lipid (dry weight %)	6.27 ± 1.47	6.30 ± 1.34	0.001	0.976
Ash (dry weight %)	13.00 ± 1.16*	9.40 ± 0.82	19.253	0.012
Carbohydrates (dry weight %)	31.21 ± 4.87	39.51 ± 6.07	3.356	0.141

Mean ± standard deviation, $n = 20$

* Statistical significance was set up at ($P < 0.05$) (one-way ANOVA)

Table 2 *Mytilus galloprovincialis.* Parameter of shell shape (length (L), width (W) and height (H)) regression equations used to relate shell length (mm) to shell height (mm) or shell width (mm), and shell width (mm) to shell height (mm) of different mussel products (experimental vs. commercial); equation fitted to $\log_{10} X = \log_{10} a + b \log_{10} Y$, where *a* intercept and *b* slope, and X and Y shell measurements. Error of correlation coefficient (*r*), number of specimens (*n*) used to derive equation are also given

Allometric relation	Mussel product		*B*	SE	*t*	Sig	95% CI		R^2	*N*
							Upper	Lower		
Log H/Log L	Experimental	Intercept	–0.185	0.224	–0.823	0.421	–	0.287	0.773	20
		Slope	0.952	0.122	7.834	0.000	0.697	1.208		
	Commercial	Intercept	0.390	0.241	1.620	0.123	–0.116	0.897	0.552	20
		Slope	0.616	0.131	4.709	0.000	0.341	0.891		
Log W/Log L	Experimental	Intercept	–0.874	0.284	–3.075	0.007	–	–	0.882	20
		Slope	1.225	0.154	7.951	0.000	0.901	1.549		
	Commercial	Intercept	–0.501	0.243	–2.066	0.053	–1.011	0.008	0.873	20
		Slope	1.000	0.132	7.595	0.000	0.724	1.277		
Log H/Log W	Experimental	Intercept	0.735	0.161	4.557	0.000	0.396	1.073	0.775	20
		Slope	0.604	0.116	5.197	0.000	0.360	0.848		
	Commercial	Intercept	0.790	0.149	5.287	0.000	0.476	1.104	0.757	20
		Slope	0.548	0.111	4.917	0.000	0.314	0.782		

Table 3 *Mytilus galloprovincialis.* Mussel product type effects on shell length (mm) to shell width (mm) and on shell length (mm) to shell height (mm), allometric relations analyses. Regression equation exponent *b* (slope) were compared using ANCOVA (Rohlf and Sokal 1981), with α set at 0.05

Allometric relation	Source of variation	ANCOVA			R^2
		df	*F*	Sig	
Log H/Log L	Log length (covariable)	1	83.561	0.000	0.759
	Product type (factor)	1	29.992	0.000	
	Error	37			
Log W/Log L	Log length (covariable)	1	121.922	0.000	0.803
	Product type (factor)	1	19.953	0.000	
	Error	37			

Biochemical composition and fatty acids

The proximate composition of both mussel products is shown in Table 1. Moisture levels were relatively high within both samples. Mussels from commercial origin (Galician Rías) showed slightly higher moisture values (78.77% vs. 76.57%) than mussels from the open ocean conditions of the present study, yet significant differences were not detected.

Significantly lower ash levels were detected within the commercial individuals (13.00% vs. 9.40%) compared to the study ones.

Lipid contents were slightly higher in the commercial samples (6.27% vs. 6.30%), although no significant differences were found. Protein contents did not show any significant differences between mussel products. Finally, glycogen levels (71.06 ± 16.71 $mg \cdot g^{-1}$ and 100.59 ± 25.05 $mg \cdot g^{-1}$ at experimental and commercial mussels, respectively) were slightly higher in commercial samples, although differences were not statistically significant.

The FA composition of both mussel products are shown in Table 4. The saturated fatty acids (SAFA) accounted for 29% in experimental mussels and significantly lower content (25%) for commercial mussels. Saturated fatty acid predominated over the monounsaturated fatty acid (MUFA) with 12% for the experimental and significantly higher level (22%) for the commercial samples. Both mussel products showed the

Table 4 *M. galloprovincialis*. Fatty acid compositions (percentage of total fatty acids) of mussel tissue pools analyzed during the comparative study between the experimental open ocean mussels from the SE Bay of Biscay and commercial mussels obtained from the sheltered Rías from Galicia

Fatty acids	Experimental open ocean	Commercial Galician Rías
14:0	2.52 ± 0.80*	4.14 ± 0.60
15:0	0.57 ± 0.07*	0.36 ± 0.02
16:0	18.91 ± 1.29*	15.50 ± 0.43
17:0	1.90 ± 0.32*	1.12 ± 0.20
18:0	4.70 ± 0.93	3.74 ± 0.52
20:0	0.06 ± 0.05	0.02 ± 0.04
22:0	0.03 ± 0.03	0.18 ± 0.03*
∑SAFA	28.69*	25.05
14:1	0.42 ± 0.07	0.99 ± 0.17*
16:1ω7	2.40 ± 0.87	11.14 ± 1.42*
16:1ω5	0.30 ± 0.05	0.25 ± 0.02
17:1	0.19 ± 0.16	0.40 ± 0.10
18:1ω9	2.22 ± 0.68	1.14 ± 0.05
18:1ω7	1.22 ± 0.15	3.19 ± 0.07*
20:1 ω11	1.21 ± 0.23	1.32 ± 0.17
20:1ω9	3.48 ± 0.28*	1.59 ± 0.19
20:1ω7	0.47 ± 0.05	2.03 ± 0.16*
22:1	0.06 ± 0.06	0.24 ± 0.14
∑MUFA	11.98	22.28*
16:4ω3	8.32 ± 2.25	3.04 ± 4.70
18:2ω6	2.03 ± 0.32*	0.87 ± 0.02
18:3ω3	1.82 ± 0.44*	0.77 ± 0.09
18:4ω3	3.74 ± 1.23*	1.35 ± 0.24
20:2 α	2.76 ± 0.36*	1.57 ± 0.31
20:2 β	0.25 ± 0.04	1.14 ± 0.35*
20:2ω6	0.71 ± 0.12*	0.38 ± 0.03
20:4ω6	1.30 ± 0.21	2.61 ± 0.12*
20:4ω3	0.34 ± 0.13	0.64 ± 0.23
20:5ω3	9.85 ± 1.06	24.71 ± 2.21*
22:2 NMIDa	0.73 ± 0.10*	0.12 ± 0.10
22:2 NMIDb	1.22 ± 0.13	4.54 ± 0.43*
21:5ω3	1.20 ± 0.16	1.09 ± 0.06
22:4ω6	0.20 ± 0.04	1.90 ± 1.54
22:5ω6	0.61 ± 0.03*	0.15 ± 0.15
22:5ω3	1.12 ± 0.11	2.13 ± 0.21*
22:6ω3	23.13 ± 2.20*	5.66 ± 0.47
∑PUFA	59.33*	52.67
NMID	1.95	4.66*
Terrestrial (18:2ω6 + 18:3 ω3)	3.85*	1.64
Ω3/Ω6	10.21	6.67
DHA/EPA	2.35*	0.23

Values (mean ± SD, $n = 10$)

NMID Non-methylene-interrupted dienoic fatty acid

* Statistical significance was set up at ($P < 0.05$) (one-way ANOVA)

majority of fatty acids consisted of polyunsaturated fatty acids (PUFA), with significantly higher level (59%) for experimental mussels than for commercial mussels (52%).

In the case of SAFA, palmitic acid (16:0) was predominant (18.91 and 15.50%) in both products, followed by stearic acid 18:0 in the open ocean mussels (4.70%), and by myristic acid 14:0 in the commercial mussels (4.14%). Significant differences (one-way ANOVA; $P < 0.05$) in the SAFA profile (i.e., the sum of pentadecanoic 15:0, palmitic 16:0 and behenic 22:0 acids) were found between both mussel products.

Regarding MUFA, the commercial mussel from Rías displayed abundant related palmitoleic acid 16:1ω7 (11.4%). This was followed by vaccenic acid 18:1ω7 (3.19%) and eicosenoic acid 20:1ω7 (2.03%). Conversely, the observed trend of MUFAs in the open ocean experimental individuals was not equivalent; the major acid was gondoic acid 20:1ω9 with 3.48%, followed by palmitoleic acid 16:1ω7 with 2.40% and oleic acid 18:1ω9 with 2.20%. Thus, significant differences (one-way ANOVA; $P < 0.05$) were found between MUFA of both mussel products.

Regarding PUFAs, eicosapentaenoic (EPA) (20:5ω3) and docosahexaenoic (DHA) (22:6ω3) acids were the most important. Commercial mussels showed higher EPA (20:5ω3) with 24.71% and less DHA (22:6ω3) with 5.66%, whereas the experimental open ocean mussels showed less EPA (20:5ω3) content with 9.85% and higher (DHA (22:6ω3) with 23.13%. Thus, the DHA/EPA ratios were significantly different (one-way ANOVA; $P < 0.05$) between mussel products.

Non-methylene-interrupted dienoic fatty acid (i.e., 22:2 NMIDa and 22:2 NMIDb) were higher in the commercial samples (4.66%) than in the experimental ones (1.95%). Differences at NMID level were statistically significant (one-way ANOVA; $P < 0.05$) between mussel products.

Finally, terrestrial fatty acids (i.e., 18:2ω6 + 18:3 ω3) accounted for the 3.85% in experimental mussels and 1.64% in the commercial ones; differences between mussel products were statistically significant (one-way ANOVA; $P < 0.05$).

Sensory evaluation/consumer testing

The panelist male:female ratio (%) participating in the present study was 37.7:63.3. The range of ages between participants was distributed as follows: 18–30 years (13.3%); 31–40 years (40%); 41–50 years (33%); 51–60 years (13.3%). The 77% of participants were regular mussel consumers (3.3% more than once a week; 26.7% once a week; 46.7% once a month), whereas the 23.3% declared to be occasional mussel consumers.

The resulted acceptance for both mussel products was around 7 points out of 9 (Fig. 1). No significant differences (t test; $P > 0.05$) were detected for any of the sensory attributes analyzed between both products (Table 5). The results of the qualitative assessment showed that both mussel products showed good appearance and odor. The results of the qualitative assessment showed that both mussel products showed good appearance and odor. Particularly, the experimental ones showed to provide more intense sea-flavour with a slightly softer and more juicy texture. Some other aspects such as the whitish color on the muscle tissue and/or the salty taste were criticized when observed. A large proportion of the panelists (70%) expressed potential interest in purchasing the experimental product, while a small minority (17%) did not. The remaining 13% of panelists did not express a purchase preference. For the commercial mussels, 83% of the panelists showed purchase intention, while 10% showed uncertainty (or doubt) and 7% would not buy it (Fig. 2).

Discussion

All mussels consumed in the Basque region, during recent decades, are imported (Gracia 1996). Currently, the only way to cultivate mussels in the Basque Country is offshore and with submerged longline method (Azpeitia et al. 2016). The aim of the study was to compare if experimental open ocean cultured mussels of commercial size were as competitive as local market leader of the same commercial size. No significant differences were observed in shell length between both products, which validated the analysis of quality differences among mussel products as well as the analysis of consumer perception among products. Although panelists were not able to inspect mussel shells, only mussels without visible damage (open valves or broken shell) were used in the study. Direct influences on the appearance of the product should be derived from the differences found at biometric levels between both mussel (*M. galloprovincialis* Lmk.) origins. Different shell

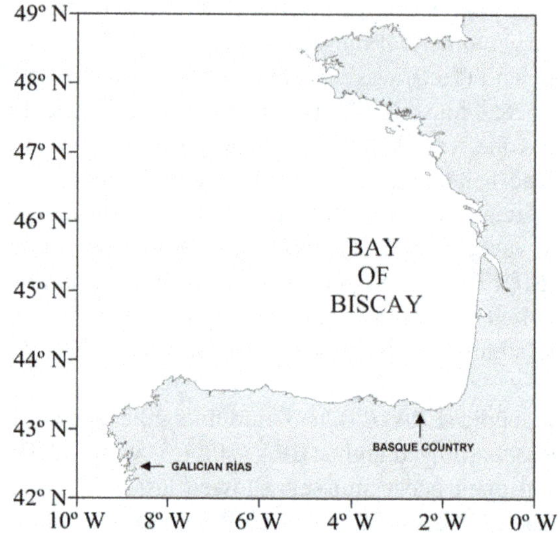

Fig. 1 Map of the Bay of Biscay showing the specific location of the open ocean experimental site at the SE Bay of Biscay (Basque Country) and the commercial mussel product from the SW Bay of Biscay (Galician Region)

Table 5 *M. galloprovincialis*. Student's *t* analysis carried out for all sensorial attributes analyzed in experimental mussel from the open ocean conditions of the SE Bay of Biscay and mussels cultured in at sheltered Rías from Galicia

	Experimental open ocean	Commercial Galician Rías	Student's *t*	
			t	*P* value
Appearance	6.467 ± 2.097	7.033 ± 1.066	1.32	0.194
Odor	6.967 ± 1.402	7.3 ± 1.088	1.029	0.308
Flavor	7.533 ± 1.106	7.4 ± 1.303	−0.427	0.671
Texture	7.033 ± 1.586	7.533 ± 0.973	1.472	0.148
Global impression	6.9 ± 1.517	7.333 ± 1.269	1.2	0.235

Values (mean ± SD, $n = 30$)

* Statistical significance was set up at ($P < 0.05$) (pairwise comparison)

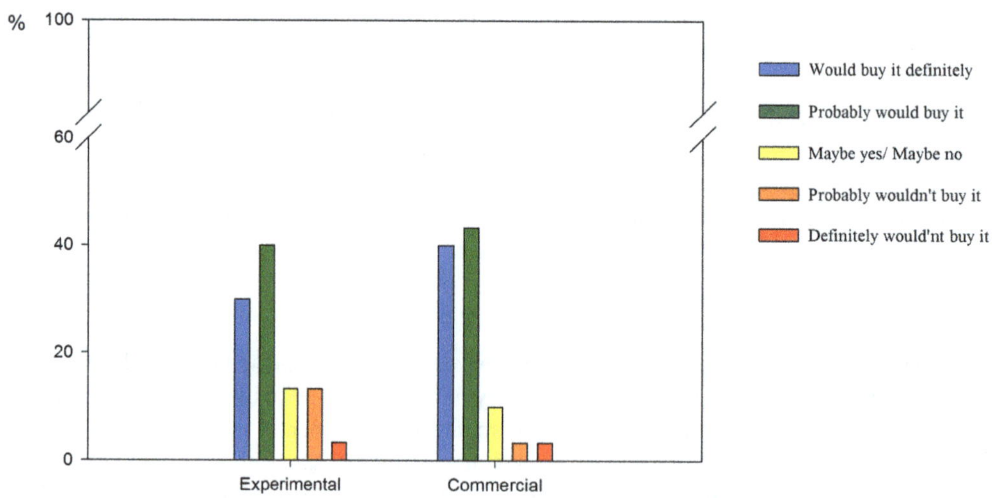

Fig. 2 *M. galloprovincialis*. Buying-intention results carried out on the comparison between mussels from the open ocean in the SE Bay of Biscay ("experimental") and from sheltered estuaries in the SW Bay of Biscay ("commercial")

shapes among mussel product were observed. Mussels cultured in Galician Rías were found to be thinner (reduced W. L^{-1} ratio) and lower (reduced H. L^{-1} ratio) than experimental open ocean mussels. Commercial mussels may have been cultured at higher densities than experimental mussels. Stocking density on experimental mussels was 681 individuals/m, whereas commercial label does not provide the consumer with those culture characteristics. Nevertheless, literature value for standard mussel density on ropes cultured in sheltered Galician Rías is between 700 and 1000 mussels·m^{-1} (Irisarri et al. 2014a). Cubillo et al. (2012) observed that mussels cultured at higher densities displayed a lower shell height to shell length ratio compared to mussels cultured at a lower density. Similarly, the author observed that mussels cultured at higher density displayed lower shell width to shell length ratio compared to those cultured at low densities.

Regarding differences in oceanographic conditions, mean current speed in the experimental site was significantly higher than values reported from previous studies carried out in Galician sheltered Rías (Pérez-Camacho et al. 2014a, b). Similarly, highly significant wave height scenarios were also found (with peaks of 796 m) at the experimental site. Indeed, only open ocean mussel growth studies of Buck and Buchholz (2004) and Langan (2000), from the German Bight and the Gulf of Maine, respectively, showed similar wave height scenarios (with peaks of 7.96 m). The mussel shells of the experimental site were characteristically round, wide and great in thickness. Although these observations were not validated through specific comparisons, they are consistent with the findings from previous studies developed with mussels from wave beaten shores (Akester and Martel, 2000, Lewis and Powel 1961, Seed 1968).

Despite the limited information available on the culture method and culture characteristic of commercial mussel product, the new product cultured at the experimental site would compete with current commercial products. No differences regarding mussel appearance of cooked samples were pointed out by the group of local panelists during the sensory test.

In this study, no relevant differences at biochemical composition level other than ash content were observed between the open ocean experimentally cultured mussels and the commercially purchased mussels from Galician Rías. Biochemical composition of mussels is affected by water temperature, nutrient availability, and the reproductive cycle of individuals (Fernández-Reiriz et al. 2007). These multiple factors change seasonally, which, in turn, cause seasonal changes in biochemical composition of mussels (De Zwaan and Zandee 1972; Dare and Edwards 1975; Ruano et al. 2012). As reported by previous authors, major differences in biochemical composition relied on: (i) feeding behavior, in terms of filtration capacity and clearance rate (Vilela 1950; Heral et al. 1980); (ii) quality and quantity of seston available in the environment; (iii) state of sexual maturity of individuals (Fernández–Reiriz et al. 2007).

This study showed similarities on proximate composition between both mussel products, consistent with previous findings from mussels from Galician waters as reported by Fuentes et al. (2009). This may be explained by similarities occurring within the main factors regulating growth and biochemical performance in both culture conditions. Both culture conditions have a close annual average in environmental conditions, the same seasonality in reproductive cycles, and/or the same genetic heritage due to geographic proximity. However, open ocean mussels exhibited a higher amount of ash when compared to commercial ones from Galician Rías. Fuentes et al. (2009) observed similar values of ash content in commercial product from Galician Rías when comparing mussel product from different Spanish regions. Higher ash content than open ocean mussels of this study were also observed in the same study in mussels from Ebro Delta and from Valencia (NE and SE Spain, respectively). Mussels cultured in the exposed environmental conditions (e.g., phytoplankton variability, strong hydrodynamics, and very marked seasonal changes in seston composition) may experience situations of stress or may have greater energetic expenditure (Azpeitia et al. 2016) than mussels cultured in sheltered conditions. Oliveira et al. (2015) found similar results when comparing open ocean mussels to sheltered cultured mussels. The authors related this difference to the hydrodynamic conditions found by mussels in the offshore culture areas, which would interfere with mussel metabolism in a set of complex interactions between temperature, food availability, growth, and reproductive cycle (Gabbot 1976). Thus, ash content is influenced by many abiotic factors such as salinity and presence of metals on the water column (Okumuş and Stirling 1998), as well as by the nutritional state and the reproductive cycle of the animal (Beninger and Lucas 1984). Regarding nutritional state, Mayzaud (1976) associated an increase in ash content of zooplankton with a state of starvation, and Wilkins (1967) observed the same in herrings. Nevertheless, this higher ash percentage may account for the decrease in other constituents (e.g., lipids, proteins) rather than an increase in the amount of ash, as data analysis was performed with relative percentages and not in absolute terms. Despite these differences in ash content, similar general properties at biochemical level should be expected when comparing the open ocean mussels from the SE Bay of Biscay to their commercial counterparts from another northern Spanish region.

Depending on the time of the year, levels of nutrients, both in Galician Rías and in the open ocean of the SE Bay of Biscay, differ. In Galician Rías, upwelling events typically take place during spring–summer. Highest chlorophyll levels have been recorded during spring (<5 $\mu g \cdot l^{-1}$)(Irisarri et al. 2014a). However, upwelling events are usually followed by thermal stratification periods during the summer which might result in a significant reduction of chlorophyll levels up to 0.4 $\mu g \cdot l^{-1}$ (Irisarri et al. 2014a). The thermocline layer prevents nutrients from fertilizing the surface and chl-a eventually becomes depleted. The author corroborated this with a negative scope for growth (SFG) observed in mussel, when abnormally high temperatures are coupled with aforementioned low energy intake in Galician Rías in summer. In the SE Bay of Biscay, there is also a thermal stratification during summer, which decreases phytoplankton biomass to lower values than those described in Galician Rías (0.2 $\mu g \cdot l^{-1}$ in surface waters) (Morán et al. 2012). However, during summer, short periods of atmospheric instability (storms) may act as modulation factors of this bloom decay in the SE Bay of Biscay. Although inputs of nutrient-rich continental waters can be observed, the intensified aforementioned current velocities may also resuspend inorganic sediments. Thus, in storm conditions, a dilution of available particulate organic matter (POM) for bivalve filter-feeders can be expected (Cranford et al. 2011; Irisarri et al. 2013; Zuñiga et al. 2014) in the experimental site.

The biochemical composition of the present contribution are consistent with previous studies reporting that mussels are high in proteins and low in calories and fat (Shahidi 2004; Fuentes et al. 2009; Gallardi et al. 2014). Indeed, some authors have reported that a daily protein intake of 0.75 g·kg body weight (which means that with 100 g of mussels for a person of 70 kg) is recommended to satisfy approximately 23% of the daily protein requirements (Garcia-Gabarra 2006; Stankovic et al. 2011; James 2013), thus demonstrating the rich nutritional value of mussels for human consumption.

In addition, it is also generally accepted that mussels represent an important source of essential fatty acids and particularly for long-chain polyunsaturated variety (PUFAs) (James 2013). In this study, not only concentration levels but also their representativeness on the profile of the major fatty acid groups (PUFA, MUFA and SAFA) is consistent with previous mussel studies (King et al. 1990; Freites et al. 2002b; Orban et al. 2002). Irisarri et al. (2014b) observed a selective accumulation of PUFA in different tissues of mussels. Mussels have a limited capability for de novo synthesis of PUFAs (Alkalani et al. 2007) and acquire them from diet. The unsaturation degree of the FA increases during the colder seasons, as mussels require these FA to maintain membrane fluidity (Hall et al. 2002). The contrary is observed in warmer seasons and thus, the PUFA level varies in opposite fashion to SAFA level. Mean seawater temperature in open ocean experimental site in summer was slightly higher (Azpeitia et al. 2016) than the typical summer temperature in Galician Rías (Pazos et al. 1997), which could have promoted a faster decrease in the unsaturation degree. This may have caused SAFA level to be higher in experimental mussel than commercial ones. Several authors have reported that higher proportions of SAFA may represent bacterial loads or organic-detritus-rich environments (Galap et al. 1999; Freites et al. 2002b). However, open ocean conditions are related to high hydrodynamic scenarios which favor marine water recycling and consequently less detritus accumulation compared to sheltered sites. Despite this higher water recycling, resuspension of material may have occurred at the experimental site. Moreover, regarding bacterial biomarker FA (15:0, 17:0, 18:1ω7 and 18:1ω7/18:1ω9 > 1; Budge et al. 2001), higher values were observed in mussels from Galician Rías compared to experimental mussels. This may indeed support the aforementioned notion that temperature differences may well lead to higher SAFA levels in experimental mussels.

Major fatty acid levels in marine bivalves are generally known for being highly dependent on biochemical and environmental conditions (Lucas and Beninger 1985; De Moreno et al. 1980; Fernández-Reiriz et al. 1989; Fuentes et al. 2009). Nevertheless, the aforementioned results are consistent with current existing knowledge on mussel biochemistry and could be of use in further research in the same field at a local level.

Palmitic acid (16:0) has been reported in numerous studies (King et al. 1990; Freites et al. 2002a; Orban et al. 2002; Alkalani et al. 2007; Fuentes et al. 2009) as the major saturated fatty acid (SAFA) present in mussels. The experimental open ocean mussels herein displayed not only the same trend, but also, consumers reported a more intense sea-flavor after tasting the product. This would be consistent with Oliveira et al. (2015), who found that palmitic acid along with DHA are mainly responsible for mussel flavour.

PUFA levels of both mussel products herein were abundant and consistent with results from previous mussel studies from other localities (King et al. 1990; Orban et al. 2002; Freites et al. 2002b; Dridi et al. 2007; Alkalani et al. 2007; Fuentes et al. 2009), where healthy mussels always contain PUFA levels near to 50% of the total fatty acids proportion. As reported by Musa-Veloso et al. (2011) a dose of 250 mg·day^{-1} of the long-chain omega-3 fatty acids EPA and DHA (PUFA) is strongly recommended for human health. In this way, although mussels are low in lipids, these lipids are of high quality and provide a healthy source of omega-3 fatty acids with numerous nutritional benefits (Shahidi 2004; Byrd-Bredbenner et al. 2009; James 2013). A high ω6/ω3 ratio has also been reported as harmful to human health (i.e., over values of 12/50; Ackman 1990; Shahidi and Miraliakbari 2004; Fuentes et al. 2009). The results on the EPA and DHA of the present study were suitable within all ranges for both types of mussels examined.

Suspension feeding bivalves feed on living and non-living available material in their habitat (i.e., generalist filter feeders) (Ward and Shumway 2004). They select particles from seston, separating the grain from the chaff, and acquire the optimum energetic budget (Ward and Shumway, 2004). Seston consists of plankton of a wide range of sizes and palatability, material resuspended from the benthos, aggregates, detritus, fecal pellets as well as microorganisms (Ward and Shumway 2004, Karayücel et al. 2010). Fatty acids (FA) have been successfully used as trophic markers of bivalves to identify food sources, habitat preferences, feeding strategies and trophic links (Kharlamenko et al. 1995, 2001; Prato et al. 2010). Several studies on mussel *Mytilus galloprovincilais* Lmk. (Freites et al. 2002a, b; Ezgeta-Balić et al. 2012; Irisarri et al. 2014a) found that variability in nutritional quality of seston was explained by the variance of FAs found in mussels. Bivalves have some capability to synthesize FAs independently of diet (Ventrella et al. 2013), but they are not capable of biosynthesizing de novo polyunsaturated FAs (PUFAs) (Fernández-Reiriz et al. 2011). In addition, fatty acids transferred from primary producers, when incorporated by bivalves, do not undergo major changes (Graeve et al. 1994; Xu and Yang 2007), and thus, fatty acid measurements are a qualitative measurement of both energy transfer and assimilation of nutrients in mussel tissue. FA signature in mussel tissues can reveal the food source of the bivalve (i.e., several weeks prior) (Ezgeta-Balić et al. 2012) and thus, certain FA and their ratios can be used as biomarkers of the diet of the mussel. With these FA ratios, the seasonal contribution of bacteria, phytoplankton classes, as well as microzooplankton to the diet of the mussel can be assessed (Budge et al. 2001; Handå et al. 2012). Such analysis is an important factor in understanding the feeding physiology of mussels (Xu and Yang 2007).

In spite of presenting healthy EPA and DHA proportions, the experimental open ocean mussels of the present study showed less EPA and higher DHA contents than the commercial ones from Galician Rías. EPA level could be related to the ingestion of dinoflagellates (Freites et al. 2002a), whereas DHA could be a consequence of consumption and accumulation of diatoms (Handå et al. 2012). Sidari et al. (1998) observed that *Mytilus galloprovincialis* Lmk. as a preference selected dinoflagellates rather than diatoms. DHA/EPA ratio was >1 in experimental mussels, whereas, it was <1 in commercial mussels. These FA trophic markers (Budge et al. 2001; Dalsgaard et al. 2003) suggest mussels from the experimental site feed on dinoflagellates. Additionally, the ratio 16:1ω7/16:0 was <1 in experimental mussels and >1 in commercial mussels. Thus, all these FA trophic markers suggest commercial mussels feed on diatoms rather than dinoflagellates (Budge et al. 2001; Dalsgaard et al. 2003). No water samples were collected for phytoplankton analysis. However, as described by Álvarez et al. (2008) in the western Iberian Peninsula (i.e., Portugal and Galicia) during spring–summer the continental shelves are strongly affected by upwelling pulses. These natural events are caused by the prevailing winds and the coastline orientation, causing an inflow of deep nutrient-rich waters which boosts phytoplankton productivity (Varela et al. 2005). Moreover, as described by Figueiras et al. (2002) and Álvarez-Salgado et al. (2008), these upwelling pulses result in the transportation of dinoflagellates to the continental shelf and leave the embayment dominated by diatoms. Similarly, Irisarri et al. (2014b) found that diatom FA biomarkers were significantly higher in mussel tissues from Galician Rías during spring–summer although those FA trophic markers were >1 throughout the entire year. Oliveira et al. (2015) also reported higher levels of EPA in mussels (*M. galloprovincialis* Lmk.) from the Galician Rías compared to individuals of the same species but different culture origin [S Atlantic Sea (Portugal)]. The authors suggested that the resultant DHA/EPA ratios would reflect the higher relative ingestion of diatoms. The contribution, however, of phytoplankton (as diatoms and dinoflagellates) to the diet (derived by 16:1ω7/16:0 and 20:5ω3/22:6ω3 ratios) was proportionally low and moderate in open ocean and commercial (*M. galloprovincialis* Lmk.) mussels, respectively. In the present study mussels were harvested in July along the north Spanish coast. There is a high likelihood that previous poor environmental conditions (Prato et al. 2010) or post-spawning stages (Garmendia et al. 2010; Ortiz-Zarragoitia et al. 2011; Múgica et al. 2015; Cuevas et al. 2015) may have affected mussels' fatty acids composition in summer.

Fatty acid trophic markers of terrestrial organic sources [i.e., $\Sigma(18{:}2\omega6 + 18{:}3\omega3)$] higher than 2.5 were described by Budge and Parrish (1998) as indicative of a terrestrial organic diet source (i.e., terrestrial plants). A significant input of terrestrial markers was observed in the present study, where the experimental open ocean mussels showed to exceed the threshold. This may be related to the aforementioned differences in seston dynamics in Galician Rías and in the SE Bay of Biscay during summer season.

Finally, the presence of 22:2NMID acid was lower in experimental mussels compared to commercial ones. Sánchez-Lazo and Martínez-Pita (2012) also reported the presence of this fatty acid, although accounting for very little proportion.

The sensory analysis herein was consistent with previous mussel studies (Kryzynowek and Wiggin 1979; Ablett et al. 1986; Gökoglu 2002). The panelists were able to determine the mussel product qualitative characteristics on appearance, odor, flavor, texture, and global perception. Both types of mussel (open ocean and commercial) reached scores of 7 over 9 points at hedonic scale for all sensory attributes. In general terms, even though no triangle test was done, the sensory analysis session was performed stably and consistently among panelists.

Regarding appearance and odor, both characteristics were scored slightly lower for the open ocean samples when compared to the commercial ones from Galician Rías. Observed discrepancies on mussel appearance could be related to inherent biological and/or operational factors (e.g., higher variability in the coloration of mussel meat from open ocean conditions; previous industrial procedures applied on the commercial samples, etc.). The results of the present study showed no significant differences in tasting levels between open ocean and commercial products. However, most panelists also identified some distinctive attributes such as intense sea-flavour, softer and more juicy textures, and salty tastes in the open ocean samples.

These results may specifically be related to the absence of previous depuration requirement on the open ocean samples as a result of having an origin with a class A production area according to EU and local public legislation. During depuration processes, mussels are fasted and animals excrete their metabolic waste products, which include salts (Lee et al. 2008). Despite depuration of commercial mussels, the organoleptic characteristics did not differ greatly between both mussel products. The results on texture, at the time of the sensory analysis, also revealed softer meat in experimental mussels. Nevertheless, differences were not statistically significant and thus it can be concluded that softer texture does not imply mussels are less desirable. It is noteworthy that desired quality among fishery products relies on regional preferences, consumers' attitudes and methods of preservation and/or consumption (Haard 1992). Both mussel products reached 7 points out of 9 in qualitative category; a positive result considering the potential of open ocean mussels of the present study to be marketable in the near future. The tested consumers also expressed an intention to purchase both products (>70%). In general terms, this prescreening hedonic analysis showed that experimental open ocean mussels were similar in attributes to those routinely imported from the Galician Rías to the Basque Country. From a marketing perspective, the mussels cultured in the open ocean of the SE Bay of Biscay may be well accepted by consumers and fulfill the minimum requirements to fit and compete in the existing local market. The results of the present study provide baseline information to validate nutritional value characteristics and consumers' acceptance of mussels (*M. galloprovincialis* Lmk.) cultured in the open ocean of the Bay of Biscay. In addition, as a commercial validation was positive, it is now worth investigating how changes in the culture method can optimize production and commercial activity of future aquaculture in the Basque Country.

Conclusion

Space available in coastal areas for aquaculture facilities is growing more and more limited and is subjected to strict jurisdictions and limited leases. In addition, the geography of the coast in many regions might not be sheltered enough for mooring rafts or longlines, while limited seston supply and/or polluted waters can also represent limiting factors for mussel coastal aquaculture. Hence, this study investigates an interesting alternative to traditional suspended mussel culture in coastal inlets.

This study is the first sensory assessment of mussels produced in open ocean from the SE Bay of Biscay. First, this new information will be valuable for the development of strategies that favor origin specifications linked to health benefits which, in turn, would contribute to consumers' decision-making processes (i.e., purchase intention and preferences). Secondly, it will help to predict cost-effectiveness during mussel production and commercialization.

The approach of this study aimed to validate consumer acceptance as a meaningful factor in market feasibility studies. In addition, it also aimed to investigate if mussels cultured in the open ocean could compete with other mussels currently available on the market. However, future market needs will ultimately decide the final viability of mussel open ocean aquaculture.

Acknowledgements We wish to thank Gaizka Bidegain, Pilar Martín, Ainhoa Bikandi, and AZTI staff for their capable assistance during the laboratory and field works. Aileen Logan and Emma Ruizdelgado critically reviewed the manuscript. This work was partially supported by the Agriculture, Fisheries and Foods Department, and also through a PhD grant by the Department of Education, Linguistic Policy and Culture of the Basque Government.

References

Ablett RF, Gould SP, Sherwood DA (1986) Frozen storage performance of cooked cultivated mussels (*Mytilus edulis* L.)—influence of ascorbic acid and chelating agents. J Food Sci 51(5):1118–1121

Ackman RG (1990) Seafood lipids and fatty acids. Food Rev Int 6(4):617–646

Akester RJ, Martel AL (2000) Shell shape, dysodont tooth morphology, and hinge-ligament thickness in the bay mussel *Mytilus trossulus* correlate with wave exposure. Can J Zool 78(2):240–253

Alkanani T, Parrish CC, Thompson RJ, McKenzie CH (2007) Role of fatty acids in cultured mussels, *Mytilus edulis*, grown in Notre Dame Bay, Newfoundland. J Exp Mar Biol Ecol 348(1):33–45

Álvarez I, Gómez-Gesteira M, DeCastro M, Dias JM (2008) Spatio temporal evolution of upwelling regime along the western coast of the Iberian Peninsula. J Geophys Res Oceans 113:14

Álvarez-Salgado XA, Labarta U, Fernández-Reiriz MJ, Figueiras FG, Rosón G, Piedracoba S, Filgueira R, Cabanas JM (2008) Renewal time and the impact of harmful algal blooms on the extensive mussel raft culture of the Iberian coastal upwelling system (SW Europe). Harmful Algae 7(6):849–855

AOAC (Association of Official Analytical Chemists) (1975) Official Methods of Analysis, 12th edn. Association of Official Analytical Chemists, Whashington, D.C

Azpeitia K, Ferrer L, Revilla M, Pagaldai J, Mendiola D (2016) Growth, biochemical profile, and fatty acid composition of mussel (*Mytilus galloprovincialis* Lmk.) cultured in the open ocean of the Bay of Biscay (northern Spain). Aquaculture 454:95–108

Beninger PG, Lucas A (1984) Seasonal variations in condition, reproductive activity, and gross biochemical composition of two species of adult clam reared in a common habitat: *tapes decussatus* L. (Jeffreys) and *Tapes philippinarum* (Adams & Reeve). J Exp Mar Biol Ecol 79(1):19–37

Borja Á, Galparsoro I, Irigoien X, Iriondo A, Menchaca I, Muxika I, Pascual M, Quincoces I, Revilla M, Germán Rodríguez J, Santurtún M, Solaun O, Uriarte A, Valencia V, Zorita I (2011) Implementation of the European Marine Strategy Framework Directive: a methodological approach for the assessment of environmental status, from the Basque Country (Bay of Biscay). Mar Pollut Bull 62:889–904

Bostock J, McAndrew B, Richards R, Jauncey K, Telfer T, Lorenzen K, Corner R (2010) Aquaculture: global status and trends. Philos Trans R Soc B Biol Sci 365(1554):2897–2912

Bower SM, McGladdery SE, Price IM (1994) Synopsis of infectious diseases and parasites of commercially exploited shellfish. Annu Rev Fish Dis 4:1–199

Bridger CJ, Costa-Pierce BA (2003) Open ocean aquaculture: from research to Commercial reality. The World Aquaculture Society, Baton Rouge, Louisiana

Buck BH (2007) Experimental trials on the feasibility of offshore seed production of the mussel *Mytilus edulis* in the German Bight: installation, technical requirements and environmental conditions. Helgol Mar Res 61(2):87–101

Buck BH, Buchholz CM (2004) The offshore-ring: a new system design for the open ocean aquaculture of macroalgae. J Appl Phycol 16(5):355–368

Buck BH, Ebeling MW, Michler-Cieluch T (2010) Mussel cultivation as a co-use in offshore wind farms: potential and economic feasibility. Aquac Econ Manag 14(4):255–281

Budge SM, Parrish CC (1998) Lipid biogeochemistry of plankton, settling matter and sediments in Trinity Bay, Newfoundland: II. Fatty acids. Organ Geochem 29:1547–1559

Budge SM, Parrish CC, Mckenzie CH (2001) Fatty acid composition of phytoplankton, settling particulate matter and sediments at a sheltered bivalve aquaculture site. Mar Chem 76(4):285–303

Bushek D, Richardson D, Bobo MY, Coen LD (2004) Quarantine of oyster shell cultch reduces the abundance of *Perkinsus marinus*. J Shellfish Res 23(2):369–374

Byrd-Bredbenner C, Moe G, Beshgetoor D, Berning J (2009) Wardlaw's perspectives in nutrition. In: Weatley CH (ed) McGraw-Hill, New York

Byron C, Link J, Costa-Pierce B, Bengtson D (2011) Calculating ecological carrying capacity of shellfish aquaculture using mass-balance modeling: Narragansett Bay, Rhode Island. Ecol Model 222(10):1743–1755

Chust G, Borja Á, Liria P, Galparsoro I, Marcos M, Caballero A, Castro R (2009) Human impacts overwhelm the effects of sea-level rise on Basque coastal habitats (N Spain) between 1954 and 2004. Estuar Coast Shelf Sci 84:453–462

Coen LD, Brumbaugh RD, Bushek D, Grizzle R, Luckenbach MW, Posey MH, Tolley S (2007) Ecosystem services related to oyster restoration. Mar Ecol Prog Ser 341:303–307

Collins M, Borja Á (2004) Preface (to: Oceanography and Marine Environment of the Basque Country). Oceanogr Mar Environ Basqu Ctry. http://eprints.soton.ac.uk/11051/

Costa-Pierce BA (2008) Epilogue: aquaculture, innovation and social transformation. Aquac Innov Soc Transf 17:315–326

Cranford PJ, Ward JE, Shumway SE (2011) Bivalve filter feeding: variability and limits of the aquaculture biofilter. Shellfish Aquac Environ 81:124

Cubillo AM, Peteiro LG, Fernández-Reiriz MJ, Labarta U (2012) Density-dependent effects on morphological plasticity of *Mytilus galloprovincialis* in suspended culture. Aquaculture 338:246–252

Cuevas N, Zorita I, Costa PM, Franco J, Larreta J (2015) Development of histopathological indices in the digestive gland and gonad of mussels: integration with contamination levels and effects of confounding factors. Aquat Toxicol 162:152–164

Dalsgaard J, John MS, Kattner G, Müller-Navarra D, Hagen W (2003) Fatty acid trophic markers in the pelagic marine environment. Adv Mar Biol 46:225–340

Dare PJ, Edwards DB (1975) Seasonal changes in flesh weight and biochemical composition of mussels (*Mytilus edulis* L.) in the Conwy Estuary, North Wales. J Exp Mar Biol Ecol 18(2):89–97

Dawes JG (2012) Do data characteristics change according to the number of scale points used? An experiment using 5 point, 7 point and 10 point scales. Int J Mark Res 51(1):2008

De Moreno JEA, Pollero RJ, Moreno VJ, Brenner RR (1980) Lipids and fatty acids of the mussel (*Mytilus platensis* d'Orbigny) from South Atlantic waters. J Exp Mar Biol Ecol 48:263–276

De Zwaan A, Zandee DI (1972) Body distribution and seasonal changes in the glycogen content of the common sea mussel *Mytilus edulis*. Comp Biochem Physiol A Physiol 43(1):53–58

Drapeau A, Comeau LA, Landry T, Stryhn H, Davidson J (2006) Association between longline design and mussel productivity in Prince Edward Island, Canada. Aquaculture 261(3):879–889

Dridi S, Romdhane MS, Elcafsi MH (2007) Seasonal variation in weight and biochemical composition of the Pacific oyster, *Crassostrea gigas*, in relation to the gametogenic cycle and environmental conditions of the Bizert lagoon, Tunisia. Aquaculture 263(1):238–248

Duarte P, Labarta U, Fernández-Reiriz MJ (2008) Modelling local food depletion effects in mussel rafts of Galician Rías. Aquaculture 274(2):300–312

Epler S, Edgar Chambres IV, Kemp KE (1998) Hedonic scales are better predictor than jus about right scales of optimal sweetness in lemonade 1. J Sens Stud 13(2):191–197

Ezgeta-Balić D, Najdek M, Peharda M, Blažina M (2012) Seasonal fatty acid profile analysis to trace origin of food sources of four commercially important bivalves. Aquaculture 334:89–100

FAO (2014) The state of world Fisheries and Aquaculture_2014 (Rome). FAO Fisheries and Aquaculture Department 176

Fernández-Reiriz MJ, Pérez-Camacho A, Ferreiro MJ, Blanco J, Planas M, Campos MJ, Labarta U (1989) Biomass production and variation in the biochemical profile (total protein, carbohydrates, RNA, lipids and fatty acids) of seven species of marine microalgae. Aquaculture 83:17–37

Fernández-Reiriz MJ, Duarte P, Labarta U (2007) Modelos de comportamiento alimentario en el mejillón de las rías de Galicia. In: Figueras AJ (ed) Biología y cultivo del mejillón, vol 26. Editorial CSIC-CSIC Press, pp 197–223

Fernández-Reiriz MJ, Pérez-Camacho A, Peteiro LG, Labarta U (2011) Growth and kinetics of lipids and fatty acids of the clam Venerupis pullastra during larval development and postlarvae. Aquac Nutr 17(1):13–23

Figueiras FG, Labarta U, Reiriz MF (2002) Coastal upwelling, primary production and mussel growth in the Rías Baixas of Galicia. In: Sustainable increase of marine harvesting: fundamental mechanisms and new concepts. Springer, Netherlands, pp 121–131

Filgueira R, Labarta U, Fernández-Reiriz MJ (2006) Flow through chamber method for clearance rate measurements in bivalves. Design and validation of individual chambers and mesocosm. Limnol Oceanogr Methods 4:284–292

Forrest BM, Keeley NB, Hopkins GA, Webb SC, Clement DM (2009) Bivalve aquaculture in estuaries: review and synthesis of oyster cultivation effects. Aquaculture 298(1):1–15

Freites L, Labarta U, Fernández-Reiriz MJ (2002a) a Evolution of fatty acid profiles of subtidal and rocky shore mussel seed (*Mytilus galloprovincialis*, Lmk.). Influence of environmental parameters. J Exp Mar Biol Ecol 268(2):185–204

Freites L, Fernández-Reiriz MJ, Labarta U (2002b) b Lipid classes of mussel seeds *Mytilus galloprovincialis* of subtidal and rocky shore origin. Aquaculture 207(1–2):97–111

Fuentes A, Fernandez-Segovia I, Escriche I, Serra JA (2009) Comparison of physico-chemical parameters and composition of mussels (*Mytilus Galloprovincialis* Lmk.) from Different Spanish Origins. Food Chem 112:295–302

Gabbot PA (1976) Chapter 8: energy metabolism. In Bayne BL (ed) Marine mussels: their ecology and physiology. International biological programme. Cambridge University Press, Cambridge, vol 10, pp 293–355

Galap C, Netchitaïlo P, Leboulenger F, Grillot JP (1999) Variations of fatty acid contents in selected tissues of the female dog cockle (*Glycymeris glycymeris* L, Mollusca, Bivalvia) during the annual cycle. Comp Biochem Physiol A122:241–254

Gallardi D, Hobbs K, Mills T, Couturier C, Parrish CC, Murray HM (2014) Effects of extended ambient live holding on cultured blue mussels (*Mytilus edulis* L.) with reference to condition index, lipid profile, glycogen content and organoleptic testing. Aquaculture 430:149–158

Galparsoro I, Borja Á, Legorburu I, Hernández C, Chust G, Liria P (2010) Morphological characteristics of the Basque continental shelf (Bay of Biscay, northern Spain); their implications for Integrated Coastal Zone Management. Geomorphology 118(3):314–329

Galparsoro I, Liria P, Legorburu I, Bald J, Chust G, Ruiz-Minguela P, Pérez G, Marqués J, Torre-Enciso Y, González M, Borja A (2012) A Marine Spatial Planning approach to select suitable areas for installing wave energy converters on the Basque continental shelf (Bay of Biscay). Coast Manag 40:1–9

García-Gabarra A (2006) Nutrient intakes: concepts and international recommendations (part two) (Review). Nutrición Hospitalaria 21(4):437–447

Garmendia L, Soto M, Cajaraville MP, Marigómez I (2010) Seasonality in cell and tissue-level biomarkers in *Mytilus galloprovincialis*: relevance for long-term pollution monitoring. Aquat Biol 9(3):203–219

Gökoglu N (2002) A descriptive method for sensory evaluation of mussels. LWT Food Sci Technol 35(7):563–567

Gosling E (2003) Bivalve Molluscs: biology, ecology and culture. Fishing News Books-Blackwell Publishing, MPG Books, Bodmin

Gracia J (1996) El sector pesquero en la historia del País Vasco: esbozo de los conocimientos actuales y problemas abiertos a la futura investigación historiográfica. Itsas Memoria Revista de Estudios Marítimos del País Vasco 1:209–214

Graeve M, Kattner G, Hagen W (1994) Diet-induced changes in the fatty acid composition of Arctic herbivorous copepods: experimental evidence of trophic markers. J Exp Mar Biol Ecol 182(1):97–110

Grienke U, Silke J, Tasdemir D (2014) Bioactive compounds from marine mussels and their effects on human health. Food Chem 142:48–60

Haard NF (1992) Control of chemical composition and food quality attributes of cultured fish. Food Res Int 25(4):289–307

Hall JM, Parrish CC, Thompson RJ (2002) Eicosapentaenoic acid regulates scallop (*Placopecten magellanicus*) membrane fluidity in response to cold. Biol Bull 202(3):201–203

Handå A, Min H, Wang X, Broch OJ, Reitan KI, Reinertsen H, Olsen Y (2012) Incorporation of fish feed and growth of blue mussels (*Mytilus edulis*) in close proximity to salmon (*Salmo salar*) aquaculture: implications for integrated multi-trophic aquaculture in Norwegian coastal waters. Aquaculture 356:328–341

Hanson SWF, Olley J (1963) Application of the Bligh and Dyer method of lipid extraction to tissue homogenates. Biochem J 89(3):101–102

Heral M, Razet D, Maestrini S, Garnier J (1980) Composition de la matière organique particulaire dans les eaux du bassin de Marennes-Oléron: apport énergétique pour la nutrition de l'huître. In: CIEM Conseil International pour l'Exploration de la mer

Hesley C (1997) Open ocean aquaculture: chartering the future of ocean farming. In: Proceedings of an international conference. UNIHI-Seagrant-CP-98–08, University of Hawaii Sea Grant College Program, Maui, 23–25 April 1997

Hickman RW (1992) Mussel cultivation. The mussel Mytilus: ecology, physiology, genetics and culture. Elsevier, New York, pp 465–510

Irisarri J, Fernández-Reiriz MJ, Robinson SM, Cranford PJ, Labarta U (2013) Absorption efficiency of mussels *Mytilus edulis* and *Mytilus galloprovincialis* cultured under integrated multi-trophic aquaculture conditions in the Bay of Fundy (Canada) and Ría Ares-Betanzos (Spain). Aquaculture 388:182–192

Irisarri J, Fernández-Reiriz MJ, De Troch M, Labarta U (2014a) Fatty acids as tracers of trophic interactions between seston, mussels and biodeposits in a coastal embayment of mussel rafts in the proximity of fish cages. Comp Biochem Physiol B Biochem Mol Biol 172:105–115

Irisarri J, Fernández-Reiriz MJ, Cranford PJ, Labarta U (2014b) b Effects of seasonal variations in phytoplankton on the bioenergetic responses of mussels (*Mytilus galloprovincialis*) held on a raft in the proximity of red sea bream (*Pagellus bogaraveo*) net-pens. Aquaculture 428:41–53

James D (2013) Risks and benefits of seafood consumption. GLOBEFISH Research Programme, FAO, Rome, p 28

Karayücel S, Çelik MY, Karayücel İ, Erik G (2010) Growth and production of raft cultivated Mediterranean mussel (Mytilus galloprovincialis Lamarck, 1819) in Sinop, Black sea. Turk J Fish Aquat Sci 10(1). doi:10.4194/trjfas.2010.0102

Kharlamenko VI, Zhukova NV, Khotimchenko SV, Svetashev VI, Kamenev GM (1995) Fatty acids as markers of food sources in a shallow-water hydrothermal ecosystem (Kraternaya Bight, Yankich Island, Kurile Islands). Mar Ecol Prog Ser 120:231–241

Kharlamenko VI, Kiyashko SI, Imbs AB, Vyshkvartzev DI (2001) Identification of food sources of invertebrates from the seagrass *Zostera marina* community using carbon and sulfur stable isotope ratio and fatty acid analyses. Mar Ecol Prog Ser 220:103–117

King PA, McGrath D, Britton W (1990) The use of artificial substrates in monitoring mussel (*Mytilus edulis* L.) settlement on an exposed rocky shore in the west of Ireland. J Mar Biol Assoc U K 70(02):371–380

Kryzynowek J, Wiggin K (1979) Seasonal variation and frozen storage stability of blue mussel (*Mytilus edulis*). J Food Sci 44(6):1644–1645

Langan R (2000) Submerged longline culture of blue mussels (*Mytilus edulis*) at an open ocean site in the Gulf of Maine. J Shellfish Res 19:575

Langan R (2001) Advances in submerged longline culture of blue mussels *Mytilis edulis* in the open ocean. In Bridger CJ, Reid TH (eds.) Open ocean aquaculture IV, symposium program and book of abstracts. Mississippi-Alabama Sea Grant Consortium, Ocean Springs, MS MASGP-01–006, St Andrews, pp 73–74

Langan R, Horton F (2003) Design, operation and economics of submerged longline mussel culture in the open ocean. Bull Aquac Assoc Can 103:11–20

LeBlanc N, Landry T, Stryhn H, Tremblay R, McNiven M, Davidson J (2005) The effect of high air and water temperature on juvenile *Mytilus edulis* in Prince Edward Island, Canada. Aquaculture 243(1):185–194

Lee R, Lovatelli A, Ababouch L (2008) Bivalve depuration: fundamental and practical aspects. Food and Agriculture Organization of the United Nations

Lewis JR, Powell HT (1961) The occurrence of curved and ungulate forms of the mussel *Mytilus edulis* L. in the British Isles and their relationship to *M. galloprovincialis* Lamarck. Proc Zool Soc Lond 137(4):583–598

Lucas A, Beninger PG (1985) The use of physiological condition indices in marine bivalve aquaculture. Aquaculture 44(3):187–200

Martínez ML, Intralawan A, Vázquez G, Pérez-Maqueo O, Sutton P, Landgrave R (2007) The coasts of our world: ecological, economic and social importance. Ecol Econ 63(2):254–272

Mayzaud P (1976) Respiration and nitrogen excretion of zooplankton. IV. The influence of starvation on the metabolism and the biochemical composition of some species. Mar Biol 37(1):47–58

Morán XAG, Barbosa AB, Revilla M, Bode A, Orive E, Soudant D (2012) Phytoplankton and microbial plankton of the Bay of Biscay and western Iberian shelf. In: O'Brien TD, Li WKW, Morán XAG (eds.) ICES Phytoplankton and Microbial Plankton Status Report 2009/2010. ICES Cooperative Research Report 313, pp 114–135

Moroño A, Arévalo F, Fernández ML, Maneiro J, Pazos Y, Salgado C, Blanco J (2003) Accumulation and transformation of DSP toxins in mussels *Mytilus galloprovincialis* during a toxic episode caused by *Dinophysis acuminata*. Aquat Toxicol 62(4):269–280

Múgica M, Sokolova IM, Izagirre U, Marigómez I (2015) Season-dependent effects of elevated temperature on stress biomarkers, energy metabolism and gamete development in mussels. Mar Environ Res 103:1–10

Musa-Veloso K, Binns MA, Kocenas A, Chung C, Rice H, Oppedal-Olsen H, Lemke S (2011) Impact of low v. moderate intakes of long-chain n-3 fatty acids on risk of coronary heart disease. Br J Nutr 106(08):1129–1141

Myrand B, Tremblay R, Sévigny JM (2009) Decreases in multi-locus heterozygosity in suspension-cultured mussels (*Mytilus edulis*) through loss of the more heterozygous individuals. Aquaculture 295(3):188–194

Okumuş I, Stirling HP (1998) Seasonal variations in the meat weight, condition index and biochemical composition of mussels (*Mytilus edulis L.)* in suspended culture in two Scottish sea lochs. Aquaculture 159(3–4):249–261

Oliveira RA, Sykes AV, Hachero-Cruzado I, Azeiteiro UM, Esteves E (2015) A sensory and nutritional comparison of mussels (*Mytilus* sp.) produced in NW Iberia and in the Armona offshore production area (Algarve, Portugal). Food Chem 168:520–528

Orban E, Di Lena G, Nevigato T, Casini I, Marzetti A, Caproni R (2002) Seasonal changes in meat content, condition index and chemical composition of Mussels (*Mytilus galloprovincialis*) cultured in two different Italian Sites. Food Chem 77(1):57–65

Ortiz-Zarragoitia M, Garmendia L, Barbero MC, Serrano T, Marigómez I, Cajaraville MP (2011) Effects of the fuel oil spilled by the Prestige tanker on reproduction parameters of wild mussel populations. J Environ Monit 13(1):84–94

Pascual M, Borja A, Eede SV, Deneudt K, Vincx M, Galparsoro I (2011) Marine biological valuation mapping of the Basque continental shelf (Bay of Biscay), within the context of marine spatial planning. Estuar Coast Shelf Sci 95(1):186–198

Pauly D (1983) Some simple methods for the assessment of tropical fish stocks. FAO Fish. Tech. Pap. 234

Pazos AJ, Román G, Acosta CP, Abad M, Sánchez JL (1997) Seasonal changes in condition and biochemical composition of the scallop *Pecten maximus* L. from suspended culture in the Ria de Arousa (Galicia, NW Spain) in relation to environmental conditions. J Exp Mar Biol Ecol 211(2):169–193

Penney RW, Hart MJ, Templeman N (2002) Comparative growth of cultured blue mussels, *Mytilus edulis*, *M. trossulus* and their hybrids, in naturally occurring mixed-species stocks. Aquac Res 33(9):693–702

Pérez-Camacho A, Gonzalez R, Fuentes J (1991) Mussel culture in Galicia (NW Spain). Aquaculture 94(2):263–278

Pérez-Camacho A, Aguiar E, Labarta U, Vinseiro V, Fernández-Reiriz MJ, Álvarez-Salgado XA (2014a) Ecosystem-based indicators as a tool for mussel culture management strategies. Ecol Ind 45:538–548

Pérez-Camacho A, Aguiar E, Labarta U, Vinseiro V, Fernández-Reiriz MJ, Álvarez-Salgado XA (2014b) Ecosystem-based indicators as a tool for mussel culture management strategies. Ecol Ind 45:538–548

Peryam DR, Girardot NF (1952) Advanced taste-test method. Food Eng 24:58–61

Peryam DR, Pilgrim FJ (1957) The hedonic scale method of measuring food preference. Food tech 11(4):9–14

Peteiro LG, Babarro JM, Labarta U, Fernández-Reiriz MJ (2006) Growth of *Mytilus galloprovincialis* after the Prestige oil spill. ICES J Mar Sci J du Cons 63(6):1005–1013

Pogoda B, Buck BH, Saborowski R, Hagen W (2013) Biochemical and elemental composition of the offshore-cultivated oysters *Ostrea edulis* and *Crassostrea gigas*. Aquaculture 400–401:53–60

Polk M (1996) Open ocean aquaculture. Proceedings of an international conference, May 8–10, 1996, Portland, Maine. In: Polk M (ed) New Hampshire-Maine Sea Grant College Program No. UNHMP-CP-SG-96-9, pp 640

Prato E, Danieli A, Maffia M, Biandolino F (2010) Lipid and fatty acid compositions of *Mytilus galloprovincialis* cultured in the Mar Grande of Taranto (Southern Italy): feeding strategies and trophic relationships. Zool Stud 49(2):211–219

R Development Core Team (2008) R: A language and environmental for statistical computing. R Foundation fro Statistical Computing, Viena, Austria. http://www.R-project.org

Rohlf FJ, Sokal RR (1981) Comparing numerical taxonomic studies. Syst Zool 30:459–490

Ruano F, Ramos P, Quaresma M, Bandarra NM, Da Fonseca IP (2012) Evolution of fatty acid profile and condition index in mollusc bivalves submitted to different depuration periods. Revista Portuguesa da Ciências Veterinárias 111(581–582):75–84

Ryan J (2004) Farming the Deep Blue. Board Iascaigh Mhara. Technical Report 82

Sánchez-Lazo C, Martínez-Pita I (2012) Biochemical and energy dynamics during larval development of the mussel *Mytilus galloprovincialis* (Lamarck, 1819). Aquaculture 358–359:71–78

Schutz HG, Cardello AV (2001) A labeled affective magnitude (LAM) scale for assessing food liking/disliking. J Sens Stud 16(2):117–159

Seed R (1968) Factors influencing shell shape in *Mytilus Edulis*. J Mar Biol Assoc U K 48(03):561–584

Seed R, Suchanek TH (1992) Population and community ecology of *Mytilus*. Mussel Mytilus Ecol Physiol Genet Cult 25:87–170

Shahidi F (2004) Functional seafood products. ACS Symp Ser 702:29–49

Shahidi F, Miraliakbari H (2004) Omega-3 (n-3) fatty acids in health and disease: part 1—cardiovascular disease and cancer. J Med Food 7(4):387–401

Shpigel M, Neori A, Popper DM, Gordin H (1993) A proposed model for "environmentally clean" land-based culture of fish, bivalves and seaweeds. Aquaculture 117(1–2):115–128

Shumway SE, Davis C, Downey R, Karney R, Kraeuter J, Parsons J, Wikfors G (2003) Shellfish aquaculture—in praise of sustainable economies and environments. World Aquac 34(4):8–10

Sidari L, Nichetto P, Cok S, Sosa S, Tubaro A, Honsell G, Della Loggia R (1998) Phytoplankton selection by mussels, and diarrhetic shellfish poisoning. Mar Biol 131(1):103–111

Smaal AC (2002) European mussel cultivation along the Atlantic coast: production status, problems and perspectives. Dev Hydrobiol 167:89–98

Soto D, Aguilar-Manjarrez J, Brugère C, Angel D, Bailey C, Black K, Wainberg A (2008) Applying an ecosystem-based approach to aquaculture: principles, scales and some management measures. Build Ecosyst Approach Aquac 14

Stanković S, Jović M, Milanov R, Joksimović D (2011) Trace elements concentrations (Zn, Cu, Pb, Cd, As and Hg) in the Mediterranean mussel (*Mytilus galloprovincialis*) and evaluation of mussel quality and possible human health risk from cultivated and wild sites of the southeastern Adriatic Sea, Montenegro. J Serb Chem Soc 76(12):1725–1737

Stickney RR (1998) Joining forces with industry—open ocean aquaculture. In: Proceedings of the third annual international conference. TAMU-SG-99-103, Corpus Christi, Texas Sea Grant College Program

Svensson S (2003) Depuration of Okadaic acid (Diarrhetic Shellfish Toxin) in mussels, *Mytilus edulis* (Linnaeus), feeding on different quantities of nontoxic algae. Aquaculture 218(1):277–291

Varela M, Prego R, Pazos Y, Moroño Á (2005) Influence of upwelling and river runoff interaction on phytoplankton assemblages in a Middle Galician Ria and Comparison with northern and southern rias (NW Iberian Peninsula). Estuar Coast Shelf Sci 64(4):721–737

Ventrella V, Pagliarani A, Nesci S, Trombetti F, Pirini M (2013) Dietary enhancement of selected fatty acid biosynthesis in the digestive gland of *Mytilus galloprovincialis* Lmk. J Agric Food Chem 61(4):973–981

Vernocchi P, Maffei M, Lanciotti R, Suzzi G, Gardini F (2007) Characterization of Mediterranean mussels (*Mytilus galloprovincialis*) harvested in Adriatic Sea (Italy). Food Control 18:1575–1583

Vilela H (1950) Vida bentónica de *Tapes decussatus* (L.). Arquivos do Museu Bocage 21:1–120

Villanueva ND, Petenate AJ, Da Silva MA (2005) Performance of the hybrid hedonic scale as compared to the traditional hedonic, self-adjusting and ranking scales. Food Qual Prefer 16(8):691–703

Waite L, Grant J, Davidson J (2005) Bay-scale spatial growth variation of mussels *Mytilus edulis* in suspended culture, Prince Edward Island, Canada. Mar Ecol Progress Ser 297:157

Ward JE, Shumway SE (2004) Separating the grain from the chaff: particle selection in suspension-and deposit-feeding bivalves. J Exp Mar Biol Ecol 300(1):83–130

Wilkins NP (1967) Starvation of the herring, *Clupea harengus* L.: survival and some gross biochemical changes. Comp Biochem Physiol 23(2):503–518

Xu Q, Yang H (2007) Food sources of three bivalves living in two habitats of Jiaozhou Bay (Qingdao, China): indicated by lipid biomarkers and stable isotope analysis. J of Shellfish Res 26(2):561–567

Zar JH (1984) Biostatistical analysis. Prentice-Hall, Englewood Cliffs, p 718

Zúñiga D, Castro CG, Aguiar E, Labarta U, Figueiras FG, Fernández-Reiriz MJ (2014) Biodeposit contribution to natural sedimentation in a suspended *Mytilus galloprovincialis* Lmk mussel farm in a Galician Ría (NW Iberian Peninsula). Aquaculture 432:311–320

18

An overview of the utilisation of microalgae biomass derived from nutrient recycling of wet market wastewater and slaughterhouse wastewater

A. Y. Maizatul · Radin Maya Saphira Radin Mohamed · Adel A. Al-Gheethi · M. K. Amir Hashim

Abstract Microalgae have high nutritional values for aquatic organisms compared to fish meal, because microalgae cells are rich in proteins, lipids, and carbohydrates. However, the high cost for the commercial production of microalgae biomass using fresh water or artificial media limits its use as fish feed. Few studies have investigated the potential of wet market wastewater and slaughterhouse wastewater for the production of microalgae biomass. Hence, this study aims to highlight the potential of these types of wastewater as an alternative superior medium for microalgae biomass as they contain high levels of nutrients required for microalgae growth. This paper focuses on the benefits of microalgae biomass produced during the phycoremediation of wet market wastewater and slaughterhouse wastewater as fish feed. The extraction techniques for lipids and proteins as well as the studies conducted on the use of microalgae biomass as fish feed were reviewed. The results showed that microalgae biomass can be used as fish feed due to feed utilisation efficiency, physiological activity, increased resistance for several diseases, improved stress response, and improved protein retention.

Keywords Nutrients · Extraction · Fish growth · Lipids · Proteins · Fish feed

Abbreviations

GHG	Greenhouse gas
BOD	Biochemical oxygen demand
COD	Chemical oxygen demand
TSS	Total suspended solids
DO	Dissolved oxygen
HPLC	High-performance liquid chromatography
GC–MS	Gas chromatography–mass spectrometry
CBB-G	Coomasie brilliant blue-G250 dye
BSA	Bovine serum albumin
TCA	Trichloroacetic acid
NIR	Near infrared spectroscopy
SFE	Supercritical fluid extraction method

A. Y. Maizatul · Radin Maya Saphira Radin Mohamed (✉) · A. A. Al-Gheethi · M. K. A. Hashim
Department of Water and Environmental Engineering, Faculty of Civil and Environmental Engineering, Micro-pollution Research Centre (MPRC), Universiti Tun Hussein Onn Malaysia, Batu Pahat, 86400 Parit Raja, Johor, Malaysia
e-mail: maya@uthm.edu.my; alisaeed@uthm.edu.my

Introduction

Microalgae are aquatic organisms with different sizes ranging between 1 μm and 2 mm (Baharuddin et al. 2016). They are photosynthetic organisms which utilise light and carbon dioxide to create biomass (Sirakov et al. 2015). However, their growth relies on environmental factors such as nutrient quality and quantity, light, pH, turbulence, salinity, and temperature (Arkronrat et al. 2016). Microalgae play a vital role in food production and the purification of contaminated environment water by assimilating nutrients from water and wastewater (Ramaraj et al. 2015; Baharuddin et al. 2016). They play important roles in the food chain, since they represent the main food for rotifers, cladocerans, fish, and shrimp larvae due to their high nutritional value which includes lipid, protein, carotene, vitamins, carbohydrates, and other essential minerals for the development of aquaculture organisms (Hakalin et al. 2014; Baharuddin et al. 2016). Among several species of microalgae, *Chlorella* sp., *Scenedesmus* sp., and *Spirulina* sp. are common in the production of animal feed supplements, fertilizers, health skin products, and other applications (Pulz and Gross 2004).

Furthermore, microalgae represent an alternative energy resource to non-renewable energy coming from fossil fuels (Makarevicienė et al. 2011; Gour et al. 2016). The importance of microalgae as an energy resource lies in the world's high energy demand which leads to the depletion of traditional energy resources such as fossil fuels which are associated with the release of harmful gases such as nitrogen oxide, carbon dioxide, carbon monoxide, and sulphur oxide into the atmosphere (Ren 2014). The accumulation of these gases in the atmosphere can lead to health problems, the acid rain phenomenon, global warming, and depletion of the ozone layer (Rulong et al. 2012). It has been found that the total annual anthropogenic greenhouse gas (GHG) emission during the period from 1970 to 2010 was 1.3%, and has increased to 2.2% in the period between 2000 and 2010 (IPCC 2014).

Interest in microalgae as biofuel feedstock can be associated with several factors such as increasing world oil prices, declining oil supplies from fossil fuels, and global warming from excessive carbon emission of vehicles (Griffiths et al. 2011). Microalgae are a good source for biofuel production due to the high content of lipids in the microalgae cells. For instance, lipids represent 22% of the total biomass yield of *Scenedesmus* sp. which are generated in the production media with low nutrient requirements (Zhang and Hong 2014). Microalgae have a higher lipid content compared to different terrestrial crops due to their simple unicellular structure and high photosynthetic efficiency for higher oil production (Griffiths et al. 2011). Table 1 below demonstrates oil yield from microalgae compared to other terrestrial plants. It can be observed that the microalgae oil yield ranges from 4.7 to 14 L/m^2/year which is higher compared to palm oil yield (0.54 L/m^2/year), Jatropha (0.19 L/m^2/year), sunflower (0.09 L/m^2/year), and soya (0.04 L/m^2/year) (Sheehan et al. 1998; Sazdanoff 2006; Mata et al. 2010).

Microalgae can grow rapidly and are potentially useful for biofuel production. Microalgae are rich in lipids and carbohydrates. They also contain excellent properties for biofuels such as biodiesel, bioethanol, and biomethane (Singh and Gu 2010). The previous studies had identified *Botryococcus* sp. which store long chain hydrocarbons in 50% of its dry biomass (Hannon et al. 2010). Besides that, microalgae cultivation for biofuel production used minimal land compared to other terrestrial plants such as canola, Jatropha, corn, and soybean but yields high oil content. According to Singh and Gu (2010), biodiesel yield from microalgae was 58,700 L ha^1 for microalgae containing only 30% oil by weight compared with 1190 L ha^1 for Canola and 1892 L ha^1 for Jatropha. Microalgae are able to produce biofuel from biomass and also remediate nutrients from wastewater. These characteristics make microalgae a potential crop which can be used to produce cost-competitive biofuels (Hannon et al. 2010).

Table 1 Production of oil from crops compared to algae

Types of crops	Oil yield (L/m^2/year)	References
Microalgae	4.7–14	Sheehan et al. (1998)
Palm	0.54	Mata et al. (2010)
Jatropha	0.19	Sazdanoff (2006)
Sunflower	0.09	
Soya	0.04	

Microalgae as a source of energy can be applied for the production of hydrogen gas, methane, and biogas (Golueke et al. 1957). Microalgae species with high oil content used for biodiesel production are Chlorophyta and Bacillariophyta (Griffiths et al. 2011). The previous studies have revealed that microalgae are able to produce biodiesel amounting to 6283 gallons/acre compared to corn (18 gallons/acre) and soybeans (48 gallons/acre), respectively (Chisti 2007; Kumar et al. 2015). Microalgae lipid comprises of triacylglycerol (TAG), phospholipid, and glycolipid. In general, the lipid in microalgae is stored in the form of TAG. Fast growing species have relatively low lipid production, but stress conditions may trigger lipid production up to 60% of cell dry weight (Griffiths et al. 2011). Jena et al. (2012) indicated that the lipid productivity of *Scenedesmus* sp. is 24.66 mg/L/day with 0.9976 g/L of biomass during the stationary growth phase. This value is higher than that recorded for *Chlorococcum* sp. and *Chlorella* sp. which are 12.44 vs. 16.11 mg/L/day of productively and 0.9922 vs. 0.9933 g/L of biomass yield, respectively. However, it has to be mentioned that the reported data about the production of biofuel from microalgae species are still based on the lab scale experiments, while reported data on the commercialisation of biofuel from microalgae biomass are still limited (Xin et al. 2011; Kumar et al. 2015). This might be due to the high production cost and the absence of effective methods to convert biomass into biofuel at a low operational cost (Chen et al. 2012).

Furthermore, one of the ways which microalgae biomass can be obtained is through the phycoremediation process of wastewater (Sethupathy et al. 2015). Phycoremediation is a specific term to represent the process of using microalgae to treat wastewater by the assimilation of nutrients such as total phosphorus (TP), total nitrogen (TN), and total organic carbon (TOC) from wastewater (Al-Darmaki et al. 2012; Jais et al. 2017). In contrast, phytoremediation is most common term used to represent a similar process conducted by plants. The efficiency of this process lies in its dual role which refers to the treatment of wastewater and the production of biomass, respectively (Pahazri et al. 2016). It has been found that the microalgae in the phycoremediation process have the capacity between 70 and 90% to remove nitrate, sulphate, and phosphate (Nandeshwar and Satpute 2014). The phycoremediation process is associated with the high production of microalgae biomass which has numerous applications (Atiku et al. 2016). Moreover, the utilisation of microalgae biomass as fish feed might be more applicable without more extractions and preparation processes, because microalgae represent natural fish food in water (Brown 2002; Guedes and Malcata 2012; Skarka 2012; Baharuddin et al. 2016).

The current review deals with the technical feasibility of phycoremediation for the treatment of wet market wastewater and slaughterhouse wastewater, the production of microalgae biomass, and the improvement of wastewater characteristics before the final disposal into the environment. The main aim of this article is to provide a summary of recent information concerning the use of microalgae biomass as fish feed. To do so, an overview of the phycoremediation process and factors affecting biomass production are presented. The authors recommend that the utilisation of microalgae biomass for aquaculture food depends on the nutrition values and the extraction process of the lipids and proteins from the biomass yield instead of the microalgae species. Moreover, the reader is strongly encouraged to refer to the original research papers for information on experimental conditions.

Characteristics of wet market wastewater and slaughterhouse wastewater

Increased urbanisation and industrialisation has placed increased water quality and quantity stress in many countries such as Malaysia (Chan 2006; Lee et al. 2015). Indeed, the main problem lies in the quantities of wastewater generated from different human activities which are discharged into natural bodies of water. These practices reduce access to clean water and increase the pollution levels of natural waters. Therefore, the management and treatment of wastewater are some of the initial steps to limit environmental pollution and maintain the river basin capacity (Lee et al. 2015).

Wastewater is a general term which refers to different types of wastewater such as industrial wastewater, hospital wastewater, domestic wastewater, public market wastewater, slaughterhouse wastewater, and grey-water. The level of organic matter for each type of wastewater is different. Therefore, the physical, chemical, and biological characteristics such as biochemical oxygen demand (BOD), chemical oxygen demand (COD), high total suspended solids (TSS), heavy metals, pH, temperature and complex mixture of fats, carbohydrates, and proteins differ (Rajakumar et al. 2011). For instance, the high concentrations of BOD and COD as well as

Table 2 Physicochemical characteristics of wastewater before and after phycoremediation by *Scenedesmus sp.* (Ajayan et al. 2016)

Characteristics	Before	After
Colour	Light brown	Colourless
Odour	Odourless	Odourless
Temperature (°C)	26.8	NR
TSS	5.6	8.3
TDS	1753	1134
TS	4528	2768
BOD	326	212
COD	872	548
DO	1.60	11.2
Nitrate	49	21.7
Phosphorus	3.90	0.10
Chromium	12.8	5.40
Copper	4.90	1.00
Lead	3.2	1.91
Zinc	7.4	2.52

All values are in mg/L except for those otherwise mentioned

infectious agents in sewage are associated with high content of organic matter. Hence, the selection of the treatment process depends on the wastewater composition. The major requirement in wastewater treatment is the removal of nutrients which would reduce microbial loads in the wastewater to acceptable limits prior to discharge and reuse (Al-Gheethi et al. 2013). Ajayan et al. (2016) studied the phycoremediation of wastewater by *Scenedesmus* sp. and the composition of wastewater was found to have improved in terms of colour, TDS, TS, BOD, COD, and DO (Table 2). The colour of wastewater changed from light brown to colourless, whereas the TDS value decreased from 1753 to 1134 mg/L. Then, the BOD and COD values reduced from 326 to 212 mg/L and 872 to 548 mg/L, respectively. Other than that, dissolved oxygen (DO) increased from 1.60 to 11.2 mg/L. Nutrient and heavy metals composition in tannery wastewater reduced slightly. The nitrate and phosphorus content reduced from 49 to 21.7 mg/L and 3.90 to 0.10 mg/L, respectively. On the other hand, chromium, copper, lead, and zinc reduced by about 50% after phycoremediation (Ajayan et al. 2016).

The discharge of wastewater with high organic matter contributes to the reduction of dissolved oxygen (DO) in water. Thus, fish and other aquatic biota cannot survive in water with a low DO content (Mohamed et al. 2016). Besides, wastewater has high nutrient levels such as phosphorus and nitrogen (including ammonia) that can lead to eutrophication which can be toxic to aquatic organisms. In addition, eutrophication also depletes oxygen levels, destroys nursery grounds and leads to the extinction of certain species (Al-Gheethi et al. 2015).

In terms of nutrients, it has been reported that wet market wastewater and slaughterhouse wastewater are very rich in nitrogen and phosphorus (Jais et al. 2015). Slaughterhouse wastewater originates from the washing process of meat, poultry, and fish, while wet market wastewater includes wastewater generated from the washing of fruit and vegetables in public markets (Jais et al. 2017). These types of wastewaters are characterised by the presence of high levels of nitrogen, phosphorus, COD, BOD, suspended solids, fats, oils, and grease compared to residential wastewater (Santos and Robbins 2004). The differences between chemical and physical characteristics of slaughterhouse wastewater and wet market wastewater are illustrated in Table 3. It can be noted that the pH of wet market wastewater ranged from 5.6 to 7.5 (Zulkifli et al. 2011; Omar et al. 2016), while the pH of slaughterhouse wastewater ranged from pH 6.5 and 7.6 (Rajakumar et al. 2011; Bustillo-Lecompte and Mehrvar 2015). Moreover, the slaughterhouse wastewater has higher turbidity compared to wet market wastewater (275 vs. 74.9 mg/L, respectively) (Jais et al. 2015; Bustillo-Lecompte and Mehrvar 2015). In terms of BOD and COD concentrations, it was reported that these parameters in the slaughterhouse wastewater are more than the parameters in wet market wastewater. According to Li et al. (2008), BOD and COD in the slaughterhouse wastewater ranged between 750 and 2895 ± 585 and between

Table 3 Characteristic of wet market and slaughterhouse wastewater

Parameter*	Wet market wastewater				Slaughterhouse wastewater			
	Jais et al. (2015)	Zulkifli et al. (2011)	Noori et al. (2016)	Omar et al. (2016)	Li et al. (2008)	Bustillo-Lecompte et al. (2015)	Budiyono et al. (2011)	Rajakumar et al. (2011)
pH	5.9–6.1	5.6–5.8	5.5–6.5	7.0–7.5	NR	6.5	7.19 ± 0.06	7–7.6
Turbidity	57.1–74.9	N/A	NR	NR	NR	275	NR	NR
BOD	85.39–92.62	71–122	NR	NR	2895 ± 585	2649	1873 ± 421	750–1890
COD	448.2–464.19	381–560	1150 ± 5	1708	4672 ± 952	5577	3756 ± 687	3000–4800
TSS	131.9–133.7	60–122	NR	140	1403 ± 596	3092	1171 ± 311	300–950
SO_4	32.3 ± 0.78	NR	NR	NR	NR	NR	NR	NR
Cl	32 ± 0.69	NR	NR	NR	NR	NR	NR	NR
TN	36.9 ± 0.5	30.3–37.3	NR	288	356 ± 46	156	212 ± 106	16–165
TP	1.61 ± 0.13	0–22.2	NR	66	29 ± 10	42.8	NR	16–32
TOC	118.67 ± 2.89	NR	NR	NR	NR	862	NR	NR
Oil and grease	5.22 ± 0.07	13–43	NR	NR	NR	NR	NR	800–1385
TS	NR	NR	NR	NR	NR	NR	NR	400–3900
TVS	NR	NR	NR	NR	NR	NR	NR	800–1800
VFA	NR	NR	NR	NR	NR	NR	NR	250–540
Alkalinity ($CaCO_3$)	NR	NR	NR	NR	NR	NR	NR	600–1340
TKN	NR	NR	NR	NR	NR	NR	NR	109–325
Protein	NR	NR	265 ± 12	NR	NR	NR	1303 ± 653	580–1000

* All parameters are expressed as mg L^{-1} except for Turbidity (NTU) and pH. Biochemical oxygen demand (BOD); Chemical Oxygen Demand (COD); high Total suspended solids (TSS); Total Organic Carbon (TOC); Total nitrogen (TN); Total phosphorus (TP); volatile fatty acids (VFA); TKN–Total Kjeldahl Nitrogen (TKN); Sulphate (SO_4); Total Chlorine (Cl); Total Volatile Solids (TVS); Total Solids (TS); NR (Non-reported)

3000 and 5577 mg/L, respectively, whereas they ranged between 71 and 122 mg/L and between 381 and 1708 mg/L, respectively, in the wet market wastewater. The high concentrations of BOD and COD in slaughterhouse wastewater might be due to the presence of blood and complex mixtures of fats, proteins, and fibres which contribute to the increase of organic matter (Rajakumar et al. 2011; Kundu et al. 2013). The high concentrations of nutrients in slaughterhouse wastewater might provide a more suitable environment for microalgae growth. Nevertheless, the similarity between slaughterhouse wastewater and wet market wastewater lies in the level of nutrients which is within the range required for microalgae growth. Therefore, both types of wastewaters contribute to the contamination of natural water systems and eutrophication (Atiku et al. 2016; Pahazri et al. 2016).

The increasing production and associated consumption of meat and poultry had increased the discharge of wastewater. Between 2005 and 2013, the consumption rate increased by more than 60% for beef which increased from 138,980 to 201,556 tons, whereas the consumption rate for chicken increased by more than 50% which rose from 785,660 to 1,390,660 tons. The huge quantities of these foods are correlated with huge amounts of wastewater generated. Consequently, the discharge of these wastes into the environment lead to the heavy pollution of natural waters and the occurrence of eutrophication (Bello and Oyedemi 2009; Kundu et al.2013).

Phycoremediation of wet market and slaughterhouse wastewater and the production of microalgae biomass

Several treatment technologies are used for the treatment of wastewater (Table 4). The proper treatment technologies of slaughterhouse wastewater have the potential to produce high-quality wastewater that meets the international standards for the disposal process. Anaerobic treatment in upflow anaerobic sludge blanket

Table 4 Different technologies used to treat slaughterhouse wastewater

Technology	Characteristic of slaughterhouse wastewater	Efficiency	References
Anaerobic treatment in up flow anaerobic sludge blanket (UASB) and aerobic filter	High organic content with average COD of 8000 mg/L with 70% proteins. TSS represent 15 and 30% of COD	COD removal is 90%	Ruiz et al. (1997)
Anaerobic sequencing batch reactors	Total chemical oxygen demand (TCOD) ranged 6908 to 11500 mg/L. Approximately 50% in the form of suspended solids (SS)	COD reduction is 90% within 2 days of hydraulic retention time	Masse and Masse (2000)
Moving bed sequencing batch reactor	COD, BOD, and suspended solids in the range of 4700–8000 mg/L, respectively	COD and BOD removal efficiency is greater than 80 and 90%, respectively. TKN removal efficiency of 86–93%	Sombatsompop et al. (2011)
Fixed bed sequencing batch reactor (FBSBR)	The wastewater has COD loadings in the range of 0.5–1.5 kg COD/m^3 per day	COD, TN, and Phosphorus removal efficiency at range 90–96%, 60–88% and 76–90%, respectively	Rahimi et al. (2011)
Chemical coagulation and electrocoagulation techniques	COD and BOD_5 of raw wastewater in the range of 5817 ± 473 and 2543 ± 362 mg/L	Removal of COD and BOD_5 more than 99% is obtained by adding 100 mg/L PACl and applied 40 V voltage	Bazrafshan et al. (2012)

(UASB) and aerobic filter methods exhibited high efficiency in removing COD of up to 90% (Ruiz et al. 1997). The moving bed sequencing batch reactor had been reported as an effective method for COD, BOD, and TKN removal at 80, 90, and 86–93%, respectively (Sombatsompop et al. 2011). Rahimi et al. (2011) studied the treatment of slaughterhouse wastewater with a fixed bed sequencing batch reactor (FBSBR). The study showed that COD, TN, and TP removal efficiency ranged between 90 and 96%, 60 and 88%, and 76 and 90%, respectively. The electrocoagulation and chemical coagulation techniques for slaughterhouse wastewater treatment had successfully removed COD and BOD_5 by more than 99% after adding 100 mg/L PACl and applying 40 V voltage (Bazrafshan et al. 2012).

Most technologies such as anaerobic treatment as well as chemical coagulation and electrocoagulation techniques are effective for the reduction of the main parameters of wastewater (Sombatsompop et al. 2011; Bazrafshan et al. 2012). However, the cost of the treatment technology should be considered especially in developing countries. Besides, these technologies are designed in a centralised system in which a sewerage network is needed to collect the wastewater generated from different wet markets and slaughterhouses into the central treatment plant and this represents the main challenge in developing countries. Therefore, to reduce the huge quantities of untreated wastewater discharged into the environment, an individual system fixed at each wet market and slaughterhouse location might contribute significantly to the requirements. Phycoremediation might be an alternative technology, because it is cheaper and more effective for removing nutrients compared to other methods (Kwarciak-Kozłowska et al. 2014; Sethupathy et al. 2015; Wurochekke et al. 2016). The phycoremediation process is a microalgae-based technology where microalgae are inoculated into wastewater which acts a culture medium. The microalgae cell growth in the wastewater leads to the removal of nutrients via the assimilation mechanism (Kwarciak-Kozłowska et al. 2014; Satpal and Khambete 2016). Besides, phycoremediation of wastewater using microalgae produces large amounts of biomass which have recently been used for several applications such as animal and fish feeds, fertilizers, as well as biofuel (Al-Darmaki et al. 2012).

Many microalgae species have been used in the phycoremediation process of wastewater. *Scenedesmus* sp. is one of the most common microalgae used, because it has high potential to tolerate acidic eutrophic water conditions and a wide range of temperature for growth (20–36 °C). Therefore, the lipid and carbohydrate contents in their cells are normally high (Xin et al. 2011; Cassidy 2011; Baharuddin et al. 2016). Microalgae have the ability to grow in an environment with pH ranging from 7 to 9.5 at which the microalgae exhibit a higher efficiency in capturing CO_2 in the atmosphere which then induces the high production of biomass (Liu et al. 2005; Zang et al. 2011). However, chlorophyll content in microalgae decreases when the pH value increases from pH 8.5 to 9.5. The decrease in chlorophyll content is associated with low microalgae activity and the low removal of nutrients from wastewater. Rai et al. (2015) revealed that the maximum production of the biomass was recorded at pH 7 (1.3 g L^{-1}), while the lowest production was noted at pH 8 (0.9 g L^{-1}). Therefore, the pH of wastewater during the phycoremediation process needs to be within the optimal pH range. This can be achieved using a continuous operating system where new wastewater is added to renew microalgae growth and to provide more nutrients. Moreover, pH factor is one of the many factors, which has more influence on microalgae growth. For example, temperature is a vital factor which has a linear relationship with microalgae growth, because it is able to modify growth rates and bio-product productivity (Huang et al. 2008; Ras et al. 2013; Minhas et al. 2016). Effects of temperature include cellular chemical composition, rate of photosynthesis, nutrient uptake, CO_2 fixation, and respiration intensity (Tan et al. 2009). *Scenedesmus* sp. grows within a temperature ranging between 10 to 40 °C. To counteract temperature fluctuation, the microalgae cell will undergo the shrinking process (Ras et al. 2013). *Spirulina plantesis* exhibited maximum growth at 35 °C (0.091 doubling/day), while it dropped to 0.041 doubling/day at 40 °C. The variations in the surrounding temperature of the microalgae production medium might also affect the chlorophyll and protein content (Cassidy 2011; Xin et al. 2011). At high temperature, the chlorophyll and protein content in the microalgae cell is reduced, but the carotenoids, saccharides (sugars), and lipids will increase (Cassidy 2011). Microalgae growth rates are inversely proportional to the cell size. The cell sizes increases at low temperature but shrinks at high temperature (Chen et al. 2012). At a lower temperature (4 °C or below), photosynthesis is completely inhibited (Takemura et al. 1985).

Another factor which affects microalgae growth and cell content is light (Kumar et al. 2011; Pérez-Pazos and Fernández-Izquierdo 2011). The variation in light intensity and photoperiod might influence lipid production in the microalgae cells, because lipoid production is associated with carbon content (Pérez-Pazos and Fernández-Izquierdo 2011). Microalgae use carbon at low light intensity to synthesise amino acids, whereas under saturated light, the microalgae cells produce sugars, lipids, and starch (Pérez-Pazos and Fernández-Izquierdo 2011). Rai et al. (2015) estimated biomass and lipid production under different light and dark periods. The study revealed that the maximum biomass (0.54 g/L) and lipid yield (0.079 g/L) were recorded when microalgae were exposed to light for 24 h, whereas the minimum production amounted to 0.25 and 0.04 g/L for biomass and lipids, respectively, with 8 h of incubation under light. However, excessive light exposure can cause light prohibition in microalgae (Ren 2014). Kumar et al. (2011) found that the maximum light intensity to produce carotenoid in *Spirulina platensis* was 3500 lx, while the highest amount of biomass was obtained at 2000 lx (0.71 g/L).

In general, wastewater contains a high level of nutrients (nitrogen, phosphorus, and carbon) and organic matter which act as elements to increase microalgae biomass (Riano et al. 2016). The availability of nutrients in wastewater such as nitrogen can improve the production of biomass. Riano et al. (2016) indicated that *C.sorokiniana*biomass produced 3.0–4.5% dry matter from nitrogen uptakes, whereas the nitrogen content in biomass cells ranged from 27 to 40%. The nitrogen levels in wastewater affect microalgae lipid production and growth rates by 1–10% (Minhas et al. 2016) as high concentrations of nitrogen might increase the quantity of microalgae biomass. In contrast, the production of lipid and carbohydrates increases when there is nitrogen deficiency. Li et al. (2008) reported that *Neochloris* sp. and *Nannochloropsis* sp. had increased lipid yield under conditions of nitrogen deficiency, while *Tetraselmis* sp. and *Scenedesmus* sp. had an increased production of carbohydrates under similar conditions (Minhas et al. 2016). Hence, to optimise the production of biomass alongside with high lipid yields, microalgae should be cultivated in nitrogen rich conditions for high production of biomass yield and then transferred to low nitrogen conditions for lipid production.

Carbon source also has an effect on microalgae biomass. Makareviciene et al. (2011) reported that the biomass yield of *Chlorella* sp. and *Scenedesmus* sp. increased simultaneously with the increase of CO_2 concentrations. The study revealed that at a carbon dioxide concentration of 24%, the growth of microalgae reached a maximum biomass of 0.2 and 0.12 g/L, respectively, for *Chlorella* sp. and *Scenedesmus* sp.

Phosphorous essential for the production of algal biomass is needed in quantities of approximately 0.03–0.06% in the medium to sustain algal growth (Hannon et al. 2010). Phosphorus is present in wastewater as orthophosphate (PO_4^{3-}) which is optimal for the production of microalgae biomass (Solovchenko et al. 2016). Rasala and Mayfield (2015) reported that phosphorus uptake by microalgae from wastewater is stored in cells in the form of polyphosphate. Mulbry et al. (2008) revealed that microalgae have a high efficiency in removing phosphorus from wastewater as much as 70 to 90%. Microalgae take up inorganic phosphate from wastewater via luxury uptake in which algae cells absorb much more phosphate from water than what is necessary for growth (Solovchenko et al. 2016). Microalgae growth and phosphate uptake from wastewater are linearly proportional to biomass yield. However, the phosphorus uptake from wastewater can be limited by factors such as light (self-shading), carbon dioxide, and oxygen levels. The main pollutant which is normally used by microalgae is phosphorus. Phosphorus is released into the waterways from farm manure, fertilizers, detergents, industrial effluents, human waste, and decaying plants (Sen et al. 2013). Phosphorus is essential for the synthesis of nucleic acids, phospholipids, and phosphate esters, and microalgae can obtain phosphorus from wastewater to stimulate growth due to their role in metabolic and anabolic pathways (Pahazri et al. 2016). Jais et al. (2017) stated that dried microalgae use phosphorus to build cells via an anabolic pathway. Phosphorus levels in wastewater should be considered, because it may limit microalgae growth (Pahazri et al. 2016; Jais et al. 2017). Orthophosphate is the most preferred form of phosphorus for microalgae growth as it easily binds to iron and action for extracellular hydrolytic enzymes. Thus, wastewater with a high concentration of phosphorus and a low amount of iron can be a good medium for microalgae growth.

Extraction and determination of nutritional values of microalgae biomass

Numerous analytical techniques including chromatography, capillary electrophoresis, infrared spectroscopy, light scattering detection, and nuclear magnetic resonance spectroscopy have been used for measuring the concentration of carbohydrates in microalgae biomass yields. The easiest, low-cost method for the determination of the concentration of carbohydrates is using the calorimetric method. This method depends on the reaction between a reagent (Phenol) and hydrolysed carbohydrates which further develops a colour complex detected using an electromagnetic spectrum. However, this method is less accurate for determining carbohydrates in very low concentrations. Besides, the chemical reagents can cause serious health problems such as being corrosive to skin, eyes, and some may even affect the respiratory system (Albalasmeh et al. 2013). High-performance liquid chromatography (HPLC) is highly efficient in the determination of carbohydrates in microalgae biomass. HPLC is rapid, specific, sensitive, and precise. Gas chromatography–mass spectrometry (GC–MS) is another technique used for measuring carbohydrates in biomass yield. However, in this method, the carbohydrates should be extracted before the test. Carbohydrates can also be determined using the capillary electrophoresis method after being extracted with borates. Carbohydrates are separated in the gel based on size when voltage goes through it. The smaller the size of the particles, the further away it gets in the gel across the electric field.

Extraction of protein from microalgae is crucial to increase protein efficiency and there are several methods for microalgae extraction such as bead beating, sonication, potter homogenisation, and microwaves to disrupt the microalgae cell wall. Schwenzfeier et al. (2011) used bead milling for the extraction of proteins from *Tetraselmis* sp. The results revealed that the protein yield increased from 36 to 64%. Barbarino and Lourenco (2005) and Murphy et al. (2000) both indicated that the extraction of protein using potter homogeniser and sonication increased the yield of protein compared to non-grinding microalgae biomass. Lee et al. (2010) claimed that the microwaves method is efficient for extracting protein from *Botryococcus* sp. *C. vulgaris* and *Scenedesmus* sp.

Protein content in microalgae biomass can be quantified using the calorimetric method (Bradford 1976; Lowry et al. 1951; Smith et al. 1985) which depends on nitrogen elements (Lopez et al. 2010). Bovine serum albumin (BSA) is the most common substance which is used as a protein standard for calibration curves in the spectrophotometer (Barbarino and Lourenco 2005). The reagent substrate used in the calorimetric method is coomasie brilliant blue-G250 dye (CBB-G) which binds to amino acid residues on the microalgae strain (Bradford 1976). However, CBB-G yields a low concentration of protein compared to Lowry assay (1951) which depends on the use of Trichloroacetic acid (TCA) for protein recovery on microalgae pellets (Murphy et al. 2000). In addition, the previous studies conducted on the determination of proteins based on the usage of CBB-G in the calorimetric method in *Porphyridium* sp., *Phaeodactylum* sp, and *Dunaliella* sp. demonstrated that the protein yield from biomass may differ due to the interference of several substances such as Phenol (Murphy et al. 2000). Therefore, TCA is more accurate for the determination of proteins, since the function of TCA is to coagulate with amino acids (Rajamani and Hilda 1987). Proteins are formed by linkages of amino acids (Craig 2009). There are ten essential amino acids that cannot be synthesised by fish. These amino acids include methionine, arginine, threonine, tryptophan, histidine, isoleucine, lysine, leucine, valine, and phenylalanine, and should be supplied through the diet (Craig 2009). The fish feed prepared using plant (soybean meal) protein are typically low in methionine; therefore, extra methionine must be added to soybean meal-based diets to promote optimal growth and health (Ajani et al. 2016). Thus, it is important to know and match the protein requirements and the amino acid requirements of each of the fish species.

Rodríguez et al. (1997) studied the amino acids found in two species of microalgae (*Chlorella pyrenoisdosa* and *Chlamydomonasreinhardii*) via high-performance liquid chromatography (HPLC) using pre-column fluorescence derivatisation. The study revealed that the predominant amino acids found in *Chlorella pyrenoisdosa* were arginine, alanine, lysine, serine, and glutamic acid, whereas in *C. reinhardii,* the predominant amino acids were alanine, arginine, leucine, lysine, serine, and glutamic acid. The high content of amino acids may be explained by their roles as a nitrogen reserve and for tricarboxylic acid metabolism (Rodríguez et al. 1997). Derrien et al. (1998) determined free amino acids from five species of microalgae which are commonly used in aquaculture such as *Tetraselmis suecica*, *Skeletonema costatum*, *Chaetoceros calcitrans*, *Thalassiosira* sp., and *Isochrisis galbana.* The microalgae species belonging to *S. costatum* and *C.calcitrans* had aspartic acid as a dominant amino acid and about 10% of isoleucine. *Thalassiosira* sp.is rich in serine, glutamic acid, and tyrosine. *I. galbana* showed a completely different composition in free amino acids, with 67.9% of tyrosine followed by serine and arginine. Different cultivation conditions such as temperature, light, pH, and mixing can affect the amino acid composition of microalgae for aquaculture feed (Derrien et al. 1998).

Other than protein and amino acids, lipids are also an important component in microalgae biomass. There are several methods for lipid extraction from microalgae (Table 5). It appeared that electroporation and the pressurized solvent extraction method for lipid extraction had the highest efficiency compared to other methods such as organic solvent, bead beating, and osmotic shock method (Kumar et al. 2015). Besides efficiency rating, the energy consumption, cost, and environmental effects should also be considered. Organic solvent, pressurized solvent extraction, and bead beating methods have high energy requirement in terms of extraction, whereas electroporation and osmotic shock had less energy consumption. Usually low energy consumption methods are cost-effective methods. For instance, osmotic shock requires less energy and low operation cost. Lipid extraction is performed through the hexane Soxhlet extraction and Bligh–Dyer method of 2:1 methanol–chloroform (Bligh and Dyer 1959; Zhang and Hong 2014). However, different microalgae strains show different efficiencies in terms of lipid extraction where the Soxhlet method is normally used for high-quality lipids (fatty acid and triglycerides), whereas the Bligh–Dryer method is normally used for extracting total lipids from microalgae.

The Soxhlet extraction method (AOAC 1995) is the quantification of lipids using solid–liquid extraction in the Soxhlet apparatus. In this method, the solvent is boiled, condensed and passed through microalgae tissue several times for lipid extraction. Then, the solvent is evaporated before the fat is weighed. However, the Soxhlet method needs a large amount of solvent and it is time-consuming. Besides that, lipids can also be determined using near infrared spectroscopy (NIR) at specific wavelengths. The carbonyl group in the ester linkage of lipids is capable of absorbing infrared energy. The method is non-destructive and the microalgae biomass sample can be used for contaminant analysis. However, NIR is not widely used for lipid determination. King (2002) had demonstrated the use of the supercritical fluid extraction method (SEF) to extract lipids. The samples are extracted with liquid carbon dioxide (solvent) before the evaporation and weighing

Table 5 Different lipid extraction methods: cost and energy efficiency (Kumar et al. 2015)

Method	Efficiency rating	Cost	Energy requirement	Remarks
Organic solvent	Moderate	High	Intensive	Health and Environmental Hazard
Pressurized solvent extraction	High	High	Intensive	Health and Environmental Hazard
Bead beating	Moderate	Cost-effective	Intensive	Difficult to scale up
Electroporation	Very high	Initial investment high cost then operate lower cost	Less energy	Appears promising but need to study for pilot scale
Osmotic shock	Moderate–high	Low-cost method	Less energy	Longer treatment time >48 h

process. The SFE method is effective for lipid quantification and no organic solvent is needed for analysis, but it is an expensive method. The lipid extraction methods described by Folch et al. (1957) and Bligh and Dyer (1959) are used as standard procedures for total lipid recovery. Folch et al. (1957) described a method known as a single-step extraction method for lipid recovery. The method used chloroform: methanol (2:1) to extract lipid before washing it with 0.9% potassium chloride (KCl). Bligh and Dryer method (1959) consists of a three-step solvent extraction method for lipid quantification. The method used methanol: chloroform, followed by chloroform and water extraction. The lipid is determined by evaporation of the solvent. Both of these methods are simple and well established, but the Bligh and Dryer (1959) method can analyse microalgae biomass directly without pre-drying treatment.

Cho et al. (2012) examined various solvents to extract lipid from microalgae. The solvents include acetone, hexane, hexane:ether (1:1, v/v), ethanol, and chloroform:methanol (2:1, v/v). The results revealed that chloroform:methanol was the best solvent for lipid extraction with an efficiency of 15% (w/w) compared to 1.3, 2.1, 4.5, and 11.1% (w/w) of lipid obtained using acetone, hexane, hexane:ether (1:1, v/v), and 70% ethanol, respectively. However, using chemical solvents for the quantification of lipid has several disadvantages related to their high toxicity towards humans and the surrounding environment (Zhang and Hong 2014). In terms of the lipid contents in microalgae cells, it has been found that *Scenedesmus* sp. has the highest lipid content compared to *Chlorococcum* sp. and *Chlorella* sp. with 24% lipid per dry weight during the early stationary phase (Jena et al. 2012). Dayananda et al. (2010) reported that the lipid content from *Botryococcus* sp. extracted using chloroform: methanol (1:2) reached 14%.

In lipid extraction, fatty acid methyl ester (FAME) analysis is used to determine docosahexaenoic acid (DHA) and eicosapentaenoic acid (EPA) which are essential for fish growth. Fatty acids such as eicosapentaenoic acid (EPA: 20:5n-3) and docosahexaenoic acid (DHA: 22:6n-3) cannot be synthesised by freshwater fish and must be supplied through diet. Both of these fatty acids are necessary for metabolic functions and cellular membrane components. Prommuak et al. (2012) investigated the fatty acid profile of *Chlorella vulgaris* using GC analysis. The most abundant FAME was methyl linoleate, methyl palmitate, and a small amount of methyl oleate and stearate. These FAMEs were reported to be the most common components in crops (Prommuak et al. 2012). Microalgae are the primary producers of polyunsaturated fatty acids (PUFAs). Pieber et al. (2012) extracted PUFA from *Nannochloropsisoculata* using the pressurized fluid extraction technique (PFE) and accelerated solvent extraction (ASE) using different solvents. *N. oculata* was rich in eicosapentaenoic acid (EPA) when extracted via ethanol extraction (36 $\pm$ 4 mass%) compared to low yields from *n*-hexane extraction (6.1 $\pm$ 0.3 mass%). The EPA yield of 3.7 $\pm$ 0.1 mass% with the use of ethanol indicates that EPA production from *N. oculata* is economically beneficial.

Microalgae biomass quality as fish feed

Aquaculture refers to the production of aquatic organisms such as shrimp, fish, crustacean, and shellfish in cages. Globally, aquaculture has developed rapidly to reduce the gap between market demand and supply. Fish stock worldwide has been declining due to overfishing, but the demand for food has been steadily increasing (Focardi et al. 2005). Therefore, aquaculture development is an alternative way to fulfil the demand for seafood worldwide. The sustainability of the aquaculture industry depends on several factors such as wild stock and fisheries production (Focardi et al. 2005). The aquaculture industry relies on juvenile fish being caught from the wild to supply stock rather than using hatcheries which cause the bycatch of other species. It is reported that 15–20 billion fry of other species such as finfish and shellfish in Honduras, Indian, and Bangladesh has been left aside after accidentally being caught from the wild (Islam et al. 2004; Sarkar and Bhattacharya 2003).

The increasing demand for seafood has led to the intensive development of aquaculture industries all over the world (Gao et al. 2016). However, the development of the aquaculture industry leads to increasing feeding cost and wastewater problems caused by fish food residue and fish faeces (Roy et al. 2011; Gao et al. 2016). High nutrient content such as carbon, phosphorus, and nitrogen from uneaten fish food and faeces leads to eutrophication and the outbreak of diseases. However, the improvement of feed quality can help reduce nutrient pollutants in water columns (Focardi et al. 2005; Mente et al. 2006). High level of nutrients in water columns from fish farming may create favourable conditions for the growth of dinoflagellates (Mente et al.

2006) and lead to harmful algal blooms (HAB) which can deplete oxygen level, reduce light penetration, and release toxins associated with paralytic shellfish poisoning. The previous studies by Joseph et al. (1988) and Michels et al. (2014) had turned problems into opportunities in which integrated aquaculture systems were used to convert waste into algal biomass. The cultivation of microalgae in aquaculture wastewater has been reported where the microalgae may help eliminate excess nutrients from wastewater and produce biomass which can be converted into potential bio-products (Joseph et al. 1988; Michels et al. 2014).

The high feeding cost of wild catch for the production of fish oil and fish meal had caused a decline in the aquaculture industry. Besides that, fishmeal is expensive (Abdulrahman and Ameen 2014). Thus, the replacement of fish meal using microalgae as a protein source may help to resolve problems existing within the aquaculture industry. Microalgae incur low cost compared to fish meal (Abdulrahman and Ameen 2014). Microalgae biomass has been used widely to substitute the traditional fish meal due to their positive effects on fish growth, feeding efficiency, nutrient composition, and fish body development (Roy et al. 2011). Microalgae also have high nutritional values such as protein, carbohydrates, lipid, and trace elements such as vitamins which are considered important for aquaculture (Guedes and Malcata 2012, Skarka 2012). Microalgae are used as a live feed for all growth stages of bivalve molluscs such as oysters, scallops, clams and mussels, crustaceans, and some fish species (Brown 2002). The nutritional value of microalgae is due to the presence of essential macro and micronutrients in their cells (Guedes and Malcata 2012). It has been reported that microalgae cells have 30–40% of protein, 10–20% of lipid content, and 5–15% of carbohydrates (Brown 2002). Therefore, microalgae have high potential to be used as an alternative protein source for fish feed (Sirakov et al. 2015). The composition of microalgae biomass in terms of nutritional value: protein, carbohydrates, and lipid, is illustrated in Table 6. Based on the data presented in the table, it can be seen that *Spirulina maxima* has the highest protein percentage (60–71%), followed by *C. vulgaris* (51–58%) and *S. obliquus* (50–56%). In contrast, *Dunaliellasalina* has the highest carbohydrate content (32%) followed by *S.dimorphus* (21–52%) and *Tetraselmismaculata* (15%). *S.dimorphus* has the highest lipid composition (16–40%) followed by *B.braunii* (Tartiel 2005; Um and Kim 2009).

Several studies have been conducted on microalgae in fish diets. The utilisation of microalgae as fish feed has induced the growth rate of fish, increased feeding utilisation efficiency and physiological activity, increased the resistance for several diseases, improved stress response, and improved protein retention (Sirakov et al. 2015). Algae meal has been also recommended as a good feed complement for counteracting intestinal inflammation produced by soybean meal and is used as a binding agent for aqua feed palletisation (Hashim and Saat 1992; Grammes et al. 2013). Microalgae have been proven as an alternative for fish feed which can reduce cost but, at the same time, provide high protein content to the consumer (Sirakov et al. 2015).

Microalgae biomass used as fish meal for improving fish growth rate is related mainly to the high content of carbohydrates, lipids, and proteins. The carbohydrates in microalgae are available in the form of starch which is easily metabolised by fish (Choix et al. 2012). Carbohydrates represent the main energy source for animals

Table 6 Composition of microalgae biomass in term of nutritional value: protein, carbohydrate, and lipid (Tartiel 2005; Um and Kim 2009)

Microalgae species	Protein	Carbohydrate	Lipid
Botryococcus braunii	40	2	33
Chlorella sp.	46.7	11.6	14.8
C. vulgaris	51–58	12–17	14–22
Dunaliella Salina	57	32	6
D. bioculata	49	4	8
Scenedesmus sp.	52.3	10.06	12.2
S. dimorphus	8–18	21–52	16–40
S. obliquus	50–56	10–17	12–14
Spirulina maxima	60–71	13–16	6–7
S. platensis	42–63	8–14	4–9
Tetraselmis maculata	52	15	3

including fish. Besides that, the production of astaxanthin from microalgae has been used to enhance colouration for salmon and rainbow trout (Skarka 2012). Common species which produce astaxanthin are *Haematococcus* sp., *Chlorella* sp., *Chlorococcum* sp, and *Xanthophyllomyces* sp.

The development of the aquaculture industry in Malaysia requires the consumption of fish meal and fish oil. However, fish meal and fish oil are highly dependent on wild catch fisheries as a source of fish food. The production of aqua feeds is essential as a source of dietary lipids (fish oil) and proteins (fish meal). Dietary protein levels for different species vary from 30 to 57%, and 5 to 8% for dietary lipid levels (Mente et al. 2006). Other critical problems related to the wild catch of fisheries include high cost and limited stock. To reduce the pressure on the demand for fishmeal and fish oil, it needs to move towards sustainably produced plant-based feeds. Table 7 shows the trend of fishmeal and fish oil consumption in the aquaculture industry for three products: salmonid, shrimp, and marine finfish. It is clear that there was an increase in fishmeal and fish oil consumption which was 343,000 tonnes of fishmeal (1992) to 789,000 tonnes fishmeal for salmonid, 232,000 tonnes to 670,000 tonnes fishmeal for shrimp, and 180,000 tonnes to 590,000 tonnes for marine finfish (Tacon et al. 2006).

The substitution of fishmeal with proteins from plants is becoming a necessity, because high-quality plant proteins are readily available. It also involves low-cost fluctuation and sustainable production (Mente et al. 2006). Currently, the replacement of fishmeal is constrained by anti-nutritional factors that cause adverse effects on fish health. For instance, carbohydrate content in soya was found to be less suitable for rainbow trouts (*Oncorhynchus mykiss*) as it can cause prolonged postprandial hyperglycaemia (Panserat and Kaushik 2002). Therefore, in-depth research is needed to understand the dietary requirements of fish for the development of diets that can replace fishmeal as the major source of dietary protein for farmed fish

Microalgae is seen as a fish feed alternative due to their low production cost, rapid growth rates, abundance in stock, and high nutritional value. It was found that microalgae biomass have successfully been used as fish feed for Tilapia (*Oreochromisniloticus)* and provided sufficient protein to enhance fish weight (Badwy et al. 2008; Vizcaíno et al. 2014; Norambuena et al. 2015).

Roy et al. (2011) examined the use of microalgae biomass as fish feed. The study formulated algae mix feeds and the feed contained 10.32% of crude lipid, 24.29 kJ/g of gross energy, 7.08% of calcium, and 1.46% of phosphorus. Feeding fish (*Oreochromis* sp.) with a mixed algal diet had increased the survival rates of fish by 99.7% compared to the conventional feeds (96.5%). It has also increased fish body weight from 1.85 to 7.96 g (303.27%) compared to 1.85–6.5 g for the conventional feed. Abdulrahman and Ameen (2014) had formulated a diet for carps (*Carpinuscarpio*) using *Spirulina* sp. in five different dried microalgae compositions: 0% (T1), 5% (T2), 10% (T3), 15% (T4), and 20% (T5). The study was conducted to determine the weight gain and survival rates of carp feed on *Spirulina* sp. The results found that carps fed with 10% of dried *Spirulina* sp. (T3) had a higher body weight (16.593 g) compared to other replacements T1, T2, T4, and T5 which resulted in body weight measuring 8.375, 12.663, 13.000, and 15.033 g, respectively. Besides that, the carp showed a higher survival rate for T2 (92.857%) and T4 (92.857%).

Radhakrishna et al. (2015) investigated the effects of replacement for 25, 50, 75, and 100% of *C. vulgaris* biomass on the growth performance, energy utilisation, and digestive enzyme of freshwater prawn during the post-larvae stage (*Macrobrachiumrosenbergii*). From the study, it was reported that 50% of fishmeal replaced with *C. vulgaris* enhanced the growth performance (1.252 ± 0.04 g/day), survival rate (93.33 ± 2.50%), and feeding rate efficiency (2.54 ± 0.22%). It was concluded that *C. vulgaris* was easily absorbed and digested by animals. Microalgae applications as aquaculture feed are presented in Table 8 which summarised microalgae potential as aquaculture feed. The highest growth rate of Nile Tilapia had been reported by Attalla and Mikhail

Table 7 Estimated use of fishmeal and fish oil for three types of aquaculture products (Adapted from Tacon et al. 2006)

Aquaculture product	1992 (tonnes)		2003 (tonnes)	
	Fishmeal	Fish oil	Fishmeal	Fish oil
Salmonid	343,000	107,700	789,000	535,000
Shrimp	232,000	27,800	670,000	58,300
Marine finfish	180,000	36,000	590, 000	110,600

Table 8 Microalgae applications as aquaculture feeds

Microalgae species	Types of fish/aquaculture	Growth rate (%)	References
Chlorella spp.	Nile Tilapia	50	Badwy et al. (2008)
Scenedesmus spp.		50	
Dunaliella spp.	Nile Tilapia	50–75	Attalla and Mikhail (2008)
Schizochytrium sp.	Channel catfish	2	Li et al. (2009)
Spirulina	Mekong Giant Catfish	5–10	Tongsiri et al. (2010)
Tetrasel missuecica	Sea Bass	10–20	Tulli et al. (2012)
Spirulina platensis	Angel fish	5–20	Mahsa et al. (2013)
Arthrospira plantesis	Whiteleg Shrimp	25–75	Macias-Sancho et al. (2014)
Scenedesmus almeriensis	Gilthead seabream	12–39	Vizcaíno et al. (2014)
Entomoneis spp.	Juvenile Atlantic Salmon	2.5	Norambuena et al. (2015)

(2008) followed by Badwy et al. (2008). Attalla and Mikhail (2008) used *Dunaliella* spp. as feeds and achieved 50–75% growth rates for Nile tilapia. Badwy et al. (2008) studied two types of microalgae species known as *Chlorella* spp. and *Scenedesmus* spp. The studied reported a growth rate of 50% for Nile tilapia when microalgae were used as fish feed. The lowest growth rates were reported by Li et al. (2009) for Channel catfish feeding on *Schizochytrium* sp. (2%), followed by the growth rate of juvenile Atlantic Salmon feeding on *Entomoneis spp.* (2.5%) (Norambuena et al. 2015).

Conclusion

This review has attempted to give an overview of the potential of microalgae in the phycoremediation of wet market wastewater and slaughterhouse wastewater as well as the use of microalgae in the production of microalgae biomass. It is hoped that the reader can obtain a clearer idea of the characteristics of these types of wastewater and the treatment processes used for these types of wastewater before they are finally discharged into the environment. The potential of wet market wastewater and slaughterhouse wastewater as a production media of microalgae biomass as well as the potential of biomass yields as fish feed have been explored in this paper. Therefore, these aspects need to be investigated further to promote the use of microalgae as aquaculture feed and their role in improving the characteristics and quality of fish feeding on biomass. Microalgae exhibit high effectiveness as an alternative nutrient source due to high levels of proteins, lipids, and carbohydrates. Moreover, the characteristics of wastewater are important for phycoremediation efficiency and biomass production. Wet market wastewater and slaughterhouse wastewater with a high content of total nitrogen and phosphorus are more suitable for algae growth and biomass yield production. Therefore, the phycoremediation process of wet market wastewater and slaughterhouse wastewater shows promise for the future of aquaculture. Currently, common treatment methods aim to reduce the main pollutants from waste such as sewage. However, wet market wastewater and slaughterhouse wastewater are quite different from sewage or other types of industrial wastewater which are rich in pathogens and heavy metals. Therefore, it represents a very good resource for biomass yields. All future research studies should also look into recycling and reusing treated wastewater, so that there will be minimal or zero net sludge generation.

Acknowledgements The authors gratefully acknowledge the Ministry of Higher Education of Malaysia for the research project financial support under fundamental research Grant scheme (FRGS) vot No. 1453. We also thankful to get a support from ORICC UTHM through research grants STG Vot U659 and IGSP Vot U682.

Author contributions MAY was the main researcher. The authors MRMSR, A-GAA, and AHMK have contributed significantly in this paper, collecting and analysing the information, writing, and editing the manuscript and approved the final version.

Conflict of interest The authors declared that they have no conflict of interest.

References

Abdulrahman NM, Ameen JH (2014) Replacement of fishmeal with microalgae spirulina on common carp weight gain, meat and sensitive composition and survival. Pak J Nut 13(2):93–98

Ajani EK, Orisasona O, Omitoyin BO, Osho EF (2016) Total replacement of fishmeal by soybean meal with or without methionine fortification in the diets of Nile tilapia, *Oreochromis niloticus*. J Fish Aquat Sci 11:238–243

Ajayan KV, Selvaraju M, Unnikannan P, Sruthi P (2016) Phycoremediation of tannery wastewater using microalgae *Scenedesmus* species. Int J Phytorem 17(10):1–36. doi:10.1080/15226514.2014.989313

Albalasmeh AA, Berhe AA, Ghezzehei TA (2013) A new method for rapid determination of carbohydrates and total carbon concentrations using UV spectrophotometry. Carbohyd Poly 97:253–261

Al-Darmaki A, Govindrajan L, Talebi S, Al-Rajhi S, Al-Barwani T, Al-Bulashi Z (2012) Cultivation and characterization of microalgae for wastewater treatment. Proceedings of the World Congress on Engineering 2012 vol I. WCE 2012, London, 4–6 July 2012

Al-Gheethi AA, Norli I, Lalung J, Azieda T, Ab. Kadir MO (2013) Reduction of faecal indicators and elimination of pathogens from sewage treated effluents by heat treatment. Cas J Appl Sci Res 2(2):29–45

Al-Gheethi AA, Mohamed RMSR, Afaiz Ab. Rahman M, Mas Rahayu J, Amir HK (2015) Treatment of wastewater from car washes using natural coagulation and filtration system. In: International Conference on Sustainable Environment and Water Research (ICSEWR2015), 25–26 Oct 2015, Johor Baru

AOAC (1995) Animal Feed. Association of Official Analytical Chemists, Chapter 4 in Official Methods of Analysis. 16th edition. AOAC International, Arlington, VI, USA. p 30

Arkronrat W, Deemark P, Oniam V (2016) Growth performance and proximate composition of mixed cultures of marine microalgae (*Nannochloropsis* sp. & *Tetraselmis* sp.) with monocultures. Songklanakarin. J Sci Technol 38(1):1–5

Atiku A, Mohamed RMSR, Al-Gheethi AA, Wurochekke AA, Kassim Amir H (2016) Harvesting microalgae biomass from the phycoremediation process of greywater. Environ Sci Poll Res 23:24624–24641

Attalla RF, Mikhail SK (2008) Effect of replacement of fish meal protein with boiled full fat soybean seeds and dried algae on growth performance, nutrient utilization and some blood parameters of Nile tilapia, Egypt. J Aquat Biol Fish 12(4):41–60

Badwy TM, Ibrahim EM, Zeinhom MM (2008) Partial replacement of fish meal with dried microalga (*Chlorella* spp. and *Scenedesmus* spp.) in Nile Tilapia (*Oreochromis Niloticus*) Diets. In: 8th International Symposium on Tilapia in Aquaculture, pp 801–811

Baharuddin NN, Azizi NS, Sohif HN, Karim WA, Al-Obaidi JR, Basiran MN (2016) Marine microalgae flocculation using plant: the case of *Nannochloropsis oculata* and *Moringa oleifera*. Pak J Bot 48(2):831–840

Barbarino E, Lourenco SO (2005) An evaluation of methods for extraction and quantification of protein from marine macro and microalgae. J Appl Phycol 17:447–460

Bazrafshan E, Mostafapour FK, Farzadkia M, Ownagh KA, Mahvi AH (2012) Slaughterhouse wastewater treatment by combined chemical coagulation and electrocoagulation process. PLoS One. doi:10.1371/journal.pone.0040108

Bello YO, Oyedemi DT (2009) Impact of abattoir activities and management in residential neighbourhoods: a case study of Ogbomoso, Nigeria. J Soc Sci 19:121–127

Bligh EG, Dyer WJ (1959) A rapid method of total lipid extraction and purification. Can J Biochem Physiol 37(8):911–917

Bradford MM (1976) A rapid and sensitive method for the quantitation of microgram quantities of protein utilizing the principle of protein dye binding. Anal Biochem 72:248–254

Brown MR (2002) Nutritional value of microalgae for aquaculture. In: Cruz-Suarez LE, Ricque-Marie D, Tapia-Salazar M, Gaxiola-Cortés MG, Simoes N (eds.) Advances in Aquaculture Nutrition VI. Proceedings of the VI International Symposium on Nutrition Aquaculture, Cancun (**3–6 Sep 2002**)

Budiyono B, Widiasa IN, Johari S, Sunarso S (2011) Study on slaughterhouse waste potency and characteristic for biogas production. Int J Water Res 1(2):4–7

Bustillo-Lecompte CF, Mehrvar M (2015) Slaughterhouse wastewater characteristics, treatment, and management in the meat processing industry: a review on trends and advances. J Environ Manage 161:287–302

Cassidy KO (2011) Evaluating algal growth at different temperatures. Theses and Dissertations–biosystems and agricultural engineering. Paper 3

Chan NW (2006) A Comparative study of water resources usage by households in Georgetown-Malaysia and Pattaya, Thailand. Iran J Environ Health Sci Eng 3(4):223–228

Chen SY, Pan LY, Hong MJ, Lee AC (2012) The effect of temperature on the growth of and ammonia uptake by marine microalgae. Bot Stud 53:125–133

Chisti Y (2007) Biodiesel from microalgae. Biotechnol Adv 25:294–306

Cho SC, Choi WY, Oh SH, Lee CG, Seo YC, Kim JS, Song CH, Kim GV, Lee SY, Kang DH, Lee HY (2012) enhancement of lipid extraction from marine microalga, scenedesmus associated with high-pressure homogenization process. J Biomed Biotechnol. doi:10.1155/2012/359432

Choix FJ, De-Bashana LE, Bashan Y (2012) Enhanced accumulation of starch and total carbohydrates in alginate-immobilized *Chlorella spp.* induced by *Azospirillum brasilense*: I. Autotrophic conditions. J Enzy Microb Technol 51:294–299

Craig S (2009) Understanding fish nutrition, feeds and feeding. Virginia cooperative extension, Virginia Tech, Virginia State University. Publication no 420-256.1-4. https://pubs.ext.vt.edu/420/420-256/420-256.html. Accessed 11 March 2017

Dayananda C, Kumudha A, Sarada R, Ravishankar GA (2010) Isolation, characterization and outdoor cultivation of green microalgae *Botryococcus* sp. J Sci Res Essays 5(17):2497–2505

Derrien A, Coiffard LJM, Coiffard C, Roeck-Holtzhauer YD (1998) Free amino acid analysis of five microalgae. J Appl Phycol 10:131–134

Focardi S, Corsi I, Franchi E (2005) Safety issues and sustainable development of European aquaculture: new tools for environmentally sound aquaculture. Aquacult Int 13:3–17

Folch J, Lees M, Stanley GHS (1957) A simple method for the isolation and purification of total lipides from animal tissues. J Biol Chem 226:497–509

Gao F, Li C, Yang ZH, Zeng GM, Feng LJ, Liu JZ, Liu M, Cai HW (2016) Continuous microalgae cultivation in aquaculture wastewater by a membrane photobioreactor for biomass production and nutrients removal. Ecol Eng 92:55–61

Golueke CG, Oswald WJ, Gotaas HB (1957) Anaerobic digestion of algae. Appl Environ Microbiol 5(1):47–55

Gour RS, Chawla A, Singh H, Chauhan RS, Kant A (2016) Characterization and screening of native *Scenedesmus* sp. isolates suitable for biofuel feedstock. PLoS One 11(5):1–16. doi:10.1371/journal.pone.0155321

Grammes F, Reveco FE, Romarheim OH, Landsverk T, Mydland LT, Øverland M (2013) Candida utilis and Chlorella vulgaris Counteract Intestinal Inflammation in Atlantic Salmon (Salmo salar L.). PLoS ONE 8(12):e83213. doi:10.1371/journal.pone.0083213

Griffiths MJ, Dicks RG, Richardson C, Harrison STL (2011) Advantages and challenges of microalgae as a source of oil for biodiesel. Biodiesel - feedstocks and processing technologies, In Tech, pp 1–25. http://www.intechopen.com/books/biodiesel-feedstocks-and-processing-technologies/advantages-and-challenges-of-microalgae-as-a-source-of-oil-for-biodiesel. Accessed 13 July 2016

Guedes AC, Malcata FX (2012) Nutritional value and uses of microalgae in aquaculture, aquaculture. In: Dr. Zainal Muchlisin (ed.), ISBN: 978-953-307-974-5, InTech. http://www.intechopen.com/books/aquaculture/nutritional-value-and-uses-of-microalgae-in-aquaculture

Hakalin NLS, Paz AP, Aranda DAG, Moraes LMP (2014) Enhancement of cell growth and lipid content of a freshwater microalga *Scenedesmus* sp. by optimizing nitrogen, phosphorus and vitamin concentrations for biodiesel production. Nat Sci 6:1044–1054

Hannon M, Gimpel J, Tran M, Rasala B, Mayfield S (2010) Biofuels from algae:challenges and potential. Biofuels 1(5):763–784

Hashim R, Saat MAM (1992) The utilization of seaweed meals as binding agents in pelleted feeds for snakehead (Channa striatus) fry and their effects on growth. Aquaculture 108(3–4):299–308

Huang Y, Chen M, Liu D (2008) Effect of nitrogen, phosphorus, light formation and disappearance and water temperature on the of blue–green algae bloom. J Northwest Sci Technol Univ Agric For (Nature Science Edition) 36(9):93–100

Islam MS, Wahab MA, Tanaka M (2004) Seed supply for coastal brackish water shrimp farming: environmental impacts and sustainability. Mar Pollut Bull 48(1–2):7–11

Jais NM, Mohamed RMSR, Apandi WAWM, Peralta HMM (2015) Removal of nutrients and selected heavy metals in wet market wastewater by using microalgae *Scenedesmus* sp. Appl Mech Mater 773–774:1210–1214

Jais NM, Mohamed RMSR, Al-Gheethi AA, Hashim Amir (2017) Dual role of phycoremediation of wet market wastewater for nutrients and heavy metals removal and microalgae biomass production. Clean Technol Environ Policy 19(1):37–52

Jena J, Nayak M, Panda HS, Pradhan N, Sarika C, Panda PK, Rao BVSK, Prasad RBN, Sukla LB (2012) Microalgae of Odisha Coast as a potential source of biodiesel production. World Environ 2(1):11–16

Joseph CW, Raymond PG, John RB (1988) Photobioreactor design: mixing, carbon utilization, and oxygen accumulation. Biotechnol Bioeng 31:336–344

King JW (2002) Supercritical fluid technology for lipid extraction, fractionation and reactions. In: Kuo TM, Gardner H (eds) Lipid biotechnology. Marcel Dekker Inc., New York, pp 663–687

Kumar RR, Rao PH, Arumugam M (2015) Lipid extraction methods from microalgae: a comprehensive review. Front Energy Res. doi:10.3389/fenrg.2014.00061

Kumar M, Kulshreshtha J, Singh GP (2011) Growth and biopigment accumulation of cyanobacterium Spirulina platensis at different light intensities and temperature. Brazilian J Microbiol 42:1128–1135

Kundu P, Debsarkar A, Mukherjee S (2013) Treatment of slaughter house wastewater in a sequencing batch reactor performance evaluation and biodegradation kinetics. Biomed Res Int. doi:10.1155/2013/134872

Kwarciak-Kozłowska A, Sławik-Dembiczak L, Bańka B (2014) Phycoremediation of wastewater: heavy metal and nutrient removal processes. Environ Prot Nat Resour 25(4):51–54

Lee JY, Jun S, Ahn CY, Oh HM (2010) Comparison of several methods for effective lipid extraction from microalgae. J Bioresour Technol 101:75–77

Lee KE, Mokhtar M, Mohd Hanafiah M, Abdul Halim A, Badusah J (2015) Rainwater harvesting as an alternative water resource in Malaysia: potential, policies and development. J Clean Prod. doi:10.1016/j.jclepro.2016.03.060

Li JP, Healy MG, Zhan XM, Rodgers M (2008) Nutrient removal from slaughterhouse wastewater in an intermittently aerated sequencing batch reactor. Biores Technol 99(16):7644–7650

Li MH, Robinson EH, Tucker CS, Manning BB, Khoo L (2009) Effects of dried algae Schizochytrium sp., a rich source of docosahexaenoic acid, on growth, fatty acid composition, and sensory quality of channel catfish Ictalurus punctatus. Aquaculture 292:232–236

Liu C, Jin X, Sun L (2005) Effects of pH oil growth and species changes of algae in freshwater. J Agro Environ Sci 24(2):294–298

Lopez CVG, Garcia MDCC, Fernandez FGA, Bustos CS, Chisti Y, Sevilla JMF (2010) Protein Measurements of microalgal and cynobacterial biomass. Biores Technol 101:7587–7591

Lowry OH, Rosebrough NJ, Farr AL, Randall RJ (1951) Protein measurement with the Folin phenol reagent. J Biol Chem 193:265–275

Macias-Sancho J, Poersch LH, Bauer W, Romano LA, Wasielesky W, Tesser MB (2014) Fishmeal substitution with *Arthrospira* (Spirulina platensis) in a practical diet for *Litopenaeus vannamei*: effects on growth and immunological parameters. Aquaculture 426–427:120–125

Mahsa B, Moosavi Javadi, Montajami Salar (2013) Assesment the effect of Spirulina plantesis feed on growth performance and survival rate in Angel Fish (*Pterophyllum scalare*). J Fish Int 8(3–6):74–77

Makareviciene V, Andrulevičiūtė V, Skorupskaitė V, Kasperovičienė J (2011) Cultivation of microalgae *Chlorella* sp. and *Scenedesmus* sp. as a potentional biofuel feedstock. Environ Res Eng Manag 3(57):21–27

Masse DI, Masse L (2000) Treatment of slaughterhouse wastewater in anaerobic sequencing batch reactors. Can Agr Eng 42(3):131–137

Mata TM, Martins AA, Caetano NS (2010) Microalgae for biodiesel production and other applications: a review. Renew Sustain Energy Rev 14(1):217–232

Mente E, Pierce GJ, Santos MB, Neolitou C (2006) Effect of feed and feeding in the culture of salmonids on the marine aquatic environment: a synthesis for European aquaculture. Aquacult Int 14:499–522

Michels MH, Vaskoska M, Vermuë MH, Wijffels RH (2014) Growth of *Tetraselmis suecica* in a tubular photobioreactor on wastewater from a fish farm. Water Res 65:290–296

Minhas AK, Hodgson P, Barrow CJ, Adholeya A (2016) A review on the assessment of stress conditions for simultaneous production of microalgal lipids and carotenoids. Front Microbiol 7:546. doi:10.3389/fmicb.2016.00546

Mohamed RMSR, Al-Gheethi AA, Jackson AM, Amir HK (2016) Multi component filters for domestic greywater treatment in village houses. J Am Water Works Assoc 108(7):405–414

Mulbry W, Kondrad S, Pizarro C, Kebede-Westhead E (2008) Treatment of dairy manure effluent using freshwater algae: algal productivity and recovery of manure nutrients using pilot-scale algal turf scrubbers. Bioresour Technol 99:8137–8142

Murphy CD, Moore RM, White RL (2000) Peroxidases from marine microalgae. J Appl Phycol 12:507–513

Nandeshwar SN, Satpute GD (2014) Green technical methods for treatment of waste water using microalgae and its application in the management of natural water resources—a review. Curr World Environ. doi:10.12944/CWE.9.3.34

Noori MdT, Ghangrekar MM, Mukherjee CK (2016) V2O5 microflower decorated cathode for enhancing power generation in air-cathode microbial fuel cell treating fish market wastewater. Int J Hydrog Energy 41:3638–3645

Norambuena F, Hermon K, Skrzypczyk V, Emery JA, Sharon Y, Beard A, Turchini GM (2015) Algae in fish feed: performances and fatty acid metabolism in juvenile atlantic salmon. PloS One 10(4):e0124042

Omar D, Salim MR, Salmiati (2016) Nutrient removal of greywater from wet market using sequence batch reactor. Malaysia J Anal Sci 20(1):142–148

Pahazri F, Mohamed RMS, Al-Gheethi AA, Hashim Amir (2016) Production and harvesting of microalgae biomass from wastewater, a critical review. Environ Technol Rev 5(1):39–56

Panserat S, Kaushik S (2002) Nutritional regulation of hepatic glucose metabolism in fish: example of a poor user of dietary carbohydrates, the rainbow trout. Prod Anim 15:109–117

Peiber S, Schober S, Mittelbach M (2012) Pressurized fluid extraction of polyunsaturated fatty acids from the microalga *Nannochloropsis oculata*. Biomass Bioenerg 47:474–482

Pérez-Pazos JV, Fernández-Izquierdo P (2011) Synthesis of neutral lipids in *Chlorella* sp. under different light and carbonate conditions. Sci Technol Future 4(4):47–57

Prommuak C, Pavasant P, Quitain AT, Goto M, Shotipruk A (2012) Microalgal lipid extraction and evaluation of single-step biodiesel production. Eng J 16(5):157–166

Pulz O, Gross W (2004) Valuable products from biotechnology of microalgae. Appl Microbiol Biotechnol 65:635–648

Radhakrishna S, Saravana BP, Seenivasan C, Muralisankar T (2015) Effect of dietary replacement of fishmeal with *Chlorella vulgaris* on growth performance, energy utilization and digestive enzymes in *Macrobrachium rosenbergii* postlarvae. Int J Fish Aquac 7(5):62–70

Rahimi Y, Torabian A, Mehrdadi N, Shahmoradi B (2011) Simultaneous nitrification-denitrification and phosphorus removal in a fixed bed sequencing batch reactor (FBSBR). J Haz Mat 185(2–3):852–857

Rai MP, Gautom T, Sharma N (2015) Effect of salinity, pH, light intensity on growth and lipid production of microalgae for bioenergy application. J Biol Sci 15(4):260–267

Rajakumar R, Meenambal T, Banu TR, Yeom IT (2011) Treatment of poultry slaughterhouse wastewater in up flow anaerobic filter under low up flow velocity. Int J Environ Sci Tech 8(1):149–158

Rajamani S, Hilda A (1987) Plate assay to screen fungi for proteolytic activity. Curr Sci 56:1179–1181

Ramaraj R, Tsai DDW, Chen PH (2015) Detention time study of algal biomass production with natural water medium. Chiang Mai J Sci. 42(3):549–559

Ras M, Steyer JP, Bernard O (2013) Temperature effect on microalgae: a crucial factor for outdoor production. Rev Enviro Sci BioTechnol 12(2):153–164

Rasala B, Mayfield S (2015) Photosynthetic biomanufacturing in green algae; production of recombinant proteins for industrial, nutritional, and medical uses. Photosynth Res 123:227–239

Ren TT (2014) Primary factors affecting growth of microalgae optimal light exposure duration and frequency. Graduate Theses and Dissertations. Paper 13793. Iowa State University, Ames

Riano B, Blanco S, Becares E, Garcia-Gonzalez MC (2016) Bioremediation and biomass harvesting of anaerobic digested cheese whey in microalgal-based systems for lipid production. Ecol Eng 97:40–45

Rodriguez JRB, Reina GG, Rodriguez JJS (1997) Determination of free amino acids in microalgae by high performance liquid chromatography using pre-column fluorescence derivatization. Biomed Chromatogr 11:335–336

Roy SS, Chaudhuri A, Mukherjee S, HomeChauduri S, Pal R (2011) composite algal supplementation in nutrition of *Oreochromis mossambicus*. J Algal Biom Util 2(1):10–20

Ruiz I, Veiga MC, de Santiago P, Blfizquez R (1997) Treatment of slaughterhouse wastewater in a UASB reactor and an anaerobic filter. Biores Technol 60(3):251–258

Rulong L, Wenxuan C, Bingpeng X, Xiurong K (2012) Optimization of extraction conditions for increasing microalgal lipid yield by using accelerated solvent extraction method (ASE) based on the orthogonal array design. InTech, pp 21–38

Santos CPE, Robbins DMRS (2004) Low-cost innovative solutions for treating public market wastewater in the philippines: deploying hybrid anaerobic/aerobic coco peat filtration systems. PADCO, Makati

Sarkar SK, Bhattacharya AK (2003) Conservation of biodiversity of coastal resources of Sundarbans, Northeast India: and integrated approach through environmental education. Marine Pollution Bulletin 47:260–264

Satpal, Khambete AK (2016) Waste water treatment using micro-algae—a review paper. Int J Eng Technol Manag Appl Sci. 4(2):188–192. http://www.ijetmas.com

Sazdanoff N (2006) Modelling and simulation of the algae to biodiesel fuel cycle. Honour's thesis, Ohio State University, Columbus

Schwenzfeier A, Wierenga AP, Gruppen H (2011) Isolation and characterization of soluble protein from the green microalgae *Tetraselmis* sp. J Bioresour Technol 102:9121–9127

Sen M, Alp MT, Sonmez F, Kocer MAT, Canpolat O (2013) Relationship of algae to water pollution and waste water treatment. Elshorbagy W (ed.) Water treatment, InTech, doi:10.5772/51927

Sethupathy AV, Subramanian A, Manikandan R (2015) Phyco-remediation of sewage waste water by using micro-algal strains. Int J Eng Innov Res 4(2):300–302

Sheehan J, Dunahay T, Benemann J, Roessler P (1998) A look back at the US Department of Energy's Aquatic Species Program: biodiesel from algae. In: Close-out report. National Renewable Energy Lab, Department of Energy, Golden, Colorado, Report number NREL/TP-580-24190

Singh J, Gu S (2010) Commercialization potential of microalgae for biofuels production. Renew Sustain Energy Rev 14:2586–2610

Sirakov I, Velichkova K, Stoyanova S, Staykov Y (2015) The importance of microalgae for aquaculture industry. Review. Int J Fish Stud 2(4):81–84

Skarka J (2012) Microalgae biomass potential in europe: land availability as a key issue. Technol Assess Theory Pract Des 21(1):72–80

Smith PK, Krohn RI, Hermanson GT (1985) Measurement of protein using bicinchoninic acid. Anal Biochem 150:76–85

Solovchenko A, Verschoor AM, Jablonowski ND, Nedbal L (2016) Phosphorus from wastewater to crops: an alternative path invilving microalgae. Biotechnol Adv 34:550–564

Sombatsompop K, Songpim A, Reabroi S, Inkongngam P (2011) A comparative study of sequencing batch reactor and moving bed sequencing batch reactor for piggery wastewater treatment. Maejo Int J Sci Technol 5(2):191–203

Tacon AGJ, Hasan MR, Subasinghe RP (2006) Use of fishery resources as feed inputs to aquaculture development: trends and policy implications FAO fisheries circular No. 1018. FAO, Rome, p 99

Takemura N, Iwkume T, Rusuno M (1985) Photosynthesis and primary production of *Microcystis agruginosa* in Lake Kasumigaura. J Plankton Res 7(3):303–312

Tan X, Kong F, Yu Y (2009) Effects of enhanced temperature on algae recruitment and phytoplankton community succession. China Environ Sci 29(6):578–582

Tartiel MB (2005) Physiological studies on some green algae. Ph.D. thesis, Faculty of Agriculture, Cairo University, Egypt

Tongsiri S, Mang-amphan K, Peerapornpisal Y (2010) Effect of replacing fishmeal with spirulina on growth, carcass composition and pigment of the Mekong Giant Catfish. Asian J Agr Sci 2(3):106–110

Tulli F, Chini Zittelli G, Giorgi G, Poli BM, Tibaldi E, Tredici MR (2012) On growth, feed utilization, and fillet composition of european sea bass juveniles fed organic diets. J Aquat Food Prod Technol 21(3):188–197

Um BH, Kim YS (2009) Review: a chance for Korea to advance algal-biodiesel technology. J Ind Eng Chem 15:1–7

Vizcaíno J, López G, Sáez MI, Jiménez J, Barros A, Hidalgo L (2014) Alarcón FJ (2015) Effects of the microalga *Scenedesmus almeriensis* as fishmeal alternative in diets for gilthead sea bream, *Sparus aurata*, juveniles. Aquaculture 431:34–43

Wurochekke AA, Mohamed RMS, Al-Gheethi AA, Amir HM, Matias-Peralta HM (2016) Household greywater treatment methods using natural materials and their hybrid system. J Water Health 14:914–928

Xin L, Ying HH, Ping ZY (2011) Growth and lipid accumulation of a freshwater microalga *Scenedesmus* sp. under different cultivation temperature. Biores Technol 102:3098–3102

Zang C, Huang S, Wu M (2011) Comparison of relationships between pH, dissolved oxygen and Chlorophyll *a* for aquaculture and nonaquaculture waters. Water Air Soil Poll 219(1–4):157–174

Zhang Q, Hong Y (2014) Comparison of growth and lipid accumulation properties of two oleaginous microalgae under different nutrient conditions. Front Environ Sci Eng 8(5):703–709

Zulkifli AR, Roshadah H, Tunku Khalkausar TF (2011) Control of water pollution from non-industrial premises. Department of Environment, Malaysia

19

Growth performance, feed utilisation and biological indices of Tra catfish (*Pangasianodon hypophthalmus*) cultured in net cages in pond fed diets based on locally available feed resources

Chau Thi Da · Torbjörn Lundh · Jan Erik Lindberg · Håkan Berg

Abstract The feeding experiment was conducted to evaluate the efficacy of locally available feed resources for Tra catfish (*Pangasianodon hypophthalmus*) cultured a series of 21 hapa net cages installed in the earthen pond during a 4-month period. The reference diet contained fish meal as the main crude protein (CP) source, whilst in the seven test diets 20–100 % of the fish meal CP was replaced with CP from local feed ingredients: groundnut cake (GNC), cassava leaf meal (CSLM), sweet potato leaf meal (SPLM), soybean meal (SBM), golden apple snail meal (GASM) and shrimp head meal. There were differences ($p < 0.05$) among diets in final body weight, total weight gain, daily weight gain, specific growth rate, viscera-somatic weight (VSI%), hepato-somatic index (HIS%) and intra-peritoneal fat (IPF%). The respective values for the shrimp head meal diet were numerically highest, followed in descending order by the reference, GASM, GNC, SPLM, CSLM and SBM diet. Food conversion ratio, fish survival rate, total feed intake, feed utilisation, fish fillet and kidney proportions did not differ among the reference and test diets ($p > 0.05$). However, the viscera-somatic, hepato-somatic, kidney and intra-peritoneal fat indices differed among treatments ($p < 0.05$). These data show that fish meal protein can be replaced with protein from locally available plant and animal feed ingredient resources in feed cultured in net cages in pond for Tra catfish fingerlings in effectively compromising growth performance, feed utilisation or carcass traits of fish.

Keywords Fish meal protein · Local feed resources · Tra catfish · Earthen pond-cultured · Happa-net cages

Introduction

The Tra catfish (*Pangasianodon hypophthalmus*) farming system in the Mekong Delta of Vietnam is still operated by many small-scale farmers but with increasing pressure from large-scale operators (Bui et al. 2010; Phan et al. 2009; De Silva and Phuong 2011). The catfish farming sector operates at marginal profit levels and

C. T. Da (✉)
Department of Aquaculture, Faculty of Agriculture and Natural Resources, An Giang University, 18 Ung Van Khiem Street, Dong Xuyen Wards, Long Xuyen, Vietnam
e-mail: chau.thida@gmail.com; ctda@agu.edu.vn

T. Lundh · J. E. Lindberg
Department of Animal Nutrition and Management, Swedish University of Agricultural Sciences, P. O. Box 7024, 75007 Uppsala, Sweden

H. Berg
Department of Physical Geography, Stockholm University, 106 91 Stockholm, Sweden

the long-term viability of the industry will be determined by the operation costs of grow-out farms and the farm gate price of the farmed fish (Phan et al. 2009; De Silva and Phuong 2011). A large proportion of the operation costs consists of feed costs, which are estimated to account for 70–90 % of total investment costs (Da et al. 2011; Hung et al. 2007; Phuong et al. 2007). The replacement of fish meal and trash fish (marine origin) would simply reduce the required input of the most expensive ingredients in fish feeds and thereby reduce the investment costs and increase the profit level for the farmers. This would also decrease the dependence and pressure on wild fish populations used for fishmeal. In Southern Vietnam, many potential "trash fish" species are currently used for fish sauce production and other human uses of high nutritional and economic values (Edwards et al. 2004; Loc et al. 2010).

The increasing volume of exports has to be met with an increase in the fish production, quantity and quality of feeds (De Silva and Phuong 2011). Because of its high nutritional value and palatability, fish meal and trash fish (marine origin) still remain the major dietary protein sources for striped catfish, comprising 20–60 % of the feed. However, using fish meal in fish feed is not a sustainable long-term feeding strategy (Naylor et al. 2009; FAO 2010). It is estimated that in order to produce 1.2 million tonnes of Tra catfish about 1.4 million tonnes wild-caught whole fish will be required as feed (Halls and Johns 2013; Mamat and Alfaro 2014). This reliance on wild marine fish in this aquaculture industry sector constitutes a major vulnerability of declining stocks of wild marine fish therefore more works are needed to identify the alternative fishmeal from local feed resources for this fish species (Kader et al. 2010).

Tra catfish is an omnivorous fish (Phuong 1998; Cacot and Pariselle 1999) and capable of using a wide range of protein sources efficiently. The Mekong Delta contains several locally available alternative protein feed sources that are abundant and have nutritional properties indicating that they could potentially be useful as ingredients in striped catfish feed (Da et al. 2011). Earlier studies have shown that soybean meal can be used in striped catfish feed as a replacement for fish meal without any negative effects on growth performance and carcass traits (Phumee et al. 2011; Hung et al. 2003). Da et al. (2012) also reported that earlier results of laboratory study, fish meal protein in feed for Tra catfish fingerlings can be replaced with protein from locally available feed resources *in effectively* compromising on growth performance, feed utilisation or carcass traits. However, that study and most other previous studies have evaluated that feed sources for cultured fish have been performed under controlled indoor conditions with clean water. Such an experimental model is very different from the conditions under which striped catfish are cultured on earthen pond farming systems in the Mekong Delta, using water from the rivers and canals (Da et al. 2011). The river and canal water contain dissolved nutrients and various living organisms (such as zooplankton, algae and invertebrates), which under pond conditions grow and contribute more or less to the nutrient supply of the cultured fish. Therefore, the fish response to a feed ingredient cultured in an earthen pond farming system may be different from that obtained under controlled indoor conditions with clean water. The hypothesis tested in this study was that fish growth performance and feed utilisation are better in earthen pond-cultured than indoor-cultured Tra catfish and the fish in this experiment was not feed on natural foods. This study evaluated the growth performance, feed utilisation and carcass traits of Tra catfish cultured in hapa net cages installed in earthen pond fed diets where fish meal protein was replaced with protein from groundnut cake (GNC), cassava leaf meal (CSLM), sweet potato leaf meal (SPLM), soybean meal (SBM), golden apple snail meal (GASM) and shrimp head meal (SHM).

Materials and methods

Study site

The study was carried out in an earthen pond at the Faculty of Agriculture and Natural Resources, An Giang University, An Giang province in the Mekong Delta of Vietnam.

Experimental diets

The experimental diets, consisting of one reference diet and six test diets, were formulated to meet the nutrient requirements of Tra catfish (Hung et al. 2002). The reference diet contained fish meal as the main crude protein (CP) source, whilst in the six test diets 20–100 % of the fish meal CP was replaced with CP from alternative local ingredients (Table 1).

Experimental fish

Tra catfish (*Pangasianodon hypophthalmus*) fingerlings were bought from the Research Center of Aquaculture Seed Production in An Giang province. The fish were treated with a 3 % NaCl solution for 15 min on arrival to eliminate ectoparasite infection, and were reared and quarantined in a hapa net cage (4 × 8 m × 2 m) in an earthen pond for 2 weeks to acclimatise to the hapa net cage conditions. Fish were then randomly selected from the rearing, weighed and transferred to the experimental hapa net cage (2 m × 3 m × 2 m) 1 week before the experiment commenced for acclimatisation to experimental conditions.

Experimental design

The experiment was set up as a factorial design with seven different diets fed in triplicate groups of Pangasius fingerlings reared in hapa net cages in earthen pond for 4 months. At the beginning and end of the experiment, fish were individually weighed using a digital scale. One-hundred and seventy-five homogeneous fish with an average initial body weight (BW) of 16.5 ± 0.1 g/fish were selected and distributed into each hapa net cage for each treatment.

Test feed ingredients

The chemical composition of the test feed ingredients is shown in Table 2. Groundnut cake was purchased from Vinh Long province and shrimp head meal from Kien Giang province. Soybean meal (India) and Kien Giang fish meal were purchased from the local market (AFIEX plant) in Long Xuyen city, An Giang province. Cassava leaves and sweet potato leaves (both stem and leaf) were collected during the harvest period from

Table 1 Ingredient composition of reference diet (RD) and test diets (g/kg) for Tra catfish (*Pangasianodon hypophthalmus*) fingerlings

Treatment	RD	Test diets					
		Ground-nut cake	Cassava leaf meal	Sweet potato leaf meal	Soy-bean meal	Golden apple snail meal	Shrimp head meal
Fish meal	260	195	195	208	0.0	0.0	0.0
Vegetable protein mix[a]	480	480	400	400	450	480	480
Wheat flour	200	200	200	200	200	220	260
Squid liver oil	20	20	20	20	20	20	20
Vit-min premix[b]	20	20	20	20	20	20	20
CMC[c]	20	20	20	20	20	20	20
Groundnut cake	–	65	–	–	–	–	–
Cassava leaf meal	–	–	145	–	–	–	–
Sweet potato leaf meal	–	–	–	132	–	–	–
Soybean meal	–	–	–	–	290	–	–
Golden apple snail meal	–	–	–	–	–	240	–
Shrimp head meal	–	–	–	–	–	–	200
Rate of replacement of fish meal (FM) (%)	0	25	25	20	100	100	100

[a] Commercial product used for Tra catfish feed in the Mekong Delta of Vietnam. The product is based on soybean and rice products and contains (93.2 % of DM): CP 21.2 %, EE 7.6 %, NDF 25.5 %, ash 3.1 %

[b] Vitamin and mineral premix; content per kg: vitamin A 4,000,000 UI; vitamin D3 800,000 UI; vitamin E 8500 UI; vitamin K3 750 UI; vitamin B1 375 UI; vitamin C 8750 UI; vitamin B2 1600 mg; vitamin B6 750 mg; folic acid 200 mg; vitamin B12 3000 mcg; biotin 20,000 mcg; methionine 2500 mg; Mn, Zn, Mg, K and Na 10 mg

[c] Carboxymethyl cellulose (CMC): imported from Korea

Table 2 Chemical composition (g/kg DM), gross energy (MJ/kg DM) and essential amino acid (g/kg DM) content of test ingredients

	Groundnut cake	Cassava leaf meal	Sweet potato leaf meal	Soybean meal	Golden apple snail meal	Shrimp head meal
Crude protein	316	223	166	485	564	661
Lipid	100	65	24	10	16	36
Crude fibre	41	145	14	35	10	61
NDF	342	339	397	198	239	242
Ash	55	97	170	73	118	143
Gross energy	19.5	18.9	14.6	17.8	12.3	16.6
Essential amino acids						
Arginine	27.9	11.1	7.7	24.9	32.1	38.2
Histidine	5.0	3.4	1.9	4.8	11.9	11.1
Isoleucine	12.0	10.7	7.5	18.3	18.8	20.9
Leucine	19.8	17.2	10.9	29.1	36.2	37.2
Lysine	5.2	6.7	2.6	15.1	16.3	25.5
Methionine	4.0	5.2	4.3	6.4	10.9	14.5
Phenylalanine	13.8	10.6	7.6	18.6	16.4	19.0
Threonine	6.7	8.4	5.0	14.2	11.1	9.0
Valine	13.6	11.5	8.7	18.3	20.8	19.3
Total	108.0	84.8	56.2	149.7	174.5	194.7

farms in the Tri Ton and Cho Moi districts of An Giang province, respectively. The leaves were cleaned with freshwater, sun dried for 2–3 days and then milled to a meal. Golden apple snails were purchased from farmers in Cho Moi district, An Giang province, and Tam Nong district, Dong Thap province. Only the meat of the golden apple snails was collected and it was cleaned with freshwater and sun dried for 3 days before use.

Diet preparation and feeding

The feed was produced by mixing the dry ingredients carefully and then adding squid oil and distilled water. The amount of distilled water was adjusted to get the mixture to form a stiff dough. The pelleted feed was made using an electronic meat grinder (Quoc Hung company, Vietnam), with diameter and length of pelleted feed in the range 1–2 mm. All diets were sun dried for 2–3 days and then weighed and stored in sealed plastic bags in small portions at 5 °C until use. New batches of experimental feeds were made bi-weekly. The fishes were fed by hand to apparent satiety, at a rate of about 3–5 % body weight per day, at 9.00 and 14.00 h.

Experiment system and management

The experiment was carried out in a 700 m^2 earthen pond with a depth of 2 m. A series of 21 hapa net cages with 2.0 mm mesh size were placed in the pond and were used to hold the fish. The hapa net cages were rectangular (2 × 3 m × 2 m deep) and were suspended by tying them to four melaleuca poles. One feeding sieve (feeding trap), 30 cm in diameter, was placed in each hapa net cage to retain feed and to prevent feed falling to the bottom. The feed was distributed to each feeding sieve using a small boat. All feeding sieves and happa net cages were scrubbed and cleaned monthly to remove moss and muddy slime from the mesh. The pond was prepared before the experiment by removing wastewater and pond sediments and left empty to dry for about 2 weeks. The bottom sediments of the pond were treated with 30–40 kg lime and left to dry for a week before river water was pumped into the pond. About 20 % of the water in the pond was replaced by new water from the river every 2 weeks during the experiment.

Chemical analysis

Samples of feed ingredients, diets and fish fillets, liver and kidney were analysed in duplicate using standard methods (AOAC 1997). Dry matter was determined by drying in an oven at 105 °C for 24 h. Nitrogen (N) was determined by the Kjeldahl method and crude protein (CP) was calculated as N × 6.25. Crude fat (EE) content was analysed using the Soxhlet method after acid hydrolysis of the sample. Crude fibre (CF) content was determined by standard methods (AOAC 1997) and neutral detergent fibre (NDF) was determined according to Van Soest et al. (1991). Ash content was determined by incineration in a muffle furnace at 550 °C for 12 h. Amino acid content of ingredients and diets was analysed by high-performance liquid chromatography according to Vázquez-Ortiz et al. (1995). Gross energy (MJ kg^{-1}) was determined with a bomb calorimeter (Calorimeter Parr 6300, Parr Instrument Company, Moline, IL, USA).

Water quality monitoring

Temperature (T °C) was recorded daily with a temperature meter at 8.00 and 14.00 h. Dissolved oxygen (DO mg/L), chemical oxygen demand (COD mg/L) and biochemical oxygen demand (BOD mg/L) parameters were measured twice a month; all water samples for COD and BOD were carefully taken from fish pond and filled into a 300-mL glass of biological oxygen demand (BOD). And then the water sample in glass BOD was immediately added 2 mL of manganese sulfate and was analyzed in laboratory (Stirling 1985). COD and BOD mg/L parameters were measured using the Hach Lange cuvett test method (DR2800 visual spectrophotometer, Hach Lange Gmbh, Germany).

Water samples of Nitrite nitrogen (NO_2 mg/L), nitrate nitrogen (NO_3 mg/L), total ammonia–nitrogen (TAN mg/L) parameter were collected twice a month. Water samplings were surrounded by ice and measured in laboratory using the Hach Lange cuvett test method (DR2800 visual spectrophotometer, Hach Lange Gmbh, Germany). The pH was recorded twice a month in the fish pond with a pH meter (Schott Greate, Florida, USA). Plankton species in the pond water were monitored and determined twice a month as described by Bellinger and Sigee (2010). The density of plankton was calculated by the formula: $N = (P \times C \times 100)/V$, where N is the number of plankton per litre of water in the pond, P is the number of planktonic organisms counted in different hapa net cages of different treatments, C is the volume of the plastic bottle holding the sample (100 mL) and V is the volume of water sample from each hapa net cage. The identification of zooplankton and phytoplankton species was based on Suthers and Rissik (2008).

Fish sampling for body indices calculation

Six-hundred and thirty fish (30 fish/hapa net) were randomly collected at the end of experiment. The fillet, kidney, liver, fat and carcass of fish were individually weighed using a digital scale for body indices calculation of the percentage of fillet, viscera-somatic (VSI%), hepato-somatic (HSI%), kidney and intra-peritoneal fat (IPF%) index.

Calculations

The following calculations were made:

The specific growth rate (SGR%) = $[(\ln W_f - \ln W_i)/T] \times 100$ and daily weight gain (DWG) = $(W_f - W_i)/T)$, where W_f and W_i refer to the mean final weight and the mean initial weight, respectively, and T is the feeding trial period in days.

Survival rate [(SR%) = $(TF_f/TF_i) \times 100$], where TF_f is the total number of fish at finish (harvest) and TF_i is the total number of fish at start.

Protein efficiency ratio (PER) = wet weight gain (g)/total protein intake (g).

Protein intake (PI) = feed intake (g) × percent protein in diet.

Total feed intake per fish (FI) = [total feed intake (g)/number of fish].

Feed conversion ratio (FCR) = [total feed intake (g)/total wet weight gain (g)].

Hepato-somatic index (HSI%) = [100 × (liver weight (g)/body weight (g))].

Intra-peritoneal fat (IPF%) = [100 × (intra-peritoneal fat weight (g)/body weight (g))].
Viscera-somatic weight (VSI%) = [100 × (viscera-somatic weight (g)/body weight (g))].
Kidney index (KI) = [100 × (kidney weight (g)/body weight (g))].

Statistical analysis

All data on fish growth performance, feed utilisation and carcass traits were statistically analysed by one-way analysis of variance (ANOVA), using Tukey's post hoc ANOVA test for individual comparisons ($p \leq 0.05$ level of significance). All statistical analysis was carried out using the IBM SPSS STATISTIC (Ibm 2011) programme, version 19.

Results

Chemical composition of feed ingredients and diets

The CP content of feed ingredients was highest in shrimp head meal, followed in descending order by golden apple snail meal, soybean meal, groundnut cake, cassava leaf meal and sweet potato leaf meal (Table 2).

The EE content was highest in groundnut cake, followed by cassava leaf meal and shrimp head meal, whilst the content of CF and NDF showed different patterns among feed ingredients. The highest CF and ash content were found in cassava leaf meal and shrimp head meal, whilst the highest NDF content was found in sweet potato leaf meal, groundnut cake and cassava leaf meal. The gross energy content varied between feed ingredients within the range 12.3–19.5 MJ/kg, with the lowest value in golden apple snail meal and the highest in groundnut cake. The essential amino acid (EAA) profile varied between feed ingredients (Table 2). In general, the most individual EAA was lowest in sweet potato leaf meal and cassava leaf meal. In contrast, shrimp head meal, golden apple snail meal and soybean meal were high in most EAA.

Table 3 Chemical composition (g/kg DM), and gross energy (MJ/kg DM) and amino acid (g/kg DM) content of the reference diet (RD) and test diets

Chemical composition	RD	Test diets (% of fish meal crude protein replaced)					
		Ground-nut cake (25 %)	Cassava leaf meal (25 %)	Sweet potato leaf meal (20 %)	Soybean meal (100 %)	Golden apple snail meal (100 %)	Shrimp head meal (100 %)
Crude protein	225	230	227	223	234	227	225
Lipid	45	39	44	38	35	31	35
Crude fibre	29	45	51	51	37	53	59
NDF	166	242	261	278	201	253	259
Ash	102	81	73	84	41	41	89
Gross energy	16.2	16.7	17.1	16.5	17.2	16.8	15.9
Essential amino acids							
Arginine	14.8	14.7	13.4	13.2	13.7	14.6	14.7
Histidine	6.4	5.6	5.5	5.5	3.2	4.8	4.2
Isoleucine	9.7	9.3	9.3	9.0	10.2	9.7	9.5
Leucine	17.6	16.6	16.6	16.0	16.5	17.3	16.3
Lysine	9.3	8.0	8.2	7.9	6.9	6.6	7.8
Methionine	5.9	5.3	5.4	5.4	4.0	4.9	5.3
Phenylalanine	9.8	9.7	9.5	9.1	10.6	9.6	9.6
Threonine	6.3	6.1	6.3	5.8	7.7	6.6	5.7
Valine	10.5	10.1	10.1	9.8	10.5	10.5	9.5
Total	90.3	85.4	84.3	81.7	83.3	84.6	82.6

The chemical composition, gross energy content and EAA profiles of experimental diets were similar amongst diets with 223–230 g CP/kg, 15.9–17.2 MJ GE/kg and 81.7–90.3 g total EAA/kg (Table 3).

Growth performance and feed utilisation

There were differences ($p = 0.01$–0.03) between diets in final body weight (BW), total weight gain (WG), daily weight gain (DWG) and specific growth rate (SGR) (Table 4).

The highest final BW, total WG, DWG and SGR were recorded for the shrimp head meal diet, followed in descending order by the golden apple snail meal and groundnut cake diets, the sweet potato leaf meal and cassava leaf meal diets and finally the soybean meal diet. The growth performance traits of fish on the shrimp head meal diet did not differ ($p > 0.05$) from those on the reference diet (Table 4). Fish fed shrimp head meal, groundnut cake and golden apple snail meal had a lower PI ($p > 0.23$) than fish on diets with soybean meal, cassava leaf meal or sweet potato leaf meal. There were no differences ($p > 0.05$) in FI, PER, FCR, PI between fish fed the reference diet and the diets with shrimp head meal, golden apple snail meal, groundnut cake, cassava leaf meal, sweet potato leaf meal and soybean meal (Table 4). The survival rates were high (range 90.5–96.3 %) and did not differ between treatments ($p > 0.05$).

Carcass and body indices

The proportion of fish fillet and kidney at the end of experiment ranged from (34.7–38.4 %) to (0.7–0.9 %) (wet weight basis), respectively, but there were no differences between the reference diet and the other diets ($p > 0.05$) (Table 5). The viscera-somatic index (VSI%), hepato-somatic index (HSI%) and (IPF%) index differed between treatments, with the highest VSI and HSI% ($p < 0.05$) for golden apple snail meal and shrimp head meal diets. The highest content of intra-peritoneal fat (IPF%) was recorded for the soybean meal, golden apple snail meal and shrimp head meal diets ($p < 0.001$).

The chemical composition of fillet, kidney and liver did not differ between the reference diet and the test diets ($p > 0.05$). The highest CP content was found in the fish fillet (151–171 g/kg), followed in descending order by liver (142–159 g/kg) and kidney (139–155 g/kg). The highest lipid content was found in liver (52–101 g/kg) and the lowest in kidney (range 10–36 g/kg) ($p < 0.05$). There were no differences ($p > 0.05$) in ash content between fillet, kidney and liver.

Water quality

Average temperature (°C) was 28.6 ± 0.9 (range 27.2–30.1), pH was 7.8 ± 1.0 (6.7–8.2) and DO content (mg/L) was 7.0 ± 1.0 (5.4–7.2). Total ammonia nitrogen (TAN), total nitrogen (TN) and total phosphorous (TP) were low during the experiment, with TAN (mg/L) of 0.4 ± 0.2 (0.1–0.8), TN (mg/L) of 0.8 ± 0.21 (0.5–1.0) and TP (mg/L) 0.6 ± 0.2 (0.2–0.9). Nitrite (NO_2) and nitrate (NO_3) were also low during the experiment, with NO_2 (mg/L) of 0.1 ± 0.02 (0.03–0.1) and NO_3 (mg/L) of 0.07 ± 0.01 (0.01–0.1). The chemical oxygen demand (COD) (mg/L) was 43.4 ± 12.7 (22–69) and the biochemical oxygen demand (BOD) (mg/L) was 19.8 ± 7.5 (11–32). The concentration of total suspended solids (TSS) (mg/L) was 251.2 ± 59.9 (115.6–356).

Plankton monitoring and assessment

Phytoplankton

Four phyla of micro algae: Chlorophyta (green-algae), Cyanophyta (blue-green algae), Bacillariophyta (diatom belonging to ochrophyta) and Euglenophyta (euglenoids), were recorded during the experiment (Fig. 1). Chlorophyta showed the highest proportion and accounted for 59.4 % of phytoplankton (17,622, range 7358–26,930 individuals/L), followed by Bacillariophyta with 17.4 % (5029, range 1198–10,878 individuals/L), Euglenophyta with 12.2 % (3625, range 1760–5598 individuals/L) and Cyanophyta with 11.4 % (3385, range 1043–5451 individuals/L). At least 13 families were recorded in the Chlorophyta and Cyanophyta phyla

Table 4 Growth performance and feed utilisation of Tra catfish (*Pangasianodon hypophthalmus*) fingerlings fed the reference diet (RD) and the test diets

	RD	Test diets (% of fish meal crude protein replaced)							
		Groundnut cake (25 %)	Cassava leaf meal (25 %)	Sweet potato leaf meal (20 %)	Soybean meal (100 %)	Golden apple snail meal (100 %)	Shrimp head meal (100 %)	SEM	*p* value
Initial BW (g)	16.1 ± 4.0	16.9 ± 4.0	16.2 ± 4.0	16.1 ± 4.0	16.2 ± 4.0	16.2 ± 4.0	16.2 ± 4.0	0.42	1.00
Final BW (g)	229.4 ± 73.7^{ab}	219.0 ± 81.3^{b}	177.9 ± 66.4^{c}	204.9 ± 96.6^{bc}	132.5 ± 39.6^{d}	222.3 ± 73.3^{b}	257.0 ± 83.3^{a}	7.45	0.02
WG (g)	213.3 ± 73.3^{ab}	202.1 ± 80.6^{b}	161.7 ± 65.7^{c}	188.8 ± 95.7^{bc}	116.3 ± 39.4^{d}	206.1 ± 72.9^{b}	240.8 ± 82.2^{a}	7.39	0.01
DWG (g)	1.90 ± 0.7^{ab}	1.81 ± 0.7^{b}	1.44 ± 4.0^{c}	1.69 ± 0.9^{bc}	1.04 ± 0.4^{d}	1.84 ± 0.7^{b}	2.15 ± 0.7^{a}	0.07	0.02
SGR (%)	2.35 ± 0.4^{ab}	2.28 ± 0.4^{b}	2.10 ± 0.4^{c}	2.20 ± 0.4^{bc}	1.87 ± 0.3^{d}	2.32 ± 0.3^{ab}	2.45 ± 0.3^{a}	0.04	0.03
FCR	1.66 ± 0.04	1.58 ± 0.1	1.97 ± 0.3	1.68 ± 0.5	2.10 ± 0.1	1.54 ± 0.4	1.42 ± 0.2	0.29	0.12
PER	2.96 ± 0.6	2.94 ± 0.3	2.35 ± 0.3	2.75 ± 0.8	2.69 ± 0.3	3.15 ± 0.5	3.18 ± 0.2	0.27	0.11
PI	0.73 ± 0.09	0.74 ± 0.02	0.79 ± 0.2	0.78 ± 0.8	0.84 ± 0.01	0.69 ± 0.02	0.77 ± 0.01	0.24	0.23
FI	326.1 ± 42.1	320.8 ± 7.0	347.3 ± 64.4	351.7 ± 14.0	359.1 ± 5.2	305.5 ± 0.02	340.1 ± 4.2	7.89	0.71
SR (%)	95.2	93.2	90.5	96.3	92.5	93.5	95.2	0.74	0.62

Means within rows with different superscript letters are significantly different ($p < 0.05$)

BW (g) Body weight, *WG* Weight gain, *DWG* Daily weight gain, *SGR* Specific growth rate, *FCR* Feed conversion ratio, *PER* Protein efficiency ratio, *PI* Protein intake, *FI* Total feed intake per fish, *SR* Survival ratio, *SEM* Standard error of the mean

Table 5 Percentage of fillet, viscera (internal organs) index (VSI), liver index (HIS), kidney index, fat index and chemical composition of fillet, kidney and liver in Tra catfish (*Pangasianodon hypophthalmus*) fingerlings fed the reference diet (RD) and the test diets

	RD	Test diets (% of fish meal crude protein replaced)							
		Ground-nut cake (25 %)	Cassava leaf meal (25 %)	Sweet potato leaf meal (20 %)	Soybean meal (100 %)	Golden apple snail meal (100 %)	Shrimp head meal (100 %)	SEM	*p* value
Fillet	38.4 ± 2.4	36.9 ± 3.0	36.1 ± 2.9	36.7 ± 2.6	34.7 ± 4.0	35.1 ± 8.5	37.1 ± 3.3	0.95	0.052
Internal organ (VSI)	11.0 ± 1.2^{c}	11.1 ± 1.6^{c}	10.1 ± 1.2^{c}	10.5 ± 0.9^{c}	13.1 ± 2.7^{a}	14.3 ± 1.5^{ab}	13.3 ± 2.0^{a}	0.64	0.013
Liver (HIS)	1.9 ± 0.3^{bc}	1.8 ± 0.2^{c}	1.9 ± 0.3^{bc}	1.7 ± 0.3^{c}	1.8 ± 0.3^{bc}	2.1 ± 0.3^{a}	2.0 ± 0.2^{ab}	0.05	0.012
Kidney	0.7 ± 0.2	0.7 ± 0.2	0.7 ± 0.2	0.4 ± 0.2	0.8 ± 0.3	0.9 ± 0.1	0.8 ± 0.2	0.08	0.274
Fat (IPF)	2.8 ± 1.1^{c}	3.1 ± 1.6^{bc}	1.8 ± 0.8^{d}	2.2 ± 0.8^{cd}	5.6 ± 1.7^{a}	5.3 ± 1.7^{a}	5.3 ± 1.4^{a}	0.25	0.001

Means within rows with different superscript letters are significantly different ($p < 0.05$)

SEM standard error of the mean

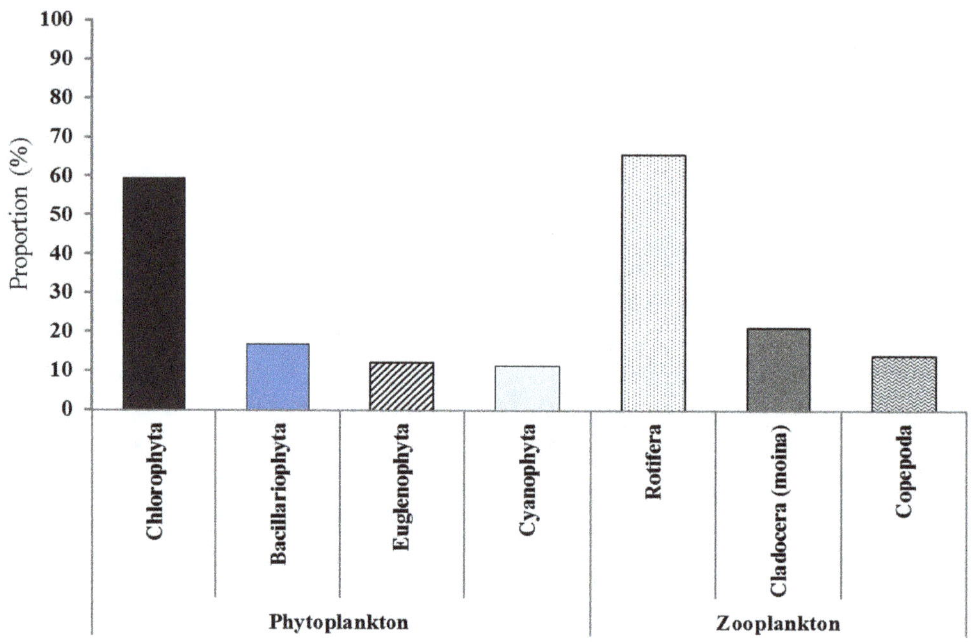

Fig. 1 Proportion (% of total) of organisms occurring in each phylum of phytoplankton and zooplankton in the pond water used for striped catfish culture over the 16-week experimental period

and there were about six families in the Bacillariophyta phylum and four families in the Euglenophyta phylum. In general, the density of phytoplankton increased as the experiment progressed (Fig. 1).

Zooplankton

Three main zooplankton groups: Copepoda, Cladocera (monia) and Rotifera were recorded during the experiment (Fig. 1). Rotifera was the most abundant group accounting for 65.1 % of zooplankton (2292, range 766–4767 individuals/mL), followed by 21.1 % for Cladocera (749, range 147–1442 individuals/mL) and 13.9 % for Copepoda (489, range 89–1442 individuals/mL). In general, the density of zooplankton decreased as the experiment progressed (Fig. 1).

Discussion

The data in the present study on Tra catfish fingerlings, cultured in hapa net cage installed in earthen pond, confirm previous laboratory studies that show that fish meal protein can be replaced with protein from locally available plant and animal ingredients in feed for striped catfish fingerlings without compromising growth performance, feed utilisation or carcass traits (Da et al. 2012). The fingerlings were fed identical diets where fish meal CP was replaced with CP from SPLM (20 % replacement), CSLM (25 % replacement), GNC (25 % replacement), SBM (100 % replacement), GAPS (100 % replacement) and SHM (100 % replacement). The WG and DWG of fish cultured in hapa net cages installed in earthen pond of current experiment were four to sixfold higher than values reported for Tra catfish and Asian catfish species cultured under other laboratory conditions (Hung et al. 2003; Phumee et al. 2011).

In general, the chemical composition and amino acid profile of test feed ingredients were in good agreement with published data (Hue et al. 2010; Phuc and Lindberg 2000, 2001; Tram et al. 2011; Nguyen et al. 2012). The average CP content of the diets studied was 22.7 % within the range (19–30 % CP) required for normal growth rate in Tra catfish fingerlings, but lower than the range required (27–32 % CP) for maximum growth rate (Hung et al. 2002). The dietary CP content was comparable to that reported for commercial Tra catfish farms, which is on average 26 % CP (range 20–30 %) for commercial feeds and 22 % CP (range 17–26 %) for farm-made feeds (Phan et al., 2009). The dietary content (g/kg DM) of arginine (13.2–14.8), histidine (3.2–6.4), leucine (16.0–17.6), methionine (4.0–5.9), phenylanine (9.1–10.6) and valine (9.5–10.5) was above the requirements of channel catfish (Wilson et al. 1980). Moreover, the dietary content (g/kg DM) of lysine (6.6–9.3), threonine (5.7–7.7) and isoleucine (9.0–10.2) met the requirements of channel catfish (Wilson 1989). Almost all EAA profiles in the diets tested in the present study were high compared with the EAA recommendations for tilapia (Jackson and Capper 1982) and common carp (Schwarz and Kirchgessner 1988).

The poorer growth performance and FCR on the soybean meal diet could be related to disturbed gut function (Nordrum et al. 2000) due to the presence of anti-nutritional factors. It could also be partly due to a lower supply of sulphur-containing amino acids in this diet than in the other diets. The CP and amino acid supply with the cassava leaf meal diet was similar to that with the other diets and the EAA profile was comparable. Although cassava leaves contain cyanogenic glycosides but the content in fresh cassava leaves is reduced by more than 80 % by sun drying and cooking (Hue et al. 2010; Phuc and Lindberg 2001), a negative impact on fish performance cannot be ruled out.

The PER was high in the present study compared with values reported for juvenile African catfish fed plant protein (Fagbenro 2004; Nyina-wamwiza et al. 2007), red-tailed catfish species (Deng et al. 2011) and juvenile Asian catfish species (Hung et al. 2004; Phumee et al. 2011). This indicates that the amino acid supply and amino acid profile in the test diets were adequate for Tra catfish requirements.

The HSI, IPF and VSI indices in present study were lower than those reported previously for Asian catfish species (Hung et al. 2003; Phumee et al. 2011). The HSI index was higher for the groundnut cake, golden apple snail meal and shrimp head meal diets than for the reference diet. Peres and Oliva-Teles (2008) demonstrated a correlation between HSI index and lysine level in the diet of turbot (*Scophthalmus maximus*) juveniles, with a linear decrease in HSI index with increased dietary lysine level. This relationship could not be confirmed in the present study. It has also been suggested that diets containing high amounts of carbohydrates should result in higher HSI index (Hernández et al. 2007; Hung et al. 2003).

The average water temperature was 28.6 °C, with a pH value of 7.8 (range 6.7–8.2) and a DO content of 7.0 (range 5.4–7.2 mg/L). The average concentrations of BOD (19.8 mg/L), COD (43.4 mg/L), TAN (0.4 mg/L), NO_2^- (0.1 mg/L) and NO_3^- (0.07 mg/L) were low compared with the limits recommended in National Standards (TCVN 5942 1995). In general, the water quality was good throughout the experiment and the values recorded can be considered low compared with the results of water quality monitoring in Tra catfish farms in the Mekong Delta (Phuong et al. 2010). The average density in water of phytoplankton was 29,661 (11,359–48,857) individuals/L and the average density of zooplankton was 3530 (1002–7651) individuals/L. This was five to threefold higher than the densities of phytoplankton and zooplankton found in river water (Phuong et al. 2010).

Conclusion

Fish meal protein in feed for Tra catfish (*Pangasianodon hypophthalmus*) fingerlings can be replaced by protein from locally available plant and animal feed ingredients in effectively compromising growth performance, feed utilisation or carcass traits of fish. However, the extent of replacement will depend on the properties of the individual feed ingredient. Further studies on grow-out fish are warranted to confirm the present findings, and to quantify the impact on fish performance and product quality traits under commercial conditions.

Acknowledgments The authors wish to thank the Sida/SAREC MEKARN ("Mekong Basin Animal Research Network") project for funding the research and a PhD scholarship for Chau Thi Da. The authors also wish to thank the laboratory staff of the National Institute of Animal Husbandry, Ministry of Agriculture and Rural Development, Vietnam, and Mr. Huynh Thanh Phuong and Mr. Le Van Kha, students at the Aquaculture Department, Faculty of Agriculture and Natural Resources, An Giang University, Vietnam, for their support and assistance during this study.

References

AOAC (1997) Animal feeds. Chapter 4. In: Cunniff PA (ed) Official methods of analysis vol 1. Association of Official Analytical Chemists International. AOAC, Arlington, pp VI, 1102

Bellinger EG, Sigee DC (2010) Introduction to freshwater algae, in freshwater algae: identification and use as bioindicators. John Wiley & Sons, Ltd, Chichester, UK, pp 1–285. doi:10.1002/9780470689554.ch1

Bui TM, Lam PT, Ingram BA, Thuy NTT, Gooley GJ, Hao NV, Phuong NT, De Silva SS (2010) Seed production practices of striped catfish (*Pangasianodon hypophthalmus*) in the Mekong Delta region, Vietnam. Aquaculture 306:92–100

Cacot P, Pariselle A (1999) Description of the sexual cycle related to the environment and set up of the artificial propagation in *Pangasius bocourti* (Sauvage 1880) and *Pangasius hypophthalmus* (Sauvage 1878) reared in floating cages and in ponds in the Mekong Delta. In: Legendre M, Pariselle A (eds) The biological diversity and aquaculture of Clariid and Pangasiid catfishes in South East Asia. Proceedings of the mid-term workshop of the 'Catfish Asia Project', 11–15 May 1998. Cantho, Vietnam, pp 71–89

Da CT, Hung LT, Berg H, Lindberg JE, Lundh T (2011) Evaluation of potential feed sources, and technical and economic considerations of small-scale commercial striped catfish (*Pangasius hypothalamus*) pond farming systems in the Mekong Delta of Vietnam. Aquac Res 44:427–438. doi:10.1111/j.1365-2109.2011.03048.x

Da CT, Lundh T, Lindberg JE (2012) Evaluation of local feed resources as alternatives to fish meal in terms of growth performance, feed utilisation and biological indices of striped catfish (*Pangasianodon hypophthalmus*) fingerlings. Aquaculture 364–365:150–156

De Silva SS, Phuong NT (2011) Striped catfish farming in the Mekong Delta, Vietnam: a tumultuous path to a global success. Rev Aquac 3:45–73

Deng J, Zhang X, Bi B, Kong L, Kang B (2011) Dietary protein requirement of juvenile Asian red-tailed catfish (*Hemibagrus wyckioides*). Anim Feed Sci Technol 170:231–238

Edwards P, Tuan LA, Allan GL (2004) A survey of marine trash fish and fish meal as aquaculture feed ingredients in Vietnam. Australian Centre for International Agricultural Research. Working Paper No. 57, pp 1–56

Fagbenro OA (2004) Soybean meal replacement by roquette (*Eruca sativa Miller*) seed meal as protein feedstuff in diets for African Catfish (*Clarias gariepinus*, Burchell 1822), fingerlings. Aquac Res 35:917–923

FAO (2010) Fisheries and Aquaculture. Food and Agriculture Organization (FAO) of United Nations. http://www.eoearth.org/article/Food_and_Agriculture_Organization_(FAO). Accessed 19 June 2011

Halls A, Johns M (2013) Assessment of the vulnerability of the Mekong Delta Pangasius catfish industry to development and climate change in the Lower Mekong Basin. Report prepared for the Sustainable Fisheries Partnership

Hernández MD, Martínez FJ, Jover M, García García B (2007) Effects of partial replacement of fish meal by soybean meal in sharpsnout seabream (*Diplodus puntazzo*) diet. Aquaculture 263:159–167

Hue KT, Van DTT, Ledin I, Spörndly E, Wredle E (2010) Effect of feeding fresh, wilted and sun-dried foliage from cassava (*Manihot esculenta Crantz*) on the performance of lambs and their intake of hydrogen cyanide. Livest Sci 131:155–161

Hung LT, Liem PT, Tu HT, Mariojouls C (2002) Comparing growth and protein requirements for fingerlings of three catfish of the Mekong River (*Pangasius bocourti, Pangagasius hypothalmus* and *Pangasius conchophilus*). J Aquac Trop 17:325–335

Hung LT, Suhenda N, Slembrouck J, Lazard J, Moreau Y (2003) Comparison of starch utilization in fingerlings of two Asian catfishes from the Mekong River (*Pangasius bocourti* Sauvage, 1880, *Pangasius hypophthalmus* Sauvage, 1878). Aquac Nutr 9:215–222

Hung LT, Suhenda N, Slembrouck J, Lazard J, Moreau Y (2004) Comparison of dietary protein and energy utilization in three Asian catfishes (*Pangasius bocourti, P. hypophthalmus and P. djambal*). Aquac Nutr 10:317–326

Hung LT, Truc LTT, Huy HPV (2007) Case study on the use of farm-made feeds and commercially formulated pellets for pangasiid catfish culture in the Mekong Delta, Viet Nam: Study and analysis of feeds and fertilizers for sustainable aquaculture development. Food and Agriculture Organization of United Nattion—FAO, Fisheries Technical Paper No. 497, pp 363–377

IBM SPSS STATISTIC (2011) IBM SPSS STATISTIC program, version 19 statistical software packages. IBM Corporation, New York

Jackson AJ, Capper BS (1982) Investigations into the requirements of the tilapia (*Sarotherodon mossambicus*) for dietary methionine, lysine and arginine in semi-synthetic diets. Aquaculture 29:289–297

Kader MA, Koshio S, Ishikawa M, Yokoyama S, Bulbul M (2010) Supplemental effects of some crude ingredients in improving nutritive values of low fishmeal diets for red sea bream, Pagrus major. Aquaculture 308:136–144

Loc V, Bush S, Sinh L, Khiem N (2010) High and low value fish chains in the Mekong Delta: challenges for livelihoods and governance. Environ Dev Sustain 12:889–908

Mamat NZ, Alfaro AC (2014) Evaluation of microalgal and formulated diets for the culture of the New Zealand pipi clam *Paphies australis*. Int Aquat Res 6:57. doi:10.1007/s40071-014-0057-7

Naylor RL, Hardy RW, Bureau DP, Chiu A, Elliott M (2009) Feeding aquaculture in an era of finite resources. Proc Natl Acad Sci USA 106:15103–15110

Nguyen THL, Ngoan LD, Bosch G, Verstegen MWA, Hendriks WH (2012) Ileal and total tract apparent crude protein and amino acid digestibility of ensiled and dried cassava leaves and sweet potato vines in growing pigs. Anim Feed Sci Technol 172:171–179

Nordrum S, Bakke-McKellep AM, Krogdahl Å, Buddington RK (2000) Effects of soybean meal and salinity on intestinal transport of nutrients in Atlantic salmon (*Salmo salar L.*) and rainbow trout (*Oncorhynchus mykiss*). Comp Biochem Physiol Part B 125:317–335

Nyina-wamwiza L, Wathelet B, Kestemont P (2007) Potential of local agricultural by-products for the rearing of African catfish (*Clarias gariepinus*) in Rwanda: effects on growth, feed utilization and body composition. Aquac Res 38:206–214

Peres H, Oliva-Teles A (2008) Lysine requirement and efficiency of lysine utilization in turbot (*Scophthalmus maximus*) juveniles. Aquaculture 275:283–290

Phan TL, Tam BM, Thuy NTT, Geoff GJ, Brett IA, Hao NV, Phuong NT, Silva SSD (2009) Current status of farming practices of striped catfish, *Pangasianodon hypophthalmus* in the Mekong Delta, Vietnam. Aquaculture 296:227–236

Phuc BHN, Lindberg JE (2000) Ileal and total tract digestibility in growing pigs given cassava root meal diets with inclusion of cassava leaves, leucaena leaves and groundnut foliage. Br Soc Anim Sci 17:301–308

Phuc BHN, Lindberg JE (2001) Ileal apparent digestibility of amino acids in growing pigs given a cassava root meal diet with inclusion of cassava leaves, leucaena leaves and groundnut foliage. Anim Sci 72:511–517

Phumee P, Wei WY, Ramachandran S, Hashim R (2011) Evaluation of soybean meal in the formulated diets for juvenile *Pangasianodon hypophthalmus* (Sauvage, 1878). Aquac Nutr 17:214–222

Phuong NT (1998) Pangasius catfish cage aquaculture in the Mekong Delta, Vietnam: current status and study for feeding improvement. College of Aquaculture and Fisheries, Can Tho University, Vietnam, Doctoral thesis no. 1999 (Library of Can Tho University), p 29

Phuong NT, Sinh LX, Thinh NQ (2007) Economics of aquaculture feeding practices: Vietnam. FAO Fish Tech Pap Rome 505:183–205

Phuong NT, Ut VN, Tung VT, Hang NTT, Lien NTK, Oanh DTH, Huong DTT, Morales EJ (2010) Water quality monitoring in striped catfish (Pangasianodom hypophthalmus) farms in the Mekong Delta, Vietnam. College of Aquacultrue and Fisheries, Can Tho University, Vietnam, Sustainable Fisheries Partnership, pp 1–19. http://media.sustainablefish.org/Final%20report_Water%20Quality2009_final.pdf. Accessed 29 May 2012

Schwarz FJ, Kirchgessner M (1988) Amino acid composition of carp (*Cyprinus carpio* L.) with varying protein and energy supplies. Aquaculture 72:307–317

Stirling HP (1985) Chemical and biological methods of water analysis for aquaculturists. Institute of Aquaculture, University of Stirling, Stirling, p 117

Suthers IM, Rissik D (2008) Plankton: a guide to their ecology and monitoring for water quality. CSIRO Publishing, Collingwood, pp 1–273

TCVN 5942 (1995) National standard (TCVN 5942-1995). Ministry of Fisheries, Hanoi. Sector Standard (TCVN 5942)

Tram NDQ, Ngoan LD, Hung LT, Lindberg JE (2011) A comparative study on the apparent digestibility of selected feedstuffs in Hybride catfish (*Claria macrocephalus* × *Clarias gariepinus*) and Nile tilapia (*Oreochromis niloticus*). Aquac Nutr 17:636–643

Van Soest PJ, Robertson JB, Lewis BA (1991) Method of dietary fiber, neutral detergent fiber, and nonstarch polysaccharides in relation to animal nutrition. J Dairy Sci 74:3583–4597

Vázquez-Ortiz FA, Caire G, Huguere-Ciapara I, Hernández G (1995) High-performance liquid chromatographic determination of free amino acid in shrimp. J Liq Chromatogr 18:2059–2068

Wilson RP (1989) Amino acids and proteins. In: Halver JE (ed) Fish nutrition, 2nd edn. Academic Press, New York, pp 111–151

Wilson RP, Poe WE, Robinson EH (1980) Leucine, isoleucine, valine and histidine requirements of fingerling channel catfish. J Nutr 110:6273

Permissions

All chapters in this book were first published in IAR, by Springer; hereby published with permission under the Creative Commons Attribution License or equivalent. Every chapter published in this book has been scrutinized by our experts. Their significance has been extensively debated. The topics covered herein carry significant findings which will fuel the growth of the discipline. They may even be implemented as practical applications or may be referred to as a beginning point for another development.

The contributors of this book come from diverse backgrounds, making this book a truly international effort. This book will bring forth new frontiers with its revolutionizing research information and detailed analysis of the nascent developments around the world.

We would like to thank all the contributing authors for lending their expertise to make the book truly unique. They have played a crucial role in the development of this book. Without their invaluable contributions this book wouldn't have been possible. They have made vital efforts to compile up to date information on the varied aspects of this subject to make this book a valuable addition to the collection of many professionals and students.

This book was conceptualized with the vision of imparting up-to-date information and advanced data in this field. To ensure the same, a matchless editorial board was set up. Every individual on the board went through rigorous rounds of assessment to prove their worth. After which they invested a large part of their time researching and compiling the most relevant data for our readers.

The editorial board has been involved in producing this book since its inception. They have spent rigorous hours researching and exploring the diverse topics which have resulted in the successful publishing of this book. They have passed on their knowledge of decades through this book. To expedite this challenging task, the publisher supported the team at every step. A small team of assistant editors was also appointed to further simplify the editing procedure and attain best results for the readers.

Apart from the editorial board, the designing team has also invested a significant amount of their time in understanding the subject and creating the most relevant covers. They scrutinized every image to scout for the most suitable representation of the subject and create an appropriate cover for the book.

The publishing team has been an ardent support to the editorial, designing and production team. Their endless efforts to recruit the best for this project, has resulted in the accomplishment of this book. They are a veteran in the field of academics and their pool of knowledge is as vast as their experience in printing. Their expertise and guidance has proved useful at every step. Their uncompromising quality standards have made this book an exceptional effort. Their encouragement from time to time has been an inspiration for everyone.

The publisher and the editorial board hope that this book will prove to be a valuable piece of knowledge for researchers, students, practitioners and scholars across the globe.

List of Contributors

Stephan S. W. Ende, Vanessa Fuchs, Annabel Schuhn, Christiane von der Marwitz, Andrea Wirtz, Joachim Henjes and Matthew Slater
Alfred Wegener Institute, Helmholtz Centre for Polar and Marine Research, Am Handelshafen 12, 27570 Bremerhaven, Germany

José Fernando M. Gonçalves, Mariana Hinzmann, Jorge Machado and Sandra Araújo
Interdisciplinary Centre of Marine and Environmental Research, CIIMAR-UP, University of Porto, Terminal de Cruzeiros do Porto de Leixões, Av. General Norton de Matos s/n, 4450-208 Matosinhos, Portugal
ICBAS-UP, Abel Salazar Institute of Biomedical Sciences, University of Porto, Rua Jorge Viterbo Ferreira n.8 228, 4050-313 Porto, Portugal

Rodrigo O. A. Ozório
Interdisciplinary Centre of Marine and Environmental Research, CIIMAR-UP, University of Porto, Terminal de Cruzeiros do Porto de Leixões, Av. General Norton de Matos s/n, 4450-208 Matosinhos, Portugal

José Carlos Oliveira
CHO-Hospital Centre of Porto, Lg. Prof. Abel Salazar, 4099-001 Porto, Portugal

Adnan Hussain Gora
Division of Marine Biotechnology, Chennai Research Centre of ICAR-Central Marine Fisheries Research Institute, Chennai, India
Chennai Research Centre of ICAR-Central Marine Fisheries Research Institute, 75, Santhome High Road, R.A. Puram, Chennai 600028, India

Saima Rehman
Division of Marine Biotechnology, Chennai Research Centre of ICAR-Central Marine Fisheries Research Institute, Chennai, India

Narottam Prasad Sahu, Sujata Sahoo and Showkat Ahmad Dar
Division of Fish Nutrition, Biochemistry and Physiology, Central Institute of Fisheries Education, Mumbai, India

Irshad Ahmad
Division of Aquaculture, Central Institute of Fisheries Education, Mumbai, India

Deepak Agarwal
Division of Fish Genetics and Biotechnology, Central Institute of Fisheries Education, Mumbai, India

F. G. Irungu and A. K. Faraj
Department of Dairy and Food Science and Technology, Egerton University, Egerton, Kenya

C. M. Mutungi
Department of Dairy and Food Science and Technology, Egerton University, Egerton, Kenya
International Centre of Insect Physiology and Ecology, Nairobi, Kenya
International Institute of Tropical Agriculture (IITA), Plot NO. 25, Mikocheni Light Industrial Area, Dar es Salaam, Tanzania

C. Tanga, S. Ekesi and K. K. M. Fiaboe
International Centre of Insect Physiology and Ecology, Nairobi, Kenya

H. Affognon
International Crops Research Institute for the Semi-Arid Tropics, BP 320, Bamako, Mali

D. Nakimbugwe
Department of Food Technology and Nutrition, School of Food Technology, Nutrition and Bio-Engineering, Makerere University, Kampala, Uganda

Kumar Katya, M. Z. S. Borsra, Dev Ganesan, Giva Kuppusamy, Max Herriman and Sayed Azam Ali
Crops for The Future, Jalan Broga, 43500 Kuala Lumpur, Malaysia

Luís F. Pereira and Rodrigo O. A. Ozório
CIIMAR-Centro Interdisciplinar de Investigação Marinha e Ambiental, Universidade do Porto, Terminal de Cruzeiros do Porto de Leixões, Avenida General Norton de Matos s/n, 4450-208 Matosinhos, Portugal

Andrew Salter
School of Biosciences, University of Nottingham, Nottingham, UK

Gaspar Lobo, José F. M. Gonçalves and Maria J. Peixoto
CIIMAR-Centro Interdisciplinar de Investigação Marinha e Ambiental, Universidade do Porto, Terminal de Cruzeiros do Porto de Leixões, Avenida General Norton de Matos s/n, 4450-208 Matosinhos, Portugal
ICBAS-Instituto de Ciências Biomédicas de Abel Salazar, Universidade do Porto, Rua de Jorge Viterbo Ferreira 228,4050-313 Porto, Portugal

Caroline Nebo and Maria Célia Portella
Faculdade de Ciências Agrárias e Veterinárias, Universidade Estadual Paulista (Unesp), Jaboticabal, SP 14884-900, Brazil

Ken Overturf
USDA-ARS, Hagerman Fish Culture Experiment Station, 3059F National Fish Hatchery Rd, Hagerman, ID 83332, USA

Andreas Brezas
University of Idaho, Hagerman Fish Culture Experiment Station, 3059F National Fish Hatchery Rd, Hagerman, ID 83332, USA

Maeli Dal-Pai-Silva
Universidade Estadual Paulista (Unesp), Instituto de Biociências, Botucatu, SP 18618-000, Brazil

Wing-Keong Ng and Chia-Ling Lim
Fish Nutrition Laboratory, School of Biological Sciences, Universiti Sains Malaysia, 11800 USM, Penang, Malaysia

Nicholas Romano
Fish Nutrition Laboratory, School of Biological Sciences, Universiti Sains Malaysia, 11800 USM, Penang, Malaysia
Department of Aquaculture, Faculty of Agriculture, Universiti Putra Malaysia, Serdang, Selangor, Malaysia

Beng-Chu Kua
National Fish Health Research Centre, Batu Maung, Penang, Malaysia

Norhayati Yusuf
School of Science and Technology, Universiti Malaysia Terengganu, 21030 Kuala Terengganu, Malaysia

Inna Petrova, Ignat Tolstorebrov and Trygve Magne Eikevik
Department of Energy and Process Engineering, Norwegian University of Science and Technology, 7491 Trondheim, Norway

Norazira Abdu Rahman
School of Fisheries and Aquaculture Sciences, Universiti Malaysia Terengganu, 21030 Kuala Terengganu, Malaysia

Helena Khatoon and Noor Azilah Haris
Institute of Tropical Aquaculture, Universiti Malaysia Terengganu, 21030 Kuala Terengganu, Malaysia

Sanjoy Banerjee
Institute of Bioscience, Universiti Putra Malaysia, Serdang, Malaysia

Fathurrahman Lananan
East Coast Environmental Research Institute, Sultan Zainal Abidin University, Gong Badak Campus, 21300 Kuala Terengganu, Terengganu, Malaysia

Katayama Tomoyo
Graduate School of Agricultural and Life Sciences, The University of Tokyo, 1-1-1, Yayoi, Bunkyo-ku, Tokyo 113-8657, Japan

Weiwei Song, Chunlin Wang, Ronghua Li and Changkao Mu
Key Laboratory of the Ministry of Education for Applied Marine Biotechnology, School of Marine, Ningbo University, Fenghua Rd. 818, Ningbo 315211, Zhejiang, People's Republic of China
Collaborative Innovation Center for Zhejiang Marine High-Efficiency and Healthy Aquaculture, Ningbo University, Ningbo, People's Republic of China

Jingjing Wang, Wei Zhao, Yajie Liang, Yifan Han, Xingbiao Wang, Xiaoxia Zhang, Xiaomei Ma and Zhiyong Huang
Tianjin Key Laboratory of Industrial Biological Systems and Process Engineering, Tianjin Institute of Industrial Biotechnology, Chinese Academy of Sciences, No. 32, West 7th Road, Tianjin Airport Economic Area, Tianjin 300308, China

Demin Zhang
School of Marine Sciences, Ningbo University, Ningbo 315211, China

Suthajaree Yenmak
Department of Aquaculture, Faculty of Fisheries, Kasetsart University, Bangkok 10900, Thailand
Center for Advanced Study in Agriculture and Food, Institute for Advanced Studies, Kasetsart University, Bangkok 10900, Thailand

Weerakit Joerakate and Supawadee Poompuang
Department of Aquaculture, Faculty of Fisheries, Kasetsart University, Bangkok 10900, Thailand

Fui Fui Ching, Norazmi Othman, Azaharie Anuar and Rossita Shapawi
Borneo Marine Research Institute, Universiti Malaysia Sabah, Jln UMS, 88450 Kota Kinabalu, Sabah, Malaysia

Shigeharu Senoo
Aquaculture Technology and Production Center, Kindai University, 3153 Shirahama, Wakayama 649-2211, Japan

Morgane Paul Magouana Anvo
Laboratoire d'Hydrobiologie et d'Eco-technologie des eaux, UFR, Biosciences, Université Félix Houphouët-Boigny, 22 BP 582, Abidjan 22, Côte d'Ivoire
Unité de Recherche en Aquaculture et Biodiversité Aquatique/Laboratoire d'Etude et de Recherche sur les Ressources Naturelles et Sciences de l'Environnement (UR-ABAQ/LERNSE), Université Polytechnique de Bobo-Dioulasso, 01 BP 1091, Bobo-Dioulasso 01, Burkina Faso

Benie´ Rose Danielle Aboua and Essetchi Paul Kouamelan
Laboratoire d'Hydrobiologie et d'Eco-technologie des eaux, UFR
Biosciences, Université Félix Houphouët-Boigny, 22 BP 582, Abidjan 22, Côte d'Ivoire

Rokyatou Sissao, Chantal Yvette Zoungrana-Kaboré and Aboubacar Toguyéni
Unité de Recherche en Aquaculture et Biodiversité Aquatique/Laboratoire d'Etude et de Recherche sur les Ressources Naturelles et Sciences de l'Environnement (UR-ABAQ/LERNSE), Université Polytechnique de Bobo-Dioulasso, 01 BP 1091, Bobo-Dioulasso 01, Burkina Faso

Athanase Kraidy Otchoumou
Laboratoire de biotechnologie, UFR Biosciences, Université Félix Houphouët-Boigny, 22 BP 582, Abidjan 22, Côte d'Ivoire

Yaa Tiwaah Amoah, Mohammad Moniruzzaman, Seunghan Lee, Jinho Bae, Seonghun Won, Minji Seong and Sungchul C. Bai
Department of Marine Bio-materials and Aquaculture/Feeds and Foods Nutrition Research Center, Pukyong National University, Busan 608-737, Republic of Korea

K. Azpeitia and D. Mendiola
Marine Research Division, AZTI-Tecnalia, Herrera Kaia; Portualdea, s/n, 20110 Pasaia, Spain

Y. Ríos, I. Garcia and J. Pagaldai
AZTI-Tecnalia, Astondo Bidea, Edificio 609, Parque Tecnológico de Bizkaia, 48160 Derio, Bizkaia, Spain

A. Y. Maizatul, Radin Maya Saphira Radin Mohamed, Adel A. Al-Gheethi and M. K. Amir Hashim
Department of Water and Environmental Engineering, Faculty of Civil and Environmental Engineering, Micro-pollution Research Centre (MPRC), Universiti Tun Hussein Onn Malaysia, Batu Pahat, 86400 Parit Raja, Johor, Malaysia

K. Azpeitia, M. Revilla and D. Mendiola
AZTI-Marine Research, Herrera Kaia, Portualdea z/g, 20110 Pasaia, Spain

M. Ortiz-Zarragoitia
Plentzia Marine Station (PiE), University of the Basque Country (UPV/EHU), Areatza z/g, 48620 Plentzia, Basque Country, Spain

Chau Thi Da
Department of Aquaculture, Faculty of Agriculture and Natural Resources, An Giang University, 18 Ung Van Khiem Street, Dong Xuyen Wards, Long Xuyen, Vietnam

Torbjörn Lundh and Jan Erik Lindberg
Department of Animal Nutrition and Management, Swedish University of Agricultural Sciences, 75007 Uppsala, Sweden

Håkan Berg
Department of Physical Geography, Stockholm University, 106 91 Stockholm, Sweden

Index

www.ingramcontent.com/pod-product-compliance
Lightning Source LLC
LaVergne TN
LVHW080913210926
848485LV00004B/644